INVENTING
OURSELVES
OUT OF JOBS?

Studies in Industry and Society
Philip B. Scranton, Series Editor

Published with the assistance of the Hagley Museum and Library

Related Titles in the Series:
David A. Hounshell, *From the American System to Mass Production, 1800–1932: The Development of Manufacturing Technology in the United States*
JoAnne Yates, *Control through Communication: The Rise of System in American Management*
James P. Kraft, *Stage to Studio: Musicians and the Sound Revolution, 1890–1950*
Lindy Biggs, *The Rational Factory: Architecture, Technology, and Work in America's Age of Mass Production*
Pamela Walker Laird, *Advertising Progress: American Business and the Rise of Consumer Marketing*

INVENTING OURSELVES OUT OF JOBS?

AMERICA'S DEBATE OVER TECHNOLOGICAL UNEMPLOYMENT 1929–1981

Amy Sue Bix

The Johns Hopkins University Press

BALTIMORE AND LONDON

The Johns Hopkins University Press
2715 North Charles Street
Baltimore, Maryland 21218-4363
www.press.jhu.edu

Library of Congress Cataloging-in-Publication Data
will be found at the end of this book.
A catalog record for this book is available
from the British Library.

To my mother, Mary Bix, and
to the memory of my father, Ira Bix

CONTENTS

Acknowledgments *ix*

Prologue: Technology as Progress? *1*

1
"Economy of a Madhouse": Entering the Depression-Era
Debate over Technological Unemployment *9*

2
"Finding Jobs Faster Than Invention Can Take Them Away":
Government's Role in the Technological Unemployment Debate *43*

3
"No Power on Earth Can Stop Improved Machinery":
Labor's Concern about Displacement *80*

4
"Machinery Don't Eat": Displacement as a Theme
in Depression Culture *114*

5
"The Machine Has Been Libeled": The Business
Community's Defense *143*

6
"Innocence or Guilt of Science": Scientists and Engineers
Mobilize to Justify Mechanization *168*

7
"What Will the Smug Machine Age Do?":
Envisioning Past, Present, and Future as America
Moves from Depression to War *204*

8
"Automation Just Killed Us": The Displacement
Question in Postwar America *236*

Epilogue: Revisiting the Technological
Unemployment Debate *280*

Notes *313*

Essay on Sources *361*

Index *365*

ACKNOWLEDGMENTS

IN COMPLETING THIS WORK, I would like to express my appreciation to a number of wonderful professional colleagues and personal acquaintances. Special thanks go to Stuart Leslie of the Johns Hopkins University, who first encouraged me to investigate this topic and patiently guided me through the earliest stages of writing. I am also indebted to Sharon Kingsland, Robert Kargon, Louis Galambos, Ron Walters, and Erica Schoenberger, all of the Johns Hopkins University, who read and commented on my Ph.D. dissertation. I also owe a special thanks to my fellow graduate students in the Department of the History of Science, Technology, and Medicine during the late 1980s and early 1990s for their many expressions of friendship.

I want to thank a number of people at Iowa State University, first and foremost Alan Marcus, for his encouragement and valuable counsel. George McJimsey, David Wilson, and Andrejs Plakans reviewed drafts of this manuscript and provided important suggestions. I would further like to mention my gratitude to Hamilton Cravens, Carole Kennedy, and the other members of the history department at Iowa State for offering me the chance to work in such a productive environment. Finally, I will always be grateful for the expert assistance of Robert Brugger, Philip Scranton, one anonymous reviewer, manuscript editor Celestia Ward, and the production staff at the Johns Hopkins University Press.

I received wonderful help from the research and support staffs of the Milton S. Eisenhower Library at the Johns Hopkins University and the Parks Library at Iowa State University, especially from people in the government documents, interlibrary loan, and audio-visual departments. Archivists and staff members at the Herbert Hoover Presidential Library in West Branch,

Iowa, the Franklin Delano Roosevelt Presidential Library in Hyde Park, New York, and the National Archives in Washington, D.C., provided generous access to material.

The National Science Foundation offered financial support for this project through a fellowship for graduate study, while the National Endowment for the Humanities offered a summer stipend. The Smithsonian Institution's National Museum of American History contributed a one-year doctoral fellowship to support my work, for which I would like to especially thank Arthur Molella. I must also thank the Hoover Library and the Center for the History of Physics in the American Institute of Physics, each of which provided a grant allowing me to visit and work in their collections. Finally, financial support from Iowa State University afforded me the much-needed opportunity to pursue the final stages of writing.

On the personal level, I owe more than I can express to many, many others. The family, especially my mother, Mary Bix, has been a rock-solid source of support, and this book is really an offering of love to them. Robert Bix and Peggy Kahn acted as sympathetic comrades in academia, while Jonathan Bix always made life a little brighter. Taner Edis pulled me out of situations of computer crisis and supported my writing with patience and companionship. Mostly, this book is a tribute to my father, Ira Bix, and my oldest brother, Michael Bix, both of whose lives represented the best possible inspiration. For their models of intellectual strength and personal integrity, I owe them more thanks than one book can possibly convey.

Many people have contributed important advice throughout the course of my work, and I give them all my deepest thanks. Any errors of fact, interpretation, or writing, of course, remain entirely my responsibility.

INVENTING OURSELVES OUT OF JOBS?

Prologue

Technology as Progress?

WHEN PRESENT-DAY OBSERVERS look for historical references to technological unemployment, Luddism, the protest movement British workers mounted during the early nineteenth century's Industrial Revolution, instantly comes to mind. In fact, the term *Luddite* has become synonymous for a person who destroys, fears, or just dislikes new machinery. Many people in the United States might be surprised to find another episode of controversy over the social dimensions of mechanization much closer to home. Just sixty years ago, Americans wrestled with fears that advances in workplace technology would exact too heavy a toll on labor. The revolutionary potential of new production equipment both impressed and alarmed observers, who feared that the sheer speed of change might cause social and economic catastrophe.

During the Depression decade, 1930 to 1940, many citizens worried that stubbornly high unemployment rates signified a deep imbalance in the Machine Age system. Development of such grave misgivings brought the United States to a watershed, a time when national emergency led men and women to reconsider some fundamental assumptions about technology and progress. Soaring prosperity back in the 1920s had allowed optimists to argue that the modern world approached unprecedented glory. Miracles of scientific discovery and engineering achievement would create virtually unlimited wealth, raising life in America to the peak of civilization. The great crash of 1929 smashed such hopes and shook people's trust in the soundness of a technocentric economy.

As unemployment skyrocketed, approaching 25 percent, workers displayed a growing concern that mechanization eliminated jobs in factories, agriculture, and the service sector. Installation of dial telephone systems threatened to reduce the need for switchboard operators, while steelmakers' conversion to continuous-strip production made that industry less labor-intensive. Rail-

Fig. 1. "Sign of the Time." If one cartoon can capture the essence of the Depression's technological unemployment debate, this undoubtedly does. While the machines actually entering factories, agriculture, and stores during the 1930s in no way resembled a humanoid form, the emotional and symbolic connotations of the robot made it an ideal icon for workers' uncertainty. Illustration by Jerger, *The Unemployed*, no. 4 (1931): 18.

road unions blamed their members' job losses partly on the introduction of more powerful locomotives, and coal miners feared that the development of new loading equipment jeopardized their interests. Manufacturers realized that machines could take over the tasks of filling cereal boxes, rolling cigarettes, or even twisting pretzels. In automaking and elsewhere, technology seemed to substitute for people's skill and experience; photoelectric eyes promised to operate more efficiently and accurately than the best human inspectors. Economists reported that the spread of tractors and other farm machinery had disrupted rural labor, and sociologists predicted that the invention of a viable cotton picker would devastate Southern populations. Statisticians collected data showing that even musicians had felt the pinch, since, in the rush to offer "talking pictures," movie theaters were firing the orchestras that had provided accompaniment for silent films.

Government officials supplied evidence seeming to suggest that such incidents could not be dismissed as minor aberrations. Cases of "technological unemployment" had multiplied over recent years, as employers' quest for mechanization continued to raise the rate of labor displacement. Men could not retrain for new jobs overnight, researchers warned, and, in any case, business had not increased fast enough to reabsorb ousted workers. On numerous occasions, President Franklin Delano Roosevelt voiced concern that technological innovation would cause a chronic imbalance in the nation's labor market. Changes in production might simply outrace society's ability to keep up, making high rates of joblessness a permanent phenomenon.

The issue of technological unemployment continued to draw attention throughout the 1930s. Economists, sociologists, statisticians, political scientists, and other academics all analyzed different angles, filling professional journals and conference programs with the results of their studies. Congressmen organized hearings on the subject of mechanization and introduced various proposals for legislative action. Newspapers and magazines kept the topic in the public eye, as reporters detailed the latest innovations in production. At one union convention after another, labor representatives delivered fiery speeches condemning displacement and debated various strategies to address the danger. Talk of "machines versus men" pervaded popular culture, becoming a dominant theme in literature, radio programs, and movies. Authors of science fiction plotted doomsday scenarios in which mechanization made the human species obsolete. Writers of children's books and school texts adapted the subject for younger sensitivities. Comedians turned the problem of displacement into bitter humor, while illustrators and cartoonists created a new visual vocabulary to symbolize the social cost of workplace change.

Leaders of the nation's corporate, scientific, and technical communities viewed such events with deep dismay. Those professionals had a lot invested in Machine Age development; from their perspective, talk of displacement seemed an ungrateful and unpatriotic attempt to make inventors a scapegoat for disaster. Quickly moving to defend themselves, prominent business executives argued that no one had really proved any link between mechanization and labor elimination. Classical economic theory demonstrated that innovation should cause nothing more than temporary and local job loss, while stimulating genuine economic progress and opening new work opportunities in the long run. Advocates mounted public relations campaigns insisting that, far from undermining well-being, research and development clearly advanced civilization by expanding production. Depression-era workers worried that the modern economy had spun out of control, but to men of business, science, and engineering, new workplace equipment still offered the ultimate ideal of efficiency and power.

Questions about the human effect of workplace change struck at the heart of a conversation about how technology fit into modern life. Mounting levels of personal distress imbued talk of displacement with a sudden emotional urgency. In weighing the implications of mechanization, people struggled with different ways of interpreting their country's past, understanding its present, and envisioning its future. Especially since the mid-nineteenth century, Americans had come to prize technology as a symbol of national progress. The connection of the Central Pacific and Union Pacific lines at Promontory Point, Utah, in 1869 realized the dream that railroads would conquer the West. By taming nature, technology had broken a new frontier and paved the way for settlement, commerce, and civilization. Seven years later, Philadelphia's Centennial Exposition elicited equally great pride, as crowds clustered around the enormous Corliss steam engine driving rows of exhibits in Machinery Hall. The 1883 dedication of the Brooklyn Bridge marked a triumph of engineering, while electrification literally changed the appearance of everyday life. Cities from New York to Muncie, Indiana, competed to light up shopping districts, while spectacular displays of electricity occasioned exuberant flourishes of rhetoric. The first decades of the twentieth century witnessed the development and popularization of automobiles, airplanes, radio, and new types of home appliances. New industries grew out of each invention, and assembly line reorganization of work allowed manufacturers to turn out more goods faster than ever. Such advances made it tempting to believe that modern science and technology provided the key to continual prosperity.

The economic crunch of the thirties forced men and women to reevaluate assumptions about the historical meaning of technological change. Stories began circulating that some Roman emperor or England's Queen Elizabeth I had moved to protect workers by limiting the pace of mechanization. Depression-era fears encouraged social scientists to revisit the history of the English Luddite revolt and to reassess the economic theories of David Ricardo, John Stuart Mill, and other authorities. Many concluded that the twentieth century represented a break with precedent, a time when threatening dimensions of technical advance had come to the fore. Scientists and engineers had accelerated research and invention to an unbelievable pace, conjuring up machines capable of taking over more and more human roles. Perhaps the relatively slow pace of life in previous centuries had given people a chance to keep up, but the breakneck speed of modern innovation now appeared to make that impossible. In the 1920s, mechanizing production had seemed to guarantee prosperity; during the 1930s, people feared that changing workplace technology might become America's social and economic downfall.

Since the 1970s, historians and sociologists have devoted substantial attention to the subject of how technological changes have affected labor relations in coal mining, longshoring, secretarial work, and other specific occupations. A number of academics have suggested that many businessmen came to value mechanization not only as a way of increasing output but also as a method of exerting more control over the workforce. By incorporating aspects of production tasks into equipment, managers could reduce the value of workers' skill and experience. Such studies have yielded new insight into how employers have, over time, used technology to reshape the labor process in their own interest.[1]

To date, however, scholars have largely overlooked the importance of the twentieth century's broader debate over technological unemployment as a social and economic flashpoint. The subject tends to fall into the cracks between disciplines. By concentrating so hard on exploring the nature of *machine development*, historians of technology often end up relegating the labor-related *consequences of machine introduction* to secondary importance. Sociologists focus on dimensions of control and conflict *on the job*, while neglecting the issue of how changing technology complicated prospects of *keeping the job*. To the extent that general labor histories mention displacement, they usually treat it only as a side issue in fights over working conditions and the length of the workweek. Such approaches wind up overlooking one simple fact, that twentieth-century economic uncertainty made the subject of mechanization and jobs into a vital question in its own right. A few historians have examined cases of how specific devices, such as cigar-rolling machines and coal loaders, affected particular groups of American workers. While valuable in their detail, such assessments risk ignoring the whole cultural context of such disputes. The technological unemployment issue became vital to Depression-era thought precisely because concern stretched across occupational lines to become a nationwide controversy.[2]

Throughout the thirties, Americans found it difficult to establish any definitive numbers on how many jobs had been created by mechanization and how many had been lost. Such economic problems proved inherently complicated. Many technological changes might occur simultaneously, having varied effects on different segments of the workforce. Industries and occupations came under stress for many reasons, making it difficult to isolate the impact of job loss resulting specifically from the introduction of new equipment. In the first years of the Depression, the government's inadequate mechanisms for collecting economic data forced officials to admit they could only guess at the numbers on displacement. Even by the late thirties, after the Roosevelt administration had organized special efforts to gather statistics, the exact relationship between mechanization and employment remained a matter for fundamental dispute. Each side of the controversy, both those expressing and those dismissing concern, could cite evidence designed to bolster its argument. Patterns of production and labor movement appeared quite different, depending on the particular industries, parameters, or time periods selected for comparison. Mechanization aroused substantial alarm, but it still represented only one among many dimensions of uncertainty during the Depression. Bank failures, international monetary chaos, cyclical unemployment, and many other sources of economic complexity each caused dismay.

At bottom, the discussion of technological unemployment drew so much attention and remained so controversial precisely because it was more than just a quarrel over economic statistics. The debate over displacement represented Americans' agonized attempt to come to terms with twentieth-century stress. It underlined the search to understand how the modern workplace had changed, the feeling that new production technology had altered the entire balance of economic and social life. It raised difficult questions of policy and ethics. Did modern society bear a moral or practical obligation to find some way of helping those who had been hurt, even temporarily, by the adoption of new equipment? Could displaced men and women be retrained for new opportunities, and, if so, who should underwrite the cost of such an operation—the federal government, individual states, specific companies, the business community as a whole, or labor unions? In which direction was the country heading—toward a future ripe for industrial expansion and the promise of new consumer wonders or toward an ongoing Machine Age employment crisis?

Given the inherent difficulties of determining exactly how every form of workplace change affected employment patterns, this book does not attempt to add up numbers to resolve the still tricky question of how much technological unemployment *really* existed in the 1930s. Rather, it details how the complexity of technological economics stimulated an intense, ongoing dialogue about the social implications of modernization. Impressions matter; people acted and reacted based on their *perceptions* of mechanization. This work follows the social course of that debate and tells the intellectual history of an idea, the notion that Americans must come to grips with the human consequences of introducing increasingly powerful machines into the workplace.

It seems safe to assume that concern about displacement was not merely a mass delusion. Intelligent observers, from presidents and respected economists to ordinary working Americans, presented good evidence that the adoption of new equipment had created real problems, at least in some occupational sectors. Even the strongest proponents of mechanization might admit that change resulted in *some* displacement; the disagreement came over its severity and significance. Economic theory treated job loss and job creation as impersonal events, removing the debate to a level of abstraction. In real life, those Americans immediately affected could not simply assume that, under a free market system, everything would turn out for the best. The Depression did not permit people the luxury to trust in reassuring words about long-term expansion, since even a brief period without work could exact a deep financial and emotional toll on vulnerable families.

It is worth noting that Depression-era episodes of fear about machines and job loss were not limited to the United States. Theoretical study of the relationship between technological innovation and employment stemmed from deep roots in European classical economics, and American social scientists continually referred back to the work of Jean-Baptiste Say, Jean Charles Sismondi, David Ricardo, and John Stuart Mill. Still more important, European nations, especially Britain and Germany, joined in agonizing over the evidence of rising labor displacement. Laws passed in Danzig in 1933 required factory owners to obtain government permission before adopting new machines. In subsequent months, the Nazi Party banned engineering rationalization procedures that reduced employment, declaring, "Never again must the worker be replaced by a machine." Concern ran along parallel lines in different places, and yet the United States managed to make the issue into something uniquely American. Discussion about workplace change became entwined with particular musings about the meaning of American history, the western frontier, and a sense of national destiny. For that reason, I concentrate in this book on the history of technological unemployment talk in the United States.[3]

The criticism of workplace mechanization during the Depression drew men and women into philosophical debates about the balance between work and leisure, the relationship between production and consumption, and the relative importance of material possessions in life. Such arguments offered a wide scope for thought, raising issues that could not be resolved easily. World War II cut off the discussion of such topics by restoring employment, but the restoration of peace brought back questions about labor displacement with renewed force. Postwar industry and engineering efforts articulated an ambitious ideal of the automatic factory and the automatic office, a vision of substituting machines for even more human functions. Powerful currents of concern have continued to reappear, as economic cycles tested Americans' attitudes toward workplace change. Through the 1990s, the country kept returning to the unsettling issue of technological unemployment, as concerned observers reflected, often passionately, on the meaning of "progress" in the twentieth century.

1

"Economy of a Madhouse"

Entering the Depression-Era Debate over Technological Unemployment

THE SIGNIFICANCE OF THE CONCERN during the Depression about machines and jobs lay in that era's contrast to the technologically secure 1920s. The United States moved almost overnight from a nation confident of Machine Age prosperity to a country considering the possibility of complete economic collapse. Americans who had gloried in the wonders of modernization suddenly began to ask, Had the innovations that seemed to guarantee progress actually hastened a future of devastating job loss? Technological unemployment did not represent an entirely novel question. For more than a century, Western economists had been thinking about the interaction of technical change, wealth, and work. The Great Depression, however, raised such questions to unprecedented urgency. It took the matter beyond the abstract theorizing of economic textbooks, raising it into an urgent public awareness. The displacement issue underlined how much the rapid evolution of production machinery had come to determine prospects for both individuals and the nation.

1920s Confidence in Technology as Prosperity and Progress

In the 1920s, many Americans could feel confident in linking technological change to progress. Material improvements in everyday life seemed to strengthen the promise of prosperity. Automobiles especially seemed to validate faith in technology as an all-powerful force moving society toward wealth and triumph. Since 1908, Henry Ford's Model T had offered quality engineering and solid construction at a reasonable price, while publicity raised his mass production system to almost mythical status. By the early 1920s, the company was producing 1.5 million vehicles per year at increasingly affordable prices; though the first runabout had sold for more than eight hundred dollars, the cost of a Ford eventually fell to less than three hundred. Those who wanted a car with more status and style than the utilitarian Model T could turn to General Motors, where Alfred Sloan introduced the concept of the annual model change and targeted each product line to consumers of different means.[1]

By the mid-1920s, such marketing successes had turned automaking into the nation's leading industry measured in terms of product value. Car manufacturing generated demand for iron, steel, rubber, and petroleum, while creating a wealth of opportunities for garages and other businesses serving the driver. Automobiles literally changed the country's landscape, reshaping the urban infrastructure and redefining shopping patterns, recreational habits, and other facets of daily life. In 1929, one resident of "Middletown" (Muncie, Indiana), asked sociologists Robert and Helen Lynd, "Why on earth do you need to study what's changing this country? I can tell you what's happening in just four letters: A-U-T-O!" When even working-class families could afford inexpensive or secondhand cars, the automobile came to embody the link between technology and prosperity.[2]

Other technological developments reinforced an impression that modern innovation heralded unparalleled national progress. As radio ownership spread to more homes, broadcasters delivered entertainment, sports, politics, religious programs, and advertising to a growing audience. Improvements in airplane design and the expansion of commercial aviation stimulated intense excitement, capped in 1927 by the heroic triumph of Charles Lindbergh's flight from New York to Paris in the *Spirit of St. Louis.* With the development of artificial materials such as rayon, marketers sold consumers on the attraction of novelty. Every year, manufacturers offered a wider selection of electrical appliances, which promised to bring luxury and modernity into the ordinary American household.[3]

Such innovations seemed to exemplify a new, systematic approach to research and development as part of the capitalist system. After General Electric opened an in-house laboratory in 1901, dozens of other companies, such as AT&T, DuPont, and General Motors, rushed to set up similar facilities. Firms recruited scientists and engineers, hoping that their professional skills would provide the employer with a competitive advantage. Researchers could improve old products and devise new ones, paving the way for business expansion. Inventions would become routine products of institutionalized research, rather than await serendipitous discovery by the stereotypical lone genius. Science, under the efficient jurisdiction of business management, promised to regularize progress.[4]

The modern world had entered a golden age of technological advance, optimists declared. The development of electricity alone had added enough prosperity to "stagger the imagination," the Commissioner of Patents told Congress. Civilization advanced primarily through "the exercise of invention," he continued, and "in the race between nations, that country which fosters most the inventive spirit is bound to win." Claiming a healthy lead in that contest, the U.S. Patent Office mounted an exhibition that portrayed six out of the ten greatest inventions from the last twenty-five years as entirely or partly American. Some boosters linked that success to the patent law itself, asserting that the U.S. government had set up a superior system for encouraging technological change. Others said that inventiveness had evolved as part of American history, that the decades spent in conquering a rugged frontier had cultivated people's energy and ingenuity. Popular accounts built Thomas Edison into the ultimate example of this drive for invention, as biographers praised his relish for hard work and his joy in meeting a challenge. According to one writer, after European scientists had supposedly declared it impossible to develop a practical electric lighting system, Edison succeeded in "a triumph of sheer will." The image-makers portrayed that undefeatable spirit as part of the national character, suggesting that all Americans could share credit for Edison's discoveries.[5]

The mythologized treatments of Edison's life during the 1920s symbolized the notion of inventive genius as a historic force. Americans prided themselves on a sophisticated faith in technology; while nineteenth-century observers had ridiculed an early locomotive for losing a race to a horse, their descendants had seen that clunky train evolve into speedy perfection. Under the title "There's Magic in the Air," a 1928 *Collier's* article indicated that Americans had come to expect miracles. "Today you can't spring anything too wild for people to believe," the author declared. "If an inventor says he has a skyrocket that will carry passengers to the moon and return, there will be a waiting line for the first ride. You don't have to be an inventor to know we're on the threshold of an age of wonders made ordinary, which will bring health, wealth and happiness to each of us."[6]

Edison himself endorsed the assumption that inventive minds would conquer all technical obstacles and solve nature's mysteries, generating a continued flow of new wonders. Though inventors had struggled with various difficulties in trying to develop a practical helicopter, Edison told *Collier's* he had no doubt they could solve those problems. Previous experience had shown that when the "need for some new thing becomes insistent, men always find it." For Edison, necessity called forth invention, following the simple law of supply and demand. Streets and railroads had become overcrowded; inevitably,

America must devise the helicopter. History seemed to define a preset path of technological advance. Science writer Waldemar Kaempffert declared that a "symphony, a poem, an invention is the product not only of a man but of its time and of the race—a product of what the Germans call *Stimmung*, a certain atmosphere" of social and psychological conditions.[7]

Optimists felt justified in expecting a brilliant future. The pace of innovation appeared to keep increasing; the number of patents granted per year had risen almost to fifty thousand, with no apparent limit in sight. Evidence suggested impressive prosperity; America's gross national product grew 39 percent through the 1920s, while manufacturing output virtually doubled. Easily available credit financing for radios, automobiles, and other items helped even families of moderate means share in this culture of material ownership. The increasing mechanization of production and improvements in industrial efficiency seemed to assure abundance. Such a world celebrated change as synonymous with progress; a November 1929 *Scientific American* article (appearing just after the great crash) declared, "Commodities become antiquated quicker today than ever before; the new is always replacing the old. Somewhere, someone has a better idea."[8]

Underneath such boasting, the 1920s had its share of economic and social trouble. Labor tensions led to bitter strikes, while declining crop prices and overproduction plagued farm communities. Though higher wages allowed many employees to profit from industrial success, prosperity also seemed to widen the gulf between the upper class and working class. Nonetheless, when Republican nominee Herbert Hoover promised in 1928 that Americans would "soon be in sight of the day when poverty will be banished from this nation," his words seemed to herald a future in which national well-being justified faith in technological progress.[9]

Even in the prosperous 1920s, some people questioned the assumption that innovation proved universally beneficial, looking specifically at how new forms of mechanization affected workers. George Barnett, a statistics professor at the Johns Hopkins University, examined how technological changes in production had affected employment in printing, bottlemaking, and stonecutting. Between 1895 and 1915, he wrote, adoption of increasingly fast stone-planing machines, along with pneumatic tools and other new devices, had displaced about 50 percent of the nation's stonecutters. Bottlemaking companies had switched from semiautomatic machines to fully automatic technology after 1905, with devastating effects on labor; one operator using the Owens bottle machine could reportedly equal the production of eighteen traditional glassblowers. Unions had occasionally tried to impose labor rules that restricted the introduction of new technology, but Barnett warned that such attempts at minimizing displacement tended to fail or even backfire. His book left a fatalistic impression that workers had little recourse when competing with powerful new machines; the stonecutters and glassblowers had lost their battle. While Barnett detailed the pain and frustration experienced by displaced workers, he nonetheless concluded that mechanization had added to overall wealth, bringing long-term benefits to labor as a whole. He interpreted technological unemployment as a phenomenon affecting only workers in isolated industries. Such pain seemed an unfortunate but finite price for the country to pay; an unlucky few had to suffer in the course of progress. His 1926 book gave no indication that, as technological changes in production spread to other occupations, more American workers might face job loss.[10]

Economic writer and consumer advocate Stuart Chase did extend talk about previous episodes of labor displacement to warn of potential trouble ahead. His book *Prosperity: Fact or Myth*, completed just before the stock market debacle in October 1929, cautioned that structural economic weaknesses lurked beneath recent growth. Chase worried especially that technological changes had started coming so rapidly as to pose a serious threat. Mechanization had already begun displacing factory workers, railroad men, and miners faster than these laborers could find new jobs, Chase warned, and, since 1920, the country's net technological unemployment had risen by 650,000. Until the United States corrected that dangerous trend, he concluded, it could not be considered truly prosperous; an accelerating problem of displacement could undermine the entire nation's economic prospects.[11]

Appealing directly to labor forces, the Industrial Workers of the World sounded a more apocalyptic note in the 1925 booklet, *Unemployment and the Machine*. Author J. A. MacDonald warned that since technological improvements in production tended to decrease the amount of labor needed for a set output, mechanization promised to make factory owners rich by throwing men off the payroll. To illustrate this "paradox of the machine," MacDonald noted that the average number of workers employed in iron and steel blast furnaces had declined 2.1 percent between 1900 and 1910, even as horsepower per worker rose 42.2 percent. Given that managers always felt pressure to produce more with fewer employees, businesses would soon turn to hydroelectric or even nuclear energy, he warned. That new power would only intensify displacement, and unless labor organization succeeded in winning a six-hour workday, there would be no "limits to the unemployment of the future." Pointed graphics on the booklet cover showed huge mechanical hands pushing humans off a cliff into an "unemployment dump." An illustration inside depicted a huge industrial complex with only two humans visible; the caption read, "The modern creed: 'Plenty of machines and very few men.'" MacDonald's overwrought prose prophesied days when displacement would drive women into prostitution, men into crime and drug use.[12]

Other segments of American labor spoke about mechanization in a less melodramatic but still grave fashion. In the months preceding the crash, union representatives explicitly voiced their concern to President Herbert Hoover. In June 1929, the Iowa State Federation of Labor sent a telegram requesting the White House to convene a special economic conference "to forestall what we foresee to be a disastrous development threatening our entire economic structure." The group especially urged Hoover to think about providing relief to the "growing army of the unemployed due to the encroachments of the machine age." The president of the Iowa Federation, J. C. Lewis, insisted that the group did not oppose scientific advances and labor-saving technology per se. Rather than trying to "hamper or impede" technological change, he indicated, the union just wanted some protection.[13]

Through the summer of 1929, labor conventions in other states followed the Iowa Federation in stressing that rapid mechanization might have unfortunate social consequences. Frank Weber, general secretary of Milwaukee's Federated Trades Council, told Hoover that in the future, private business would not be able to absorb the ranks of workers displaced by technological changes in production. Representatives of particular occupations wrote to the president describing the difficulties mechanization was causing among their ranks. One member of the International Brotherhood of Paper Makers reported that, whereas paper mills had formerly employed four or more men per ton of newsprint manufactured, bigger and faster machinery allowed newer plants to turn out over one ton of paper for every worker. That declining labor ratio, the union complained, confronted its members with an "almost insurmountable" problem. While labor spokesmen judged the potential impact of new technology on jobs sufficiently serious to require attention at the highest levels of government, letters written before the crash still tended to refer to widespread displacement as a problem that might materialize at some unspecified date. The Iowa Federation spoke of being able to "foresee" trouble that the United States could yet "forestall." The papermakers' complaints suggested machines were already causing stress for workers in certain occupations, but prosperity encouraged confidence that the nation might yet avoid mass displacement.[14]

Within that optimistic context, critics derided expressions of concern about job loss as completely unfounded. After the 1928 American Federation of Labor convention, at which president William Green had described machine-related displacement as among "the most important problems affecting labor today," newspaper editorials condemned the union for attacking the very devices that had given workers higher living standards and exciting new consumer goods. The *Omaha World-Herald* acknowledged that mechanization meant "violent upsets are temporarily occasioned in particular fields" but maintained that "as soon as the industrial adjustments are made, everybody is better off than before," including the affected workers. When Green cautioned that rapid mechanization might produce a "human scrap-heap," he just appeared ungrateful for capitalism's gains, unwilling to acknowledge how much technology had contributed to modern well-being. The issue of unemployment simply did not appear urgent as long as most Americans were still caught up in the dream of prosperity. Even the national AFL journal had to admit in July 1929 that, despite "persistent reports" of new production equipment eliminating jobs, the United States still appeared economically strong overall.[15]

The 1929 government report *Recent Economic Changes in the United States* reinforced this sense of security with an assessment that officially tied technological change to prosperity. Thanks to recent scientific and engineering advances, the analysis declared, the national economy had enjoyed "striking" gains in man-hour productivity, which in turn sparked "an accelerated cycle of production-consumption." That speeding-up of economic activity had recently roused some concern about displacement, the report noted, since "unemployment can arise as a result of industrial efficiency as well as of inefficiency." In the end, the committee judged that mechanization caused only "temporary hardship" for labor. The strong economy of the 1920s had driven up demand for new products and services, Dexter Kimball argued, so most workers forced out of shrinking fields could find new employment in auto repair, white-collar work, movie theaters, and radio. Without the creation of such new job opportunities, he acknowledged, the recent pace of displacement could have generated "critical unemployment," but as long as economic prospects remained favorable, mechanization should cause no alarm.[16]

Even temporary unemployment could mean distress for millions of families through no fault of their own, the authors of *Recent Economic Changes in the United States* observed, but such pain would be worthwhile as long as it assured the nation at large of "real and permanent" progress. Old jobs inevitably disappeared as technology evolved, but history had shown that such events would not prove catastrophic. Contrary to Luddite predictions, textile machines had not destroyed British society; the United States had not experienced epidemic joblessness as railroads replaced horse-drawn carriages. Short-term labor elimination invariably appeared as a "constant accompaniment of progress in modern industry," explained economist Wesley Mitchell, but governments had safely "left the remedy to natural forces." Americans could trust that innovation ultimately generated enough economic growth to absorb displaced labor, said the *Recent Economic Changes* report, preventing any serious consequences.[17]

Onset of the Depression

As long as people continued to associate technology with prosperity, warnings about machine-related unemployment could be ignored or dismissed as the paranoia of scaremongers too stubborn to appreciate America's good fortune. Positive economic indicators of the 1920s seemed to herald the unlimited triumph of Machine Age expansion, firmly established on a foundation of scientific and engineering advance. It was tempting to assume that the United States possessed the blueprint for continual economic and social progress, plans centering on machines for production, machines for consumption, and machines for making jobs. Once the nation's favorable economic conditions turned into national catastrophe, such confidence could no longer be maintained so readily, giving sudden credibility to concerns about vanishing work.

Signs of economic downturn had in fact been apparent since at least early 1928, when slowdowns in consumer spending and construction led to cuts in industrial production and subsequent layoffs. With the October 1929 stock market crash, economic decline accelerated into catastrophe. By 1933, American manufacturing had fallen 54 percent from its previous peak, mocking the business and engineering ideal of efficiency. Unemployment soared from under 500,000 in 1929 to approximately 15 million by early 1933. Desperately seeking to explain this national catastrophe, many Americans seized on technological change as a credible focus for blame. This change in attitudes did not take long. Starting in 1930, as unemployment climbed toward 25 percent, economists, social scientists, and other professionals directed new attention to the question of how machines affected jobs. Newspaper stories and editorials put technological unemployment into the headlines, both reflecting and stimulating public concern. Anecdotes of humans pushed out of work by machines no longer seemed like isolated incidents.

While such tales carried a certain power, the issue remained clouded with uncertainty. In the early months of the Depression, no one knew exactly how many workers had been displaced by mechanization, in part because of the absence of reliable data on overall unemployment. Criticizing the inadequate estimates of joblessness, which ranged anywhere from 2.3 million to 6.6 million people, the *New York Times* in 1930 accused Washington of neglecting its duty to provide comprehensive analysis of the country's situation. Frustrated officials acknowledged the insufficiency of economic monitoring mechanisms; Commerce Department staff member E. Dana Durand flatly declared it "impossible to make any approximately correct estimate" of unemployment "at this or any other time." Instead, for most of 1930, government experts attempted to provide a rough picture by patching together partial assessments of joblessness. Washington hurriedly arranged a special nineteen-city census, but this piecemeal approach provided little information about unemployment outside urban areas or among professionals, self-employed individuals, or those in service occupations. Inadequate knowledge about general unemployment only intensified problems of trying to figure out how much displacement machines had caused. The difficulty of resolving these basic questions soon became apparent, complicating the new Depression-era discussion.[18]

Analysts encountered real problems in attempting to untangle an apparently straightforward question of economic statistics. Even within a single industry or occupation, many technological changes occurred simultaneously: automobile manufacturers substituted spot-welding for hand riveting, automatic enameling machines for hand dipping, forging and molding machines for drop forging and hand molding, photoelectric machines for human inspectors. Similarly, railroad observers pointed to a whole series of inventions that decreased manpower needs. Thanks to automatic loading machines, paint sprayers, and electric welding equipment, train companies needed fewer employees to handle basic operations. Heavier, chemically treated rails lessened duties for maintenance and replacement crews, while electric track circuits allowed railroads to replace flagmen and manually operated gates with automatic crossings. Diesel engines eliminated jobs for firemen, and the most powerful locomotives could pull trains up to two miles long, letting companies save labor by running fewer trains with more cars. Rail union statisticians applied Interstate Commerce Commission figures to calculate that such technical innovations had displaced over 28,000 rail workers between 1923 and 1929, including 10,994 trainmen, 6,507 firemen, and 4,425 engineers. A. F. Whitney, president of the Brotherhood of Railroad Trainmen, argued in 1932 that by rushing to adopt new technologies, railroad owners had "short-sightedly sacrificed men on the altar of machine economy."[19]

Just looking at one company or even one industry could not fully answer questions about technological unemployment; multiple economic, social, and technical factors entered into the equation. Defenders argued that rather than closing down opportunities, changes in production multiplied them. As mechanization increased the efficiency of automobile manufacture or railroad operation, for instance, the resulting reduction in operating costs should translate into higher wages and lower consumer prices. Classical economics predicted that rising demand would then spur new economic activity, turning into employment growth. Moreover, even if new inventions eliminated jobs for auto-parts inspectors and railroad firemen, the national course of scientific and technical progress would create new industries and open up new work.

Such arguments contained plenty of vagueness and led to much controversy. Would reality reflect classic economic theory, or had the modern Machine Age changed the rules of the game? Did a rise in efficiency really trickle down to workers and consumers, or were ordinary people paying the price for industrial modernization? How much could Americans trust that the future would bring wondrous gains in science and engineering, yielding new material goods to create both new jobs and higher standards of living? Meanwhile, what happened to men and women displaced from older lines of work? Even if economic growth did create new jobs in different locations or other industries, even if that work offered pay and status similar to eliminated jobs, how could workers cope with the inevitable difficulties—the delay between firing and reemployment, the need for new education and retraining, the problems of moving families to new areas? Exactly how many individuals did change displace anyway? Was it a minor fragment of the overall workforce, which should present no serious national problem, or a higher number that would only increase as mechanization intensified? To answer, experts had to analyze not only economic trends but also psychological and sociological questions about shifting economic sectors, industrial growth, labor costs, consumer activity, capital investment, scientific and engineering advance, and workers' education, training, and personal behavior.

Cases of Concern: Continuous-Strip Steel Mills and Dial Telephones

In the effort to understand how mechanization affected jobs, Depression-era observers began paying attention to specific cases such as steelmaking, where industrialists had begun to adopt new continuous-strip mill systems. Under old-style production, individual workers used hand tongs to lift, stack, and move fiery hot metal slabs (weighing up to 150 lbs. each) out of open-hearth furnaces to cool on floor plates before reheating them. Men then pushed, pulled, and guided slabs through roughing and finishing machines, whose steam-powered rollers compressed the metal into thin sheets; the process required workers to handle each piece fifty times or more. In 1922, the American Rolling Mill Company constructed a plant in Ashland, Kentucky, which radically changed steelmaking by eliminating most metal-handling. While monitoring instruments and gauges, supervisors controlled conveyor belts running hot slabs continuously (at one thousand feet per minute or faster) through a series of six, eight, or ten rolling stands, producing strips thirty-six inches or more in width. Displaying his admiration for the new process, Westinghouse Electric president F. A. Merrick described the operator's station as "a sort of pulpit," the center from which one man "controls every stage from the white-hot ingot to the finished structural shape by merely moving a few levers."[20]

Construction of continuous-strip mills accelerated during the late 1920s and continued even into the Depression. Between 1924 and 1938, American steel companies built twenty-seven such facilities with increasingly impressive capabilities; such plants could produce steel strips up to ninety-one inches across. The industry prided itself on yearly increases in the amount of steel manufactured by continuous-strip methods, from 181,000 net tons in 1925 to 10 million tons in 1937, doing away with the need for hundreds of workers per ton. An assistant editor of *Steel* touted mechanization with an article headlined, "Continuous Mills Voracious in Cost, but How They Produce!"[21] The switch from hand mill to continuous-strip production sparked debate about the economic and social consequences of such innovation; as a major component of the nation's manufacturing sector, the steel industry represented a highly visible test case for the issue of labor displacement.

The spread of continuous-strip mills roused strong feelings among both those asserting and those denying existence of serious technological unemployment. Each argument was heard during the 1939–40 hearings of the Temporary National Economic Committee (TNEC), in which Congress undertook a full assessment of the "concentration of economic power" in the modern production and distribution system. That broad mandate generated hearings that encompassed a number of topics, specifically including the role of "technology in our economy." In spring 1940, TNEC members began hearing testimony from economists, social scientists, government experts, businessmen, and labor representatives on the technological unemployment question. Public hearings focused on major industries, services, and agriculture, as the committee analyzed recent technological changes, employment patterns, and economic conditions. On April 11, the first of two full days devoted to discussing steelmaking, American Rolling Mill Company president Charles Hook told Congress that continuous-strip mills had not caused major displacement or distress for workers. On April 12, Philip Murray, chair of the CIO's Steel Workers Organizing Committee (SWOC), argued precisely the opposite. Both agreed that the old-style hand-rolling mills were doomed, given the drastic way continuous-sheet rolling equipment had altered steelmaking, but they concurred on little else.

Opening his testimony, Hook promised to prove that, while new technology had ended the most back-breaking part of steelmaking, it had not reduced the total number of jobs. Though mechanization did decrease labor requirements per unit of output, he said, the changes created enough new demand for steel to make up the difference. Continuous-strip mills turned out more tonnage in wider strips than could older methods of production, while the new precision in roller design and driving motors improved fine accuracy of gauge. Those innovations raised the quality of the metal and made it more affordable, converting steel into a more attractive product. In the two decades leading up to the 1930s, automobile manufacturers had roughly doubled the amount of steel they used, substituting metal for wood in car bodies and also using it to replace the cloth top. To drive home the message of more, higher quality steel at lower prices, Hook held up an early Model T fender, showing how its rough "cornmeal" surface and light weight contrasted to a smooth, solid, new Buick fender. Appliance manufacture represented another important new market for steel; since the mid-1920s, the price of electric refrigerators had dropped 58 percent and production had risen 1,027 percent. As his industry sold more metal, Hook explained, it needed extra help in finishing, processing, shipping, management, and sales, not to mention expanding operations and building new plants. Though the number of old-style hand mill workers had fallen by 27,000 between 1929 and 1937, overall steel employment had risen from 427,000 to 544,000, a 27.4 percent gain.[22]

Hook's presentation led the committee into an involved statistical discussion, as TNEC economic consultant Dewey Anderson challenged the way Hook's employment figures had been calculated. Cutting data off at 1937 distorted the picture, Anderson said, since that year represented an artificially inflated level of steel production. Hook had talked about a net growth of 117,000 new jobs, but choosing 1938 as the standard for comparison yielded very different results. Rather than resolving the issue, numbers expanded the grounds for dispute, as the steel union accused employers of lying with statistics. The SWOC agreed that total industry employment had grown through 1937, before the full effects of automatic strip-mill production appeared. But companies had opened major new facilities in 1938 and 1939, which Hook's numbers, covering data only up to 1937, did not show. Factoring in those two years revealed a loss of 38,470 jobs, Philip Murray said, not to mention another 46,300 positions on the verge of vanishing as the industry scheduled more old hand mills for closure. Murray disputed Hook's idea that higher quality and lower prices stimulated new opportunity. Since "labor is practically eliminated on the hot-strip mills," he argued, increased demand could never create extra jobs to compensate for those lost. Human requirements in steelmaking (measured in man-hours per ton) had dropped 21 percent just in the preceding four years and 36 percent since 1923. Reportedly, it took only six workers in 1939 to manufacture the amount of steel that had required ten people sixteen years earlier.[23]

Labor and management also presented radically different interpretations of how changes in production had affected individual steelworkers. The switch from hand to continuous mills had been sufficiently gradual to cushion workers from shock, Hook assured Congress. His Middletown, Ohio, plant had made special efforts to help, warning employees well ahead of schedule about forthcoming technological changes. The firm had promised to shift displaced men to other positions whenever possible, Hook said, and had paid separation allowances averaging $530 to the 393 men it could not avoid discharging. Murray discounted the suggestion that employers took care of displaced laborers, asserting that in one instance, "3,000 workers were told to go home and never to come back." One industry vice president had allegedly admitted that companies really didn't want to transfer men from older facilities into new continuous-strip mills, saying that "a hand mill worker is used to producing from 5 to 10 tons in 8 hours, and he can't get used to seeing a thousand or more tons produced on a strip mill in the same time."[24]

To validate labor's fear of technological change and personalize the issue of displacement, Murray brought one unemployed steel worker, Michael Russell, to tell the Congressmen his story. He had been a skilled, high-wage roller for thirty-two years, Russell declared, until pressure from new continuous-strip mills led U.S. Steel to close his old hand mill in 1937 and discharge the entire 1,600-man workforce. The company had found him work in another plant, but as a low-wage helper, the job he had started in three decades before. That facility closed after another seven months, Russell testified, and he had received no offer of dismissal wages. Such instances proved that businesses did not always assist displaced workers, Murray said, calling for legislation requiring employers to give displaced men six months' notice, dismissal allowances, vocational advice, and retraining.[25]

In the end, TNEC's members sided more with Murray than with Hook in their report, believing that the adoption of continuous-strip mills ultimately cost jobs. Reports suggested that the latest facilities could turn out as much steel with 130 workers as older-style plants had done with four thousand or more. Moreover, Congressional economists doubted whether labor had received its fair share of benefits from manufacturing innovations. Hourly pay for the average man in iron and steel manufacturing rose just 9.9 percent (yielding a net 20% drop in unit labor costs) between 1923 and 1935, even as technical changes raised man-hour output by 48.2 percent. Unless such trends reversed course, the TNEC report concluded pessimistically, introduction of modern equipment would displace an increasing number of Americans, and "economic and social distress may be expected to accumulate."[26]

The SWOC regarded technological unemployment as a major issue and fought the battle in the arena of public relations. It was not enough to alert Congress to the implications of continuous-strip mills; union representative Harold Ruttenberg wrote popular articles for the *New Republic* and *Harper's* declaring that mechanization had created "85,000 Victims of Progress." Such pieces disparaged the notion that industry employment would inevitably grow as workplace advances reduced prices and increased demand. After steel companies had paid off the enormous capital costs of building strip mills, Ruttenberg argued, they simply poured the savings into higher profits. Tin-plate prices remained the same in 1937 as in 1929, $117.70 per ton, though strip-mill technology had helped reduce labor costs in production from $36 to $14 a ton. Steel companies built new operations, Ruttenberg wrote, and then simply abandoned their old facilities in communities where generations had counted on work in the mill. The SWOC pointed to New Castle, Pennsylvania, as one such steel "ghost town." U.S. Steel had closed three old plants there in the 1930s, permanently displacing 5,700 workers and forcing 64 percent of the local families onto relief.[27] Such problems were not new, but they took on increased weight during a period of national depression.

Steelmaking served as a classic case showing how machines could completely redefine production processes in heavy industry. Yet during the Depression, Americans perceived technological unemployment as a much wider phenomenon, a potential threat to labor in all economic sectors. Even white-collar workers no longer seemed safe; the force of mechanization displaced telephone operators as well as steelworkers. Up to the 1920s, telephone companies had relied on operators in neighborhood offices to field requests for calls and to connect and disconnect lines by manipulating pegs and cords. Switchboard work offered good opportunities to young ladies who had a pleasant voice, some education, and the proper class and ethnic background. In 1932, women accounted for 62 percent of AT&T's total workforce; one female personnel manager reported that the Bell System employed almost 3 percent of all working women in America outside agriculture and domestic service. As long as telephone use in both homes and businesses continued expanding, it seemed likely that many women could continue to find employment in switchboard offices.[28]

Development of the dial system gave telephone companies the ability to handle many operations mechanically. Customers could place calls themselves, and electrically driven banks of connector switches automatically linked the appropriate lines. In late 1919, the Bell System began experimenting with dial mechanisms in Virginia, a decision coming only months after the company had conceded a major victory to labor organizations. New England switchboard operators had walked out in April, demanding increased pay and bargaining recognition. Repair staff and other men from the International Brotherhood of Electrical Workers joined the strike a few days later, disrupting service across five states for five days and forcing managers to agree to settlement. The union interpreted management's subsequent introduction of dial equipment as retaliation, a strategy to avoid any further trouble with the operators.[29]

In 1922, the Bell Company began preparing to convert midtown Manhattan and New Jersey to dials, and from then on, the new system spread rapidly. By 1931, Bell had replaced almost 32 percent of its old phones with dial technology, and, by 1940, the number had risen to 60 percent. Management insisted that because of growing demand, the company had "no way to handle the telephone business without machines." The rising number of calls threatened to overwhelm even the most talented human operators, especially in busy urban locations. Wall Street set new records for telephone use during the heavy stock trading of 1928 and 1929, placing almost 200,000 calls daily and receiving thousands more. Service was expanding nationwide; between 1929 and 1940, the company added 1,400,000 phones to its system, totalling 17,000,000 sets placing 90 million calls each day. Officials declared that they could never have found enough new workers to handle that volume of activity manually. One representative even extrapolated to suggest that without dial systems, AT&T ultimately would have needed to employ every woman in America as a switchboard operator.[30]

AT&T emphasized that even with dialing, central offices still needed operators to handle information requests, customer problems, and long-distance connections. The precise number of operators affected by conversion depended on an area's proportion of routine local calls, which could be dialed direct, to long-distance and toll calls. Mechanization did not completely eliminate human beings; in fact, growing demand for phone service forced the company to add 32,000 operators between 1920 and 1930. But the U.S. Bureau of Labor Statistics estimated that without dial technology, rising phone use in the 1920s would have led Bell to hire an additional 69,421 operators, meaning that dial systems had reduced requirements for switchboard labor by 66 percent. To install the new technology, AT&T had hired 5,415 additional central office repair and maintenance personnel, but during that same period, improvements in wire cabling had eliminated 7,818 line construction jobs. On balance, the bureau concluded, technical changes in phone services had eliminated 71,824 jobs, a 33 percent labor force reduction.[31]

Since the introduction of dialing had the greatest potential effect on AT&T's

female employees, the Women's Bureau of the Department of Labor devoted special attention during the early 1930s to analyzing the impact of mechanization. Researcher Ethel Best reported that within six months of the manual-to-dial change, the telephone office in Worcester, Massachusetts, dropped 285 of its 534 operators. Planning ahead, managers had warned prospective hires about the likelihood of displacement and had entirely frozen new hiring during the half-year prior to the change (instead using operators shifted from other locations and former staff returning temporarily). Most of those laid off were those women on temporary or "occasional" status, and when the company saw that customers needed time to learn dialing techniques, it immediately rehired thirty-nine of these women. Management also succeeded in transferring 131 displaced operators to nearby cities not yet converted to dial operations.[32]

Those apparent efforts to minimize damage impressed the Bureau of Labor Statistics and the Women's Bureau, which applauded the Bell System for putting "science and forethought into its human as well as its technical planning." Those observers largely accepted the company's assertion that it had succeeded in introducing dials without causing any significant harm to labor. In truth, simply warning operators ahead of time about changes did not relieve the pain of being laid off, especially in the middle of a prolonged economic downturn. Women living in remote rural areas, in depressed cities, or in areas dominated by heavy industry had few alternative options for employment. Moreover, even workers hired on a temporary basis found it demoralizing to be replaced by a machine. One admitted that the episode had left her "heartbroken," while another told the Women's Bureau, "New inventions are good for something, but they are not working good for poor people."[33]

The telephone operators' case attracted publicity and controversy throughout the decade. At the 1940 TNEC hearings, AT&T vice president William Harrison testified that the company had limited its introduction of dials to under 20 percent of new shipments to meet its "responsibility" to the "human factor in the business." The conversion of 1,250,000 phones from manual to dial operation between 1934 and 1939 had affected eleven thousand operators, he said, and of that number, 2,500 voluntarily opted to end work. Managers found it "impossible to reassign all the operators displaced," Harrison admitted, but they had successfully transferred five thousand, and the rest had only been hired as temporary transition workers. He declared that AT&T had even spread out part-time work and created make-work jobs to support sixty thousand more employees than the business really needed.[34]

AFL organizer and former operator Rose Sullivan challenged the company's description of events. While managers claimed they needed mechanization to improve service in the face of growing demand, she insisted that dial systems actually proved slower and less accurate than manual switchboards. "All operators know that the dial is highly susceptible to error," Sullivan told Congress. Though AT&T boasted of creating technical marvels, customers who dialed too fast or too slowly might wind up with a wrong number or a false busy signal. Connecting equipment could slip into the wrong groove, and ironically, users then needed to get a human operator to correct the problem. Such difficulties had generated strong consumer resistance to the new technology, she argued. Some elderly people, the visually impaired, and non-English speakers encountered special frustrations in trying to remember numbers and master the unfamiliar procedure of dialing.[35]

Frank Jewett, president of Bell Telephone Laboratories, bragged that "no other structure thus far created by man comes so near to simulating the operation of an intelligent human being." He contended that the new dial system "performs all of the functions originally performed by the telephone operator, and many others besides." Sullivan took issue with that notion, arguing that human operators remained superior in many ways. The phone company's own advertisements had praised women who heroically remained at their posts during the 1934 New England floods, even as dial systems broke down. Skilled veterans could "give preference to the authorities, hospitals, doctors, and the most affected sections" in an emergency, while a "machine doesn't know one call from another." Human operators offered callers psychological reassurance and provided value to the community, Sullivan insisted, things an automatic connection could never provide.[36]

Indeed, during the years when AT&T was converting more and more areas to dial, its advertisements still featured drawings of attractive women seated at old-style plug switchboards. Ads praised the operator for providing "swift, skilled, courteous service," even as managers announced that humans could no longer handle the job. "The alert, friendly voice of the operator is familiar to all who use the telephone," one 1935 ad read. "Truly, the telephone operators have been called 'Weavers of Speech.'" Public relations personnel loved creating flowery, self-promoting labels for its signature operator, famously dubbing her "The Voice with the Smile."[37] Dial technology might have the technical capacity to simulate the operator's job, but it could not offer the same personal appeal.

Though officials claimed they needed technology to offer better service, Sullivan told Congress that the Bell System's real motivation for mechanization had been a desire to weaken the union. When operators threatened to take a stand on wage rates, hours, and other conditions, management had realized the advantages of substituting machines for human labor. Though executives talked about making special efforts to minimize the pain, she declared, the company had really refrained from mass layoffs only out of fear for its reputation. In boasting about how many women they had transferred to other jobs or nonmechanized offices, Bell had "fooled" observers into overlooking how many ended up with less pay or lower status. Despite publicity to the contrary, dial systems had turned telephone operators into "total victims of technology," Sullivan said. Mechanization had taken an incontrovertible toll; though AT&T had hired twenty-five to fifty thousand women per year during the early 1920s, it had not added operators on anything other than a temporary basis since then. The union did not oppose all change per se, Sullivan emphasized. Trunk lines, tandem switching, and machine ringing genuinely improved customer service. But labor could not approve of extending the dial system, she said, something which represented "the perfect example of a wasteful, expensive, inefficient, clumsy, antisocial device, being substituted for satisfactory, competent human labor."[38]

Despite such protests, telephone conversion continued to go forward, and employment experts cautioned girls to take mechanization into account when choosing an occupation. While switchboard operation had once provided jobs for literally thousands of young women each year, a 1938 Civic Education Service guide described the future employment outlook as "not particularly promising." Anticipating that expansion of the dial system would further reduce the need for operators, vocational counselors began to recommend that switchboard workers start learning other skills in their spare time. Employment counseling in the Machine Age meant preparing people for the likelihood that modernization might remove entire fields of opportunity. Women needed to face the fact that dial technology had turned switchboard work into "a blind-alley occupation."[39]

Social Scientists Assess Workplace Mechanization

Concern in specific occupations highlighted the more general fear: had technological unemployment become a new national reality, threatening to spread across all industries and economic sectors? The answer turned on fundamental questions in economics: the complex interactions of scientific and technical change, capital investment and productivity growth, job creation and job loss, labor costs and selling prices, rising and declining industries, consumerism and standards of living. Depression-era economists knew that such matters had engrossed their predecessors over the past one hundred and fifty years.

Back during the Industrial Revolution, Europeans had observed the mechanization of key manufactures such as textiles and theorized about the implications. French economist Jean-Baptiste Say's 1803 *Treatise on Political Economy* declared that "industrious human agents" must "needs be thrown out of employ" whenever new devices appeared. Though such a fact seemed "clearly objectionable," Say quickly added that economic laws could "wonderfully reduce the mischiefs" of displacement. After all, the process of manufacturing and installing new machines itself created jobs, while the gradual pace of technical change gave the government time to ease the transition by hiring displaced men for public works. Most important, Say emphasized that mechanization should actually augment employment; as output rose, prices of goods would drop, stimulating an increase in consumption which could ultimately provide jobs for more people than ever. To illustrate, Say argued that though the development of printing presses had displaced manuscript copyists, the new power of production made books far more affordable, fostering a demand that employed one hundred times as many workers as before. True, desire for certain products might not expand infinitely, but purchasers could still spend elsewhere the money they saved from mechanized production. On balance, this "Law of Markets" made innovations in production a clear economic good, Say maintained. By raising the overall wealth of society, improvements cancelled out any temporary displacement.[40]

Say's Law seemed straightforward enough, but the 1819 book *New Principles of Political Economy* challenged the belief that mechanization brought rising consumption and hence stimulated employment. Swiss economist Jean Charles Léonard de Sismondi criticized Say's example of book printing for confusing cause and effect in consumer demand, cautioning that markets could easily become saturated with goods. Development of stocking frames had initially helped poor families afford better clothing, but inherent limits on that market meant that beyond a certain point, mechanization would displace workers faster than consumption could grow, bringing "national misfortune." Say had assumed manufacturers would pass the savings from production improvements on to consumers, but Sismondi wrote that when "a hundred workmen are dismissed, that the work may be done with one," the industry's "goods are not reduced to the hundredth part of their price." Labor displacement spread in geometric progression, while prices tended to fall only in arithmetical progression, he declared. Every time change threw a breadwinner out of work, families suffered a instant drop in purchasing power, leading society toward "extremes of wretchedness."[41]

Driven by the motto, "Wealth is everything, men are absolutely nothing," business would push every worker onto the streets if they could gain even 5 percent savings, Sismondi lamented. A society that came to worship output for its own sake would wind up with a "population made expendable by the Invention of Machines." A country might as well have only one inhabitant, its ruler, who "by constantly turning a crank, might produce, through automata, all the output of England." But as much as Sismondi deplored displacement, he warned that attempting to suppress ideas would prove futile, even hazardous, if such moves placed a nation at economic disadvantage in competing with neighbors. To prevent the machine from "turning against those it ought to serve," Sismondi could only recommend reforming patent law. If all patents instantly became public property, he hoped, businessmen would become less vicious in battling for superiority. In the end, Sismondi feared, "nothing can prevent" the elimination of jobs. Displacement represented a permanent problem in the modern economy, a menace "against which the social order offers no remedy."[42]

Such fundamental disagreements about the economics of production change also troubled British economist David Ricardo, who eventually reversed his initial thinking on the subject. His 1817 version of *The Principles of Political Economy and Taxation* endorsed an optimistic faith in technical advance as "a general good." Like Say, Ricardo had felt that over the long run, growing demand should compensate for any "inconvenience" of temporary job loss due to installation of new technologies. In a revised edition five years later, Ricardo declared such assumptions "erroneous." He had become convinced that mechanization indeed proved "often very injurious" to workers (though still benefitting owner and landlord). Sudden adoption of new inventions caused the most serious problems, but Ricardo believed that manufacturers more often introduced machines gradually, which allowed a chance for economic adjustments to offset the worst effects. In any case, he agreed with Sismondi that despite the impact on labor, modern economic competition meant that society could not afford to limit or discourage innovation.[43]

Ricardo's change of heart apparently upset some fellow economists, who protested that he had "done a serious injury to the science" by endorsing any idea that technological change might cause significant problems. J. R. McCulloch, for one, continued to regard such concerns as fallacious, echoing Say's argument that mechanization actually drove employment up by making goods more affordable and so expanding consumer demand. If anything, McCulloch's 1830 work indicated, technical developments placed more pressure on the interests of capitalists than on labor. He presumed that as machines took over old jobs, most workers could adapt by learning new trades. Handloom weavers, however, deserved special assistance because mechanization hurt them so badly, a situation that made "strong claims on the public sympathy." Such exceptions could not violate the more general point, McCulloch insisted, that industrial gains tended to prove "beneficial to all classes."[44]

This discussion of how variations in invention, production, consumption, and employment affected one another absorbed the interest of John Stuart Mill, Thomas Malthus, and numerous other nineteenth-century economic thinkers. In his 1867 *Das Kapital,* Karl Marx emphasized that the sheer power of new machines could devastate employment. With a calico-printing machine, he reported, a single operative could turn out as much cloth per hour as two hundred workers had done before. In the Manchester textile industry overall, inspectors' reports showed that mechanization offered mill owners savings of 10 to 33 percent on labor. Only "apologists" such as McCulloch and Mill, secure in their "bourgeois" comforts, could dismiss the fact that such radical developments gave workers good reason to feel threatened, Marx wrote. He pilloried the "pretentious cretenism" [sic] of those who minimized the difficulties facing displaced workers by insisting that expanded production always improved life.[45]

Bringing the technological unemployment debate to the United States, Commissioner of Labor Carroll D. Wright reported in 1890 that mechanization in brickmaking, printing, boot and shoe production, and a host of other industries showed an "incredible" tendency to replace labor. He declared it impossible to count how many people had been affected. Nevertheless, Wright reassured readers that displacement manifestly had not created any economic or social crisis. After all, total American employment had risen 176.07 percent from 1860 to 1890, against a population growth of 99.16 percent. Ingenuity had improved old industries and established entirely new ones, from railroads and rubber to telegraphs and telephones, creating as many if not more jobs than had been lost. Labor had received an enormous, though not equitable, share of that new wealth, Wright concluded. Though the recent economic downturn had spread misery and tension, invention remained a "friend and not the enemy of man."[46]

Through the 1920s, neoclassical economic theory still downplayed concern over mechanization. Over time, advocates maintained, price reductions gained through efficiency should allow inherent forces of consumer demand and job growth to counter worker displacement. Critics such as Charles Gide and John Hobson responded that although employment and mechanization might balance perfectly in the abstract world of economic law, complicating factors such as varying savings rates and the growth of business monopolies tended to upset that ideal equilibrium in real life. But the prosperity of the 1920s itself seemed to validate Say's Law, suggesting that the modern economy would naturally readjust to engineering advances. Faith in Machine Age progress fostered hope that the United States had grown too successful to fear technological unemployment, but the onset of depression would reopen the old debate over machines and jobs with a vengeance.[47]

Some thinkers of the 1930s worked to reestablish confidence that economic laws tending toward natural balance would prevent displacement from developing into any long-term national crisis. University of Chicago economists Paul Douglas and Aaron Director assured Americans that, while the "gloomy prophets" might talk of job loss, the power of mechanization must inevitably translate into new work. In a competitive system, they reasoned, manufacturers should pass on savings to consumers, who invariably would be drawn by lower prices to buy more of everything (with the exception of some luxury goods, whose purchase depended on different economic and psychological conditions). Americans could use that surplus pocket money to enjoy a nice vacation, go to movies more often, or just buy extra packages of chewing gum; the beauty of capitalism lay in offering people freedom and opportunity to fulfill the acquisitive wish for more. Douglas and Director relied on this material desire and elasticity of demand to sustain job growth regardless of how much mechanization increased productivity. If Americans decided to save rather than spend, banks could distribute that extra money in loans to underwrite new business ventures. Even if producers failed to pass along to consumers the savings from mechanization, that money would then finance expansion and renovation of old facilities. Through such interactions, Douglas and Director concluded, change must ultimately enhance the national economy. For "every man laid off, a new job has been created somewhere." New openings for gas station and hotel workers could offset declining prospects for farm laborers and miners, making it "therefore clear that permanent technological unemployment is impossible."[48]

In practice, Douglas and Director acknowledged, economic relationships might not work out so neatly. Consumer behavior would not change overnight, nor could workers switch occupations instantaneously in response to mechanization. For that reason, they noted, "the ultimate benefit which flows to society in the form of higher real incomes is obtained only at the cost of great and undeserved hardship to many." To prevent such misfortune from fomenting extreme, even violent, working-class opposition to all progress, the Chicago economists recommended that government support centralized employment assistance, while businesses could provide dismissal wages and retraining. Labor itself should also bear some responsibility of preparing for displacement, of accepting some compromise in ideals. Given the possible need to move out of state to follow the changing job market, Douglas and Director advised, manual workers should not purchase homes.[49]

Such arguments about the impossibility of persistent technological unemployment would have sounded familiar to Say, but other observers in the 1930s noted that the Depression had created unusually troubling dimensions that invalidated the assumptions of "normal times." Columbia economist Rexford Tugwell worried in 1931 that the country faced special problems of "occupational obsolescence" (a term he favored over *technological unemployment,* since many displaced individuals might succeed in finding work, but only in lesser positions at lower wages). Tugwell criticized Douglas and others for too readily assuming that mechanization would support employment growth. Innovation did not necessarily translate into cheaper production, he argued, given manufacturers' overhead costs and the reality that new machines did not always live up to their promise. Even if prices did fall, increased consumption could not be guaranteed, and even if demand rose, job growth might not follow automatically. Supposing that exciting inventions and new products did come along to open up markets, they still might not bring enough work to absorb everyone displaced from previous employment. After all, new firms had the best opportunity to take advantage of the latest advances in technical efficiency, so, in emerging industries, "machines are sure to be the latest device in 'labor-saving.'"[50]

For such reasons, Tugwell concluded, occupational obsolescence posed more than temporary difficulty; the United States could not patch up the problem by improving employment offices and inching up dismissal wages. Displaced men and women would need not only help in locating new work but also public support to carry them through the transition; Tugwell advocated setting up a national system of business-funded unemployment insurance and economic planning. Anticipating objections that burdening companies with those expenses would constrain economic progress, he maintained that such a system would merely reflect a truthful assessment. Present methods of economic analysis, Tugwell argued, gave an illusory impression of progress, highlighting the savings technical innovation offered while concealing the cost. Companies could appropriate all the gains from mechanization while externalizing all the losses; it was individual workers and the general community who had to bear the price. Under honest accounting, Tugwell indicated, workers' skills should be valued as highly as machinery, and labor displacement must be factored into modern business as a capital cost.[51]

Through the 1930s, American social scientists grappled with such questions in their writings, speeches, and professional conferences, which frequently featured special paper sessions and panel discussions on the relationship between machines and work. In one 1931 American Economic Association panel, for example, the University of Minnesota's Alvin Hansen shifted blame back to labor for upsetting ideal economic relations. Unions' aggressive wage demands had encouraged employers to substitute machines for workers, he said, at a time when lower interest rates made investing in equipment unusually attractive. Hansen believed that natural forces would maintain an economic balance between labor and capital in due course, but Harvard's Sumner Slichter warned that any disturbance might make displacement "serious and chronic." He explained that the prospect of continued innovation would keep managers obsessed with finding new ways to save labor. Given that tendency, Slichter suggested, the United States should create a federal labor board to ensure that new patents contained provisions protecting employment and to encourage business to reduce working hours, retrain displaced men, and pay dismissal wages. In ensuing discussion, audience members seemed to favor Slichter's call for government intervention and national economic planning, rather than trusting in economic events to run a positive course on their own.[52]

While Douglas, Director, Tugwell, and others debated theories of mechanization and economic activity, different analysts began attempting to determine the actual extent of present-day displacement. Taking the printing industry as her case study, Columbia researcher Elizabeth Baker suggested that Say's Law had held true up to the twentieth century. Expanding demand for printed material did compensate at first for the effect of new linotype machines that allowed one man to do work previously requiring four. Since 1913, however, the use of presses with automatic paper-feeding mechanisms had skyrocketed, accounting for more than 66 percent of recent sales. Most modern printers used old-fashioned manual presses (where workers inserted one sheet at a time) only for small jobs or specialty work, such as printing on thin tissue or thick cardboard. Further altering operations, some sophisticated new presses could print two colors or even two sides of a sheet simultaneously. The economics of the printing industry had not changed as fast as production technology, Baker declared, and consumer demand could no longer make up for the impact of mechanization. Business had more than doubled from 1914 to 1927, but companies met that need with only a 25 percent increase in employment, since each individual in 1927 could turn out three-fifths more volume than in 1914. For every hundred modern presses adopted, jobs dropped 11 percent, Baker concluded, and workers would continue to suffer as more firms discarded the old labor-intensive presses.[53]

Surveying fifty-three of more than two thousand pressrooms in Manhattan, the heart of the trade, Baker found that the introduction of improved presses hit hardest at semiskilled assistants, the men responsible for jobs such as hand-feeding paper. Earlier presses had generally required two assistants to every pressman, but updated versions needed just one. Displaced assistants could not easily graduate to operating automatic machines, and even when that opportunity existed, older men often resisted learning the new skills required. Most remained out of work for some time; a few managed to pick up part-time printing tasks or odd jobs in pressrooms, while others tried chauffeuring, auto repair, and bootlegging. One man Baker interviewed had seen his working life upset twice by invention; as a pianist, the fellow had accompanied silent films until sound movies took over the theater, while his subsequent work feeding printing presses ended with the introduction of automatic models. Pressroom assistants had often engaged in labor battles over employment conditions and pay in previous decades, but technological changes completely altered the terms of the industry, forcing them into a defensive fight like nothing before. Whereas once the men had performed a vital role in the pressroom, new automatic-feeding machines punctured that sense of occupational pride. In Baker's assessment, the developments of recent decades left the printing assistant "threatened with extinction."[54]

Though semiskilled assistants suffered, innovations in printing brought a relative employment increase and gains in prestige for pressmen, skilled crafts-men, and mechanics. Although jobs for assistants had dropped 5.7 percent in the plants Baker studied, pressmen's employment had risen 7.9 percent. Yet, though new presses seemed to bring advantages for pressmen, that group expressed strong misgivings about their long-term prospects. If today's presses threatened assistants' jobs, pressmen anticipated that tomorrow's designs might render them vulnerable too, as managers sought to reduce labor needs still more. Fear of technological unemployment became contagious; as employees saw how easily companies had replaced their fellows, they instinc-tively began to question where that trend would stop.

Given evidence that mechanization had generated tension across the board, Baker concluded that such issues posed one of the most difficult challenges for modern Americans. Instead of referring to "technological unemployment," she preferred to speak of *technocultural unemployment*, a term emphasizing that the country could and should adopt a proactive stance. After all, machines did not introduce themselves into the workplace, and economic patterns of pro-duction and consumption, displacement and employment, did not exist in a vacuum. She insisted that government, businesses, and unions all bore a responsibility to establish effective policy for minimizing displacement and retraining workers. In Baker's eyes, evidence from the printing industry doc-umented the human cost of change, but other parties assessed the situation quite differently. Advocates stressed that new presses had added job oppor-tunities for pressmen and took Baker to task for making sweeping policy rec-ommendations based on an investigation of one field. A *New York Times* editorial stated, "That the automatic machine has an effect of some kind on employment can hardly be denied, but exactly what that effect is we know only in the printing industry." Though other economists had generalized about technological change and work from less evidence, the episode underlined just how controversial talk about displacement had become.[55]

In the reactivated debate on mechanization, economists and social scientists began examining specific aspects of the question, such as how factors of gender, age, and skill complicated issues of displacement and readjustment. Some worried, for example, that older workers suffered a form of double jeopardy. Age both exposed them to greater likelihood of being discharged when businesses began layoffs (whether for mechanization or other reasons) and then presented an additional obstacle in finding a new position. According to the New York State Commission on Old Age Security, 89 percent of manufacturing firms discriminated against job candidates past middle age, while 59 percent refused even to accept applications from older workers. Yale industrial engineer Elliott Dunlap Smith found it ironic that the modern emphasis on rapid technological change had started shrinking the period of employability, even as medical progress had extended man's health and lifespan. If anything, increasing mechanization should make aging less of a drawback, Smith argued. Conveyor systems eliminated heavy lifting, and automatic machines reduced the stress of industrial labor, offsetting any decline in strength and endurance among older workers.[56]

Executives often perceived young people as better suited to the Machine Age, more adaptable and more knowledgeable about new technology. Some older workers, who resented and feared the powerful changes transforming their familiar workplace, did indeed resist learning new skills or prove unable to handle strange machinery. Doctors and psychologists pointed out in the 1930s that the aging process affected everyone differently and argued that the many capable and adaptable older individuals should not be condemned on the basis of stereotypes. Employers should consider the personal strengths older staff contributed to a workplace, such as reliability and experience. But in a period when mechanization could completely revolutionize production, experience with old-style operations became a liability rather than an asset. One manager rationalized age limits by saying that most men over age forty were "trying to sell themselves on the basis of what they have accomplished in the past by using methods which were good at that time but are now obsolete." Economists, social workers, and union leaders worried that the premium on youth made older workers into special victims of technological unemployment, "eliminated as ruthlessly as the machinery whenever it begins to show wear."[57]

Observers of mechanization also began asking about its relative impact on skilled versus unskilled workers. Many innovations clearly substituted for unskilled labor. The Bureau of Labor Statistics reported, for example, that ditch-digging machines with one operator and assistant could perform as much work as forty-four hand laborers.[58] At the same time, other technological changes seemed to pose an unprecedented threat to skilled workers. Deskilling did not represent a new issue; during the Industrial Revolution, spinning mules and power looms had transformed textile-making from skilled craft into unskilled machine-tending. In Depression America, widespread joblessness lent urgency to the fear that inventors could make machines equal to even the most experienced men. One case attracting attention involved the painters who had long been a part of first the carriage and then the automobile industry, embellishing expensive vehicles with striping and other decorations. Those men had formed a production aristocracy, commanding relatively high pay and respect for their eye for detail. With the development of paint guns that could spray fine lines, even young female novices could learn striping, and the job could be completed faster with fewer people. Technology helped reduce car-striping time from seventy-two minutes to twelve, replacing twenty-two skilled men earning at least $1.50 an hour with one man and four girls averaging sixty cents an hour. Stripers lost both their job security and their personal identity as machines made their artistry irrelevant.[59]

Automobile striping seemed to illustrate a particular twentieth-century twist to the old problem of deskilling. Machines had long been able to surpass man's physical strength, but new inventions carried an even more powerful capability to beat other human abilities. Automatic dialing equipment could react faster than even the best telephone operator, at least when it functioned properly. In manufacturing, new devices incorporated sophisticated sensory capabilities that promised to supersede humans' manual dexterity and visual acuity. Cigar-inspecting machines reportedly worked more quickly and accurately than people; human sorters could distinguish only eight or nine variations in color, while the technology could differentiate among several dozen shades. Photoelectric equipment introduced at Henry Ford's River Rouge plant in 1935 allegedly eliminated eight out of ten men inspecting wrist pins and fourteen out of eighteen inspecting camshafts. Even jobs requiring a delicate, experienced touch no longer seemed safe. By 1939, one Pennsylvania bakery replaced skilled "pretzel bending artists" with machines that could knot raw dough into loops ten times faster, plus conveyor belts to move the product through ovens and cooling areas automatically.[60] When machines could outperform workers who had devoted years to perfecting their senses and skills, the threat of displacement seemed to have reached a new level.

Of course, as economists since Say and Sismondi had noted, the seriousness of technological unemployment turned not just on how many positions machines eliminated, but how many people the economy could reabsorb. While Depression-era observers had little reliable information on exact numbers of men and women thrown out of work by technological change, experts knew even less about what happened to those individuals afterward. While defenders of classical economic theory maintained that consumption should open plenty of job opportunities and thereby prevent long-term joblessness, the few available studies tracing the fate of displaced workers yielded more ambiguous evidence. One 1930 study of 1,190 Connecticut rubber workers displaced when their factories could not compete with more efficient operations showed that on average, workers lost 4.3 months to unemployment. Thirteen percent remained jobless after eleven months. Sixty-six percent of those who found new work had to accept lower pay, with skilled workers losing 28 percent more income than unskilled. Age represented a significant factor: only 61 percent of subjects over age forty-five managed to find reemployment after eleven months, versus 84 percent of those between twenty-five and thirty-five.[61] That study had been started before the Depression fully set in; it seemed likely that a worse economic climate would make statistics on reemployment even grimmer.

Some evidence suggested that, as technological change irrevocably altered labor needs in old lines of work, displaced men and women had to switch occupations to gain a decent chance at reemployment. Economist Isador Lubin's 1929 study of 754 unemployed workers (some displaced by mechanization, others by plant slowdowns or relocation) showed that 54 percent of those who managed to find new jobs had done so by completely changing fields. Lubin cited instances of garment cutters who became gas station attendants, grocery clerks, or building guards. Such cases might seem to confirm the faith that, through natural shifts in production and consumption, the economic system would absorb displaced labor. Though cutting machines had eliminated jobs in the clothing industry, the rising demand for garage and grocery services had opened new positions. And yet, such breezy confidence in the workings of the market neglected to factor in the hardship for those involved. In tough times, few could afford the luxury of turning down low-level jobs in hopes of a better future offer, but those who jumped occupations often paid a high price in lost skill, lost wages, and lost time. Lubin's study began to document how many mature and experienced workers saw little choice but to "begin all over again at the bottom and learn a new trade at lower pay."[62]

Nevertheless, the possibility of reemployment for at least some displaced workers could reinforce the hope that mechanization posed no serious danger. Classical economic theory readily meshed with modern optimism, the faith that Americans' native ingenuity would keep churning out better ideas. Innovations had, in the past, led to the creation of entirely new industries. Many citizens enduring the Depression could readily remember the days before automobiles, airplanes, radio, movies, and telephones; over just a few decades, those inventions had grown into five of the nation's biggest employers. Census Bureau figures confirmed that new product areas generated work; between 1920 and 1930, the number of chauffeurs, truck drivers, and tractor drivers in America went from 285,000 to 972,000, while auto and airplane industries helped mechanics' jobs grow from 281,000 to 628,000. Employment of harness makers, railroad firemen, and streetcar motormen had dropped in that decade, sometimes precipitously, but the popularization of cars had ensured employment in the manufacture, distribution, and sales of vehicles, accessories, and travel services.[63]

Optimists held that just as automobiles and airplanes had supported job growth over the previous years, so future discoveries would spawn new industries and prevent technological unemployment from growing unmanageable. The country still needed new inventions to meet unfulfilled consumer desires, especially in the critical area of housing. Paul Douglas told a 1932 radio audience that plans for improving construction would soon make mass production of affordable homes practical. The resulting wave of building, he promised, would not only improve living conditions for thousands of families but also bring thousands of jobs.[64] Such arguments reflected faith that a historic path of innovation would allow the economy to continue on a course of natural expansion; in essence, the United States would invent its way out of labor displacement.

More cautious observers warned against counting on vague scientific and engineering developments as a panacea. The latest model automobiles had attracted buyers during good times, but the Depression made it harder to find consumers who had discretionary funds to spare on such items. With so many families experiencing economic difficulty, companies might face obstacles getting new products established. The *New York Times* cautioned that innovations often merely redirected buying rather than adding to it; the novelty of rayon attracted consumers, but it simultaneously reduced the demand for silk and cotton goods. Moreover, the inevitable lag between the germ of an idea and its translation into successful production limited its immediate value in providing employment. Displaced workers of the 1930s could find little solace in predictions that the development of some future enterprise like television would yield thousands of jobs in decades to come. Such forecasts only made writer Silas Bent scornful; if invention could miraculously open so many opportunities, he asked in 1931, why did so many currently face hardship? "All we need do is look around for one of those new jobs—business of scurrying about with a telescope—Where are those seven million new jobs the machine has been creating?"[65]

Discussion of new inventions reflected a whole set of underlying questions about America's economic and technological future. The country had changed, Mordecai Ezekiel noted. Generations of activity in farming, cattle ranching, mining, and transportation had built wealth in the past, but the United States had run out of room for expansion. Twentieth-century inventions such as airplanes and air-conditioning might raise living standards, but they could not compare to the sheer scale of capital growth during the nineteenth century. Ezekiel feared that the closing of the frontier had cut off the nation's best source of progress and concluded, "Re-absorption of the technologically unemployed today is thus no longer an automatic process." Stuart Chase, among others, similarly worried that the end of westward movement had made prospects more troublesome for the current generation. While frontier settlers had been able to feel "unlimited optimism as to the economic future," he warned, the Depression forced modern citizens to confront disturbing new realities such as the loss of jobs to mechanization.[66]

If the closing of the frontier had started shutting off economic options, Depression-era analysts wondered, where would new jobs come from? How would patterns of employment look different in years ahead? Some economists argued that America's occupational picture had started to shift radically, moving away from agriculture and manufacturing toward professional, service, and technology-related jobs. Incomplete and ambiguous employment statistics left plenty of room for dispute about such a trend, however, and experts disagreed even more fundamentally about its implications. Douglas and Director hailed expansion of the service sector as America's economic salvation, providing new avenues of opportunity for men and women displaced elsewhere. Census figures from the 1920s had already showed increasing numbers of people working as insurance agents, entertainers, schoolteachers, gas station attendants, and hotel staff, while agricultural and manufacturing employment declined. In short, the Chicago economists argued, mechanization did not permanently subtract jobs but merely transferred them from certain areas to others. While previous generations had grown up to become garment workers or farm hands, their modern counterparts would naturally gravitate to the new openings for "movie ushers, saxophone players, and house-to-house canvassers."[67]

Skeptics doubted whether service-sector growth could automatically fend off unemployment. Even for garage or hotel work, displaced men would need to learn new skills and would be at a disadvantage competing against those with experience. Even though the ranks of movie ushers and gas pumpers had grown rapidly during the 1920s, logic suggested that those lines of employment could not infinitely absorb job-seekers. Some professions had also reached saturation point; Chase cautioned that young people just entering journalism, for example, would encounter a very tight job market. AFL leader William Green, for one, found little reassurance in the promise of new openings. After all, he said, a "man laid off in a steel mill where new machinery has just been installed cannot go tomorrow and take up work as a barber, and he certainly is not prepared for the professions."[68]

Moreover, service workers were by no means immune to technological displacement, as switchboard operators had discovered. Thanks to coin-operated vending machines, retailers could cut sales personnel and let customers serve themselves. In the food business, the popularity of Automats seemed to herald the age of self-service. Inventors seemed fixated on the dream of mechanizing service everywhere. Business, scientific, and technical periodicals of the 1930s described plans for an automatic parking garage and other new gizmos. The man who had previously created the Piggly-Wiggly grocery promoted his design for a new self-service store; customers in a "Keedoozle" shop would select items from a glass display case by turning a key. Conveyor belts would carry the desired products to the front of the store, where shoppers would total up purchases themselves. The owner would need to hire just a single cashier to accept money and a few stockers to keep shelves full, the creator promised. As it turned out, the Keedoozle did not displace hundreds of grocery clerks, since it proved a commercial failure. But just as changes over previous years had fundamentally altered labor needs in the steel industry and telephone work, the new trend in mechanization suggested an equally significant impact on future service employment.[69]

As economists of the 1930s considered such questions, they consciously placed themselves within both a community of expertise and the historical debate over technological unemployment. In their discussion of production and consumption, occupational shifts, and the nation's labor supply, Douglas, Baker, Slichter, Ezekiel, and others frequently referred to arguments that Say, Sismondi, and their other predecessors had made. More than ever before in modern times, the Depression-era crisis of unemployment underlined the inherent difficulties of untangling the complex economic relationship between technological change and work. Though the crisis focused new attention on the subject, experts could not instantaneously resolve a centuries-old issue. Throughout the decade, economists continued to argue about both the broad picture and the detailed evidence. Joining the economists, other social scientists began a protracted attempt to understand the effects of workplace mechanization. Members of the American Statistical Association wrestled with technical problems about what types of numbers, measures, and comparative indicators should be used to define economic change, to measure productivity, and to gauge employment. For a more intense look, professional societies organized special meetings devoted entirely to the matter. In 1930, the American Academy of Political and Social Science hosted a conference on jobs and technological change which attracted four hundred delegates from various organizations and colleges.[70]

University of Chicago sociologist William Fielding Ogburn, who had long expressed interest in the nature and impact of the inventive process, became a central figure in the discussion. His influential 1922 book, *Social Change with Respect to Culture and Original Nature,* had characterized science and engineering as prime forces reshaping economic behavior, family life, and other parts of the modern world's interdependent social structure. When economic institutions, government, or other actors failed to adjust to technical change in timely fashion, the subsequent "cultural lag" could cause critical "maladjustment." Continuing his work into the 1930s, Ogburn interpreted labor displacement as an especially serious example of cultural lag. The pace of change kept speeding up, he emphasized, and developments in technology could exert a tremendous price on society. Even minor inventions such as vending machines might radically reshape business, causing ongoing trouble for working people. Ogburn compared the course of American science, engineering, and invention to a "huge tidal wave" that could not possibly be restrained; the United States had no choice but to try adjusting its social mechanism to compensate. In present-day life, Ogburn wrote, "technology cracks the whip." The force of mechanization raced ahead ever faster, with the result that "institutions of society slip out of gear, and humanity suffers."[71]

Ogburn's perspective encouraged the concept that America needed to cultivate skills in social invention to match its record in developing new technology. "The mechanical inventor has given us too many hats; the social inventor has not given us too many techniques and aids for social ends," wrote Arland D. Weeks. For Ogburn and others in his circle, the first step in the process of adjustment required getting experts to identify new inventions that carried an especially great potential impact. Through the early 1930s, Ogburn, S. Colum Gilfillan, Clark Tibbits, and fellow sociologists annually published lists of the previous year's most critical discoveries. The technologies they singled out ranged widely in type, from advances in television equipment, airplane safety, and the mechanization of coal mining to the advent of night golf and the discovery of imperishable paint for decorating buildings. Their analyses covered more than just displacement; the sociologists tried to anticipate how selected discoveries might affect personal behavior, family interactions, and community life. But as concern about unemployment grew, Ogburn and Gilfillan devoted more and more attention to the question of work. The cultural lag theory of displacement proved enormously influential throughout the Depression decade, especially among federal researchers.[72]

Unsurprisingly, the man who symbolized America's faith in technology as progress, Thomas Edison, ridiculed any idea that overly rapid mechanization might contribute to social or economic problems. When an interviewer asked in 1931 whether the sheer pace of new invention threatened to make life too complex, Edison replied, "No. People will live up to it. The brain—if used—has enormous capacity. People don't begin to suspect what the mind is capable of." For Edison, people's character flaws were what had caused economic depression, and triumph of will could lead America out of it. He told a radio audience that he had witnessed many downturns over his lifetime, but the country had always emerged "stronger and more prosperous." Promising that innovation would relieve human drudgery and thereby increase happiness, Edison looked forward to the day when "every task now accomplished by human hands is turned out by some machine." He visualized having a perfectly automatic machine that took in cloth at one end, controlled its cutting, and turned out completely finished suits at the other end. Edison's death, in October 1931, came in the midst of a powerful paradigm shift in American assessments of technology, after which assumptions about mechanization and well-being would never be the same.[73]

From the early 1930s on, economists and social scientists helped focus attention on the subject of production mechanization. Exactly how many Americans had new workplace technology displaced, and how many had been reabsorbed? The answers remained subjects of basic dispute between labor and business, classical economists, and critics. Yet within just a few months of the Depression's onset, the terms of debate had fundamentally changed. While 1920s prosperity had supported confidence in Say's law and reassuring theories about economic equilibrium, Americans of the 1930s no longer took such faith for granted. As popular economic writer Stuart Chase said bluntly, facing the grim financial situation of 1931, "This is the economy of a madhouse."[74]

Many uncertainties remained. Depression-era observers often cited an 1832 comment by British thinker (and early computing-machine experimenter) Charles Babbage, who declared, "That machines do not, even at their first introduction, *invariably* throw human labour out of employment, must be admitted; and it has been maintained, by persons very competent to form an opinion . . . that they never produce that effect. The solution of this question depends on facts, which, unfortunately, have not yet been collected."[75] Such an assessment seemed equally apropos one century later; vital "facts" about machines and jobs remained open to dispute. Given the national scale of unemployment, many Americans looked to the federal government to answer their questions. During this age of crisis, the issue of technology and progress would come to command attention at the highest levels of government.

2

"Finding Jobs Faster Than Invention Can Take Them Away"

Government's Role in the Technological Unemployment Debate

D URING THE 1930s, technological unemployment came to command
national political attention. As officials began searching for ways to
address the economic and social crisis, the sheer persistence of high joblessness attached grave import to the possibility that workplace technology might
be even partly responsible. Herbert Hoover's administration tried to minimize
fears, assuring Americans that the growing mechanization of production
would inevitably create opportunity over the long run. But the election of
Franklin D. Roosevelt drastically changed the Washington line on displacement, bringing in officials who treated it as a serious problem.

Economists in the Works Progress Administration undertook intensive
research into recent production and labor trends in various industries, service sectors, and agriculture. Unemployment remained stubbornly high, even
as business started to recover from depression, suggesting the disturbing possibility that mechanization had cut into the base level of work available. WPA
officials started speaking of "permanent technological unemployment." President Roosevelt himself warned Americans in his 1940 State of the Union
address that the country must begin "finding jobs faster than invention can
take them away." Throughout the decade, members of Congress also wrestled
with the issue, debating various proposals to address the changing nature of
work in the Machine Age.

At the start of Depression, experts had few reliable statistics on overall
unemployment, much less the extent of technological displacement. Officials
simply had not made it a national priority in previous years to collect comprehensive figures on the pace of mechanization and on job growth or decline

across various fields. After 1930, as job loss deepened, the popular press, labor organizations, academic observers, and the general public began calling for better information about the unemployment situation. The *London Times* editorialized that "even the most prosperous country on earth cannot indefinitely postpone" the vital need to study the modern relationship between mechanization and work.[1]

Any intelligent discussion of technological unemployment would need to be grounded on a firm factual foundation, and through default, if nothing else, the federal government seemed to bear the obligation of supplying such statistics. Historian Charles Beard, expressing his dismay at discovering that "our best authorities have few figures at hand" on technological unemployment, tried to goad the Hoover administration into corrective action. In a 1930 letter to Lillian Gilbreth, member of the President's Emergency Committee on Employment, he asked, "How many are thrown out of employment annually by changes in machinery and processes? We do not know? What becomes of those so displaced? . . . How many are taken care of by their former employers? We do not know? How many are retrained and how? . . . What hardships are imposed on the displaced? We cannot answer." Beard called on the Hoover administration to set top economists to work immediately on the most basic problem—just deciding what data to collect.[2]

As the Depression set in, other parties joined the appeal for the government to investigate displacement. In a March 1930 letter to President Hoover, AFL leader William Green wrote that, given the likelihood that production would grow even more mechanized, American workers could end up paying an "obviously unfair" price for industrial progress. To prevent such an outcome, Green recommended that Hoover establish a federal bureau to help locate new positions for "victims of technological unemployment." Such a centralized employment agency could also gather information on changing economic patterns and job trends, Green added, data that educators could then use to redirect vocational guidance. In the White House response, Hoover declared he had been "giving a great deal of thought" to technological unemployment. He mentioned the possibility of establishing a Presidential Commission or other official body to review issues of machines and jobs, writing that the "inquiries I have been able to make show a great deal of confusion as to fact."[3]

That July, Hoover's administration created a special Advisory Committee on Employment Statistics to recommend ways of improving the Bureau of Labor Statistics' admittedly "inadequate" information-gathering processes. Hoover explicitly charged the panel with suggesting directions for inquiry on technological unemployment. University of Pennsylvania researcher Joseph Willits, P. W. Litchfield of Goodyear Tire and Rubber Company, and the AFL's

John Frey formed the subcommittee on machines and labor, whose advisors included William Green, economist Sumner Slichter, statistician Ewan Clague, and the Taylor Society's Harlow Person.

From the start, these members emphasized the inherent complexities and unpredictability of economic interactions. When an employer introduced new equipment, they noted, "displacement may occur in the plant in which the improvements occur, in a competing plant several thousand miles away, or in a plant or plants manufacturing totally different products." Economists had wrestled for decades with such "exceedingly complicated" questions, the group pointed out, but Americans could no longer afford to shrug their shoulders. The sheer pace and power of invention obligated federal officials to begin systematically analyzing the problem. The committee recommended that agencies such as the Bureau of Labor Statistics compile monthly summaries of manufacturing capacity, productivity, and jobs, broken down by industry and area. Once the surveys pinpointed the most significant changes in productivity and employment, the government could mobilize its analytical expertise to concentrate on such potential trouble spots.[4]

Members of the Hoover administration's Commerce Department joined the campaign to begin special studies. In a memo to Commerce Secretary Robert Lamont, staff member Edward Eyre Hunt posed a series of what he considered imperative questions. How rapidly did new inventions appear? How did adoption of new machines affect business activity and labor conditions "as measured by (a) the scrapping of capital goods; (b) additions of new capital goods; (c) technological unemployment; (d) opportunities for new employment?" Without such information, Hunt wrote, people could not make sense of the modern economy. If innovation had begun to generate severe job losses, he argued, the country would need to devise public or private measures "to cushion the shock." If mechanization created more positions than it destroyed, Americans still had to know "Where? How rapidly?" and whether new work meant "loss of economic status." To find out, Hunt proposed forming a special subcommittee of the Conference on Unemployment that would assess the pace of invention, analyze the application of new methods, and compare job loss and job creation across two ten-industry samples. Armed with such insights, federal experts could "construct a balance sheet of technological unemployment" versus "new opportunities."[5]

Meanwhile, Secretary Lamont and President Hoover had started talking with Dexter Kimball, dean of Cornell's College of Engineering, about arranging outside studies of machines and jobs on the government's behalf. Back in 1921, the American Engineering Council had conducted a major investigation of manufacturing efficiency which helped set the agenda for the Commerce

Department under the leadership of then-Secretary Hoover. The *Waste in Industry* report had encouraged government officials to work with company executives in a quest for greater workplace efficiency, as the way to create a better standard of living for all. In 1930, Kimball hoped that the AEC could similarly answer unresolved questions about the economic impact of mechanization to provide expert guidance through a time of turbulence. Apparently backed by interest from the Hoover administration, Kimball persuaded the council unanimously to endorse the idea of undertaking a conclusive review. Within six months, however, a disappointed Kimball had to inform Hoover that the AEC had "exhausted all possible sources of financial support" without raising the necessary $50,000 for the study. He complained, "People who could help are either not interested or do not understand." Kimball did not stop pressing the White House on the issue's importance, again telling the president, "I should like to talk to you about technological unemployment."[6]

Washington leaders continued to move toward accepting the responsibility for analyzing the consequences of workplace mechanization. In practice, specific proposals for attacking the problem made little headway. Partisan controversy and bureaucratic confusion complicated the question of whether existing agencies should collect statistics or whether politicians should create new organizations to take on the job. To make matters worse, government officials still struggled with the task of documenting the scale of unemployment in general. Until they could comprehend the nation's overall job situation, chances of mastering the complex economics of machine-related displacement remained small. In March 1931, Commissioner of Labor Statistics Ethelbert Stewart declared that he still had no reliable statistics on technological unemployment. He approximated that perhaps 125,000 Americans had been thrown out of work by changes in production equipment but quickly cautioned that even trying to estimate remained difficult.[7]

The Hoover Administration's Approach to Technological Unemployment

Though government officials realized by 1930 that they did not have necessary information, discussion of machines and unemployment could not wait. Herbert Hoover himself adopted a clear line. The president expressed sympathy for those hurt by technical change but maintained that such difficulties merely represented temporary economic maladjustment. Speaking to the 1930 AFL convention, Hoover referred to displacement as a grave problem, especially in sectors such as the coal industry. While he regretted that modernization could cause workers and their families such a misery "wholly out of place in our American system," Hoover assured his AFL audience that over the long run, the affected men would surely find work in new industries and an expanded service sector.[8]

Hoover's confidence in the benefits of mechanization plainly reflected his personal background and economic philosophy. Before he entered government, Hoover had acquired fame and fortune as a Stanford-trained engineer who skillfully managed mining operations in Australia, China, and elsewhere. In many ways, the late nineteenth century and early twentieth century represented the heyday of American engineering, when increasing educational and employment opportunities drew growing numbers of young men into the field. Conscious of their new professional status, engineers debated the broader implications of that role. Many believed that America's future strength would rest on technical knowledge, making engineers some of the most important people in society. Hoover agreed with such expectations, but he believed that prominence entailed as many responsibilities as rewards. Assuming the presidency of the newly founded Federated American Engineering Societies in 1920, Hoover argued that engineers' specialized training and practical experience prepared them for public leadership. Expertise obligated engineers to help shape a better social order, one in which technology, economic welfare, and human conditions would advance together.[9]

As head of the Commerce Department during the 1920s, Hoover put such faith in expertise into practice. He envisioned a system in which his agency's staff would offer companies advice based on the best economic and technical information. Under their sophisticated guidance, managers could share ideas about rationalizing production and otherwise cooperate on ways to improve business. That new efficiency would multiply the nation's industrial capacity and create consumer abundance. Hoover wanted to see the U.S. government encourage technical innovation without interfering in private initiative. During his tenure there, the Department of Commerce promoted the growth of radio broadcasting and commercial aviation. Forecasts for the development of television and air conditioning seemed to promise a future made prosperous through a continued flow of ideas. While such an ideal suited the 1920s, the Depression called into question the assumption of progress through technological change. Inventions and the advance of mechanization had not assured basic security for many working-class families, much less a material paradise. It looked as if the Machine Age alliance of engineering and business might have created a disaster of technological unemployment rather than a utopia of efficiency.[10]

President Hoover reacted defensively to such a notion, insisting that serious technological unemployment remained an impossibility. History and economic law dictated that inventions ultimately created more jobs than they destroyed, and so the plight of displaced workers must be merely temporary. Throughout his term, Hoover periodically asserted that the worst of the Depression had passed, a position that allowed him to downplay the issue of mechanization. Over the months, as unemployment persisted, the president blamed a number of other causes, from international economic trouble to abuses of credit and investment. He told economist Arch Shaw in February 1933 that he was not "insensible" to the "effect of labor saving devices on employment." But when compared to other factors such as the failure of the nation's financial system, Hoover continued, technical revolutions caused at most "minor notes of discord."[11]

Hoover, like many of his fellow engineers, remained committed to the faith that innovation under free market capitalism must represent a positive economic force. Pointing especially to prosperity during the years 1924–28, he told Shaw that "our system of stimulated individual effort, by its creation of enterprise" and its support of invention had produced the most and the most varied products "ever known in the history of man." The nation's advances had come thanks to the "march of labor-saving devices," Hoover wrote. Labor might occasionally encounter difficulties in trying to adapt to new equipment, but any such problems could be corrected with minor "adjustments of economic mechanism."[12]

Following the President's lead, Secretary Lamont and other administration officials argued that Depression hysteria had escalated fears of machine-based job loss out of all proportion. Julius Barnes, chairman of Hoover's National Business Survey Conference, described displacement as the price civilizations had to pay for progress, something "inherent in a fluid economic and social structure." Science and invention brought so much wealth to the United States, Barnes declared, that any unemployment "which results incidentally" should not count as a "social evil." Assistant Secretary of Commerce Julius Klein seconded the message that while change unfortunately caused some "indisputable suffering," that problem would "work out" in the natural course of business development. The many "new jobs brought into being by our steadily mounting levels of living should eventually take care of all" displaced men and women. The country's future well-being would depend more than ever on its commitment to innovation, and the greatest danger came when producers invested too little in mechanization rather than too much. Workers who feared job loss would suffer far more if American industry failed to embrace the latest technology and thereby fell behind European competitors. Half the equipment in the nation's factories had already become obsolete, Klein warned. Business owners should take advantage of current conditions to snap up new machines at bargain rates, he urged, as a good-faith gesture that would itself enhance employment.[13]

Meanwhile, to address the general issue of joblessness, the administration recruited prominent figures (especially from industry and engineering) to form the President's Emergency Committee on Employment. PECE followed Hoover's principle that any drive to restore national economic and social order should rest on voluntary local efforts rather than on a centralized federal mandate. The group encouraged communities to hold their own relief drives and called for individuals to help bring down the national unemployment rate by creating one job at a time. In a "Spruce Up Your Home" campaign, PECE distributed 1,400,000 pamphlets recommending that property owners hire unemployed manual laborers to build rock gardens and sheds. Housewives could engage jobless office workers to organize and type up family recipe files, the Women's Division recommended. As long as patriotic citizens could do their bit to help the country pull through, there would be no need to worry about such controversial subjects as mechanization.[14]

PECE and its reorganized successor, the President's Organization on Unemployment Relief (POUR) served as Hoover's most visible forum on the issue of jobs. Government officials, economists, social scientists, and members of the public wrote to express their pet theories about the causes and cures for economic distress. Some correspondents argued that Washington needed to take the problem of displacement far more seriously, lest accelerating changes in production completely upset the labor market. One observer recommended that government require employers to submit ideas for mechanization to a Presidential "National Progress-Prosperity Commission." Such an organization would have the power, "if an unemployment emergency exists," to order postponement of such plans until the economy recovered enough to give displaced workers new jobs at no loss of wages. PECE and POUR reviewed all such proposals but avoided any notion of having federal regulators veto the introduction of new equipment, even in the interest of minimizing pain to workers. Such a concept undoubtedly must have appalled committee members such as L. R. Smith, head of an auto frame company that had achieved international fame for taking mechanization to new heights.[15]

At least one PECE/POUR official did focus on displacement as a significant problem. The quest for efficiency in modern business had brought about "employment insecurity through technological improvement," said Harry Wheeler, chairman of the Special Committee on Employment Plans and Suggestions. To underscore the gravity of the threat, he cited AFL figures showing that per capita hourly production had jumped 50 percent between 1923 and 1930. Such considerable changes in labor needs must inevitably harm labor,

Wheeler concluded, unless industry began to compensate for the "time economies resulting from improved machinery" by shortening the workday. As management reduced hours, they would naturally need to hire extra personnel to fill in the gap. PECE and POUR endorsed the notion of asking employers to spread jobs among more people, hoping such measures might jump-start economic activity and revive consumer confidence. DuPont, Kellogg, Eastman Kodak, U.S. Steel, Goodyear, Westinghouse, Western Electric, General Electric, Chevrolet, and other corporations experimented with share-the-work programs in the early 1930s, with varying results. Soon, POUR had to concede that just asking firms to cut hours had not made much difference in distributing employment.[16]

Despite such problems in leaving economic recovery to voluntary cooperation, PECE and POUR simply were not geared toward a more interventionist approach. Their philosophy could not support any call for more aggressive measures to make businesses institute share-the-work plans or rethink mechanization. Instead, PECE and POUR aimed to bolster people's spirits, stressing the faith that Americans' goodwill and resourcefulness could naturally restore economic soundness. The committee could not entertain any idea that the Machine Age might have problems too deeply rooted to be solved by hiring men to build rock gardens. Nevertheless, many Americans kept appealing to the administration to consider such possibilities. In 1932, a group of six thousand unemployed New Yorkers criticized business and politicians alike for overlooking the "revolutionary fact" that the tremendous power of mechanization had made existing employment policies inadequate. Republican C. H. Christensen, mayor of Palo Alto, wrote Hoover specifically to warn that the American economy had been "jerked out of balance" by the rapid advance of manufacturing ability. As a small-business owner, Christensen had his own reasons for disliking the way big competitors introduced new machinery, but he also argued labor's case that change had thrown too many people out of work. Modern techniques of industry represented "a Frankenstein monster" that had started "devouring our civilization."[17]

The administration's failure to acknowledge such concern did not mean that the president remained blind to the distress of unemployed Americans. Rather, Hoover's background and intellectual outlook had steeped him in an idealized vision of engineering progress. His experience as Commerce Secretary had further convinced Hoover that mechanization inevitably represented a positive economic incentive—a conviction too strong for even the Depression to shake. Though willing to concede that the adoption of new equipment sometimes created a brief shortage of work in certain localities, Hoover could not accept the possibility of its having grown into anything more serious. Even after he had completed his term in office, Hoover continued to downplay any link between mechanization and job loss. In a 1936 speech to the American Society of Mechanical Engineers, Hoover did acknowledge that technological unemployment had probably risen proportionately more in the United States recently than at any other point in time. However, he continued, the historical advance of civilization should persuade people to trust that mechanization eventually led to lower prices and increased consumer demand, thereby raising living standards and creating jobs. As an ironic footnote, when Hoover built a new home in California back in the 1920s, his wife Lou Henry reportedly purchased all the latest labor-saving household devices—except for a modern refrigerator. Architect Birge Clark recalled that according to the Hoovers' son Allan, "the reason that we didn't change to electric refrigerators was that his mother knew the iceman, and she didn't want to participate in eliminating him."[18]

During the early 1930s, Washington did have one group designed to address large-scale issues like technological unemployment. Before the crash, in September 1929, Hoover had organized a special Research Committee on Social Trends to analyze complex social relationships and emerging areas of strain. Wesley Mitchell, director of the private National Bureau of Economic Research, headed the group, joined by sociologist William Ogburn, University of Chicago political scientist Charles Merriam, and others. The committee's voluminous report, *Recent Social Trends in the United States*, ranged widely over matters including urban and rural life, law and crime, recreation and arts, health, religion, racial and ethnic issues, and the status of women and children. In that broad analysis, the group referred to the substitution of machines for men as a serious problem that might worsen in the future but that represented only one among many formidable challenges of twentieth-century life.

Economists Edwin Gay of Harvard and Leo Wolman of Columbia affirmed in *Recent Social Trends* that advances in science and invention had brought about an "unusual increase" in labor output during the 1920s. The gain in per capita production had come so sharply that technological unemployment now presented one of the country's "most pressing" problems. The adoption of labor-saving equipment in the railroad industry between 1920 and 1930 had greatly increased operational efficiency but eliminated up to 535,000 workers, Wolman and Gustav Peck of the City College of New York wrote. At the national level, temporary displacement posed a "problem of increasing gravity," while the prospect of accelerating mechanization in years ahead threatened to cause a permanent absence of jobs. Sometimes new technologies could have a positive effect, Ralph Hurlin of the Russell Sage Foundation and

Meredith Givens of the Social Science Research Council noted. In steelmaking, the efficiency gained through manufacturing improvements had generated enough savings to spur market growth, which opened up opportunities for workers. But Hurlin and Givens doubted whether the steel market could sustain unlimited gains in consumption, and, when the expansion stopped, an "aggregate displacement of labor" would catch up.[19]

To explain why displacement posed more of a threat in the modern age, the committee drew on Ogburn's cultural lag theory. Reporting that the number of patents issued had jumped from 143,000 during the years 1901–5 to 219,000 during the years 1926–30, the group suggested that the future would bring ever-accelerating development in science and engineering. In contrast, government agencies and other institutions tended to resist rapid alteration, meaning that, when compared to the advance of technology, the nation's social and political context would become increasingly outdated. Such uneven rates of change could create "zones of danger," as if "parts of an automobile were operating at unsynchronized speeds." Unless the United States saw either "a speeding up of social invention or a slowing down of mechanical invention, grave maladjustments" were certain to occur.[20]

Since Americans usually felt reluctant to place constraints on scientific and technical initiative, the committee concluded, they must therefore redefine social institutions to fit the development of machines. Adopting proposals for shorter hours, unemployment insurance, and old-age pensions might cushion distress for brief periods, but such half-measures would not prevent the growing problems of displacement. The lag between technical knowledge and everyday existence would continue to widen, unless the nation embarked on a bold course of social invention. Most people, conservative by nature, instinctively shunned radical experiments in life, the report continued. Therefore, experts in government and the social sciences must assume leadership and impose comprehensive social and economic adjustments, as they had done fifteen years before in mobilizing the country for war. Adapting to the Machine Age called for "effective integration of the swiftly changing elements" in life, a job requiring "nothing short of the combined intelligence of the nation" over years to come. The committee proposed that Americans might create a National Advisory Council of politicians, businessmen, labor leaders, scientists, and "statesmen in social science," who would keep an eye on workplace trends. Their deliberations would ensure that "economic, governmental, moral and cultural arrangements should not lag too far behind" the advance of research and development.[21]

Such a scheme would effectively institutionalize Ogburn's cultural lag theory, defining displacement as a problem of social disjunction that could only be solved through professional expertise. The social scientists proposed to give themselves a central role in arranging America's future, in saving the country from its own talent for mechanization. Their prediction that technological change would risk aggravating social and economic tension did not harmonize with the Hoover presidency's efforts to shore up the trust that mechanization meant progress. By 1933, that conflict ceased to matter. Future decisions about how government should address the question of displacement would rest with the incoming Roosevelt administration.

The Roosevelt Administration Faces Concern about Machines and Jobs

At the time of Roosevelt's inauguration, widespread distress made technological unemployment appear more than ever a continuing national emergency. The Hoover administration had dismissed cases of displacement as a temporary anomaly: the nation's fundamentally healthy business system would soon correct such glitches, officials promised, validating faith in mechanization. With the political changing of the guard, Americans began hearing a radically different message. Under Franklin Delano Roosevelt, Washington leaders openly considered the possibility that current problems of job loss represented merely the tip of the iceberg. Far from automatically creating more work, they warned, future mechanization in industry, agriculture, and services might turn technological unemployment into a persistent curse of modern life.[22]

The idea of technological unemployment as a crisis remained exceedingly controversial. One of the Roosevelt administration's earliest attempts to address the issue sparked internal disagreement and public embarrassment. William Ogburn had prepared a pamphlet entitled *You and Machines* for the American Council of Education, material intended for instructional use in the Civilian Conservation Corps. CCC executive Robert Fechner reportedly found Ogburn's assessment of mechanization and jobs overly pessimistic and promptly banned it from CCC camp libraries and education programs. Ogburn protested that he remained "very optimistic regarding the future." Meanwhile, University of Buffalo economist Percy Bidwell, editor of the pamphlet series, praised *You and Machines* as a well-balanced account. Ogburn had avoided the unrealistic portrayal of technology as a source of universal blessing, Bidwell said, but he also had rejected the notion of an "unconquerable monster devouring the world."[23]

In truth, Ogburn did offer encouraging indications that new technology brought concrete social and economic benefits to ordinary Americans. When manufacturers introduced better machines, the booklet indicated, gains in efficiency translated into consumer savings. To illustrate, a cartoon showed a seesaw with a robot on one end, pushing down the "cost of goods," while a happy man rises into the air on the other side, enjoying an improved standard of living. Thanks to modern production, Ogburn wrote, a "girl behind the counter at Woolworth's" owned more silk stockings than had Queen Marie Antoinette. The accompanying drawing showed a twentieth-century worker comparing his home to Versailles and announcing, "I wouldn't trade houses with you, Marie!" In the interests of humor, Ogburn sometimes stretched the truth about machines' capabilities; cute but outlandish images showed a humanoid robot washing dishes while instructing the housewife to "run along and have fun."[24]

At the same time that Ogburn emphasized how workplace mechanization made consumers richer, he argued it "has always been true" that such improvements eliminated labor. The general economic slump still accounted for most job loss; Ogburn estimated that "not more than 1 out of every 7 persons unemployed in 1933, perhaps not even 1 in 10, had his job taken away by a machine." In sectors like agriculture, however, the damage appeared clear. A single wheat farmer using the latest tractors could, in three hours, complete work that would have taken fifty-seven hours a century before. Mechanical corn harvesters might displace hundreds of field hands, and once inventors perfected the automatic cotton picker, Americans could "bet that there will be no provision for the Negroes who will lose their jobs."[25]

Other geniuses had revolutionized brickmaking and road-laying, Ogburn told readers. Restaurants could use a machine to pour, flip, and serve pancakes automatically. Hollywood's switch to sound movies had ousted ten thousand musicians, who discovered that protesting did them "no more good than it would for farm horses to kick the tractors that replace them." True, employers did not adopt new machinery instantly; the cotton picker so far remained limited to certain Southern regions, and coal-mining machines to the East. Thousands of jobs might arise from expanding demand for printing, plus the growth of business in the radio and automobile industries. The unprecedented scale of change threatened to create numerous "pools of technological unemployment," Ogburn concluded, but intensified innovation and economic development would help such pools "disappear more quickly."[26]

The imbalance in Ogburn's pamphlet came from the visual images, which conveyed a vivid impression of machines harming labor. One drawing showed a man at a factory door reading the notice, "You're fired—new machines take your job." The artist turned the icon of a humanoid robot into a visual shorthand for displacement; cartoons showed the robot trying to snatch a hammer out of a worker's hands or shoving the fellow away from his workbench. In another picture, the robot pushes an elderly man toward the edge of a cliff. Though the text included reassurances that mechanization could generate as well as destroy opportunity, none of the illustrations depicted men working on an automobile assembly line or in a radio salesroom. Readers who only looked at the pictures would overwhelmingly associate technology with job loss.[27]

Ogburn also phrased his criticism in extreme language. Calling the machine "as dangerous as a wild animal," Ogburn announced that changes in industry endangered American workers' survival as much as saber-toothed tigers had jeopardized our prehistoric ancestors. Just as early humans had discovered how to protect themselves from predators, he suggested, twentieth-century people must learn to defend themselves against the downside of progress. Ogburn defined the problem in terms of his cultural lag theory, explaining that modern invention "comes so quickly we are unprepared." Although government and society had so far remained "always behind time in adjusting," the nation might yet master the challenge. His conclusion offered plenty of ambiguity: the last drawing showed a young man triumphantly riding on the back of a robot, like an all-American cowboy, yet the final sentences referred to the machine as "a new monster," phrasing guaranteed to prove controversial.[28]

Though interpreting the effects of technological change remained a touchy venture, President Roosevelt himself did not shy away from the subject. Speaking at a 1935 press conference, he indicated that American workers might have good reason to worry about their fate. Even if the nation could immediately restore production to the peak levels of 1929, he declared, that rebound would only supply jobs for 80 percent of the unemployed, because mechanization had so greatly increased per capita efficiency in the meantime. The president's statement echoed what some union officials and social scientists had already suggested, that the economy was losing its ability to absorb displaced workers as employers kept installing more labor-saving devices. Roosevelt's comments made front-page headlines. Business leaders sharply criticized any such notion, contending that statistics showing employment lagging behind productivity either reflected meaningless fluctuations or had been misinterpreted. *New York Times* editors chastised the president for falling into the trap laid by "calamity prophets." A calm look at economic data, they insisted, proved "plainly there is no such dire disproportion between business activity and employment."[29]

Over subsequent days, other administration officials echoed the president's disturbing assessment. "Man has been thrust aside to make way for the machines," Aubrey Williams informed Americans, meaning that joblessness was, "like the airplane, the radio, the weather and taxes, here to stay." As evidence, he observed that employment rates had failed to rebound on pace with

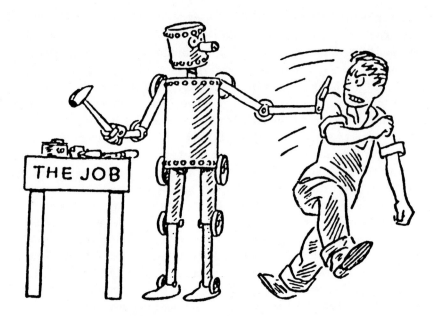

Fig. 2. Illustrations of William Ogburn's readable pamphlet on the technological unemployment issue provide plenty of food for thought. On the one hand, his robots appear rather cute, being relatively small and resembling animated tin cans. On the other hand, when shoving workers aside or hitting them in the face, the robots packed a heavy punch. Illustration by Fred Cooper, in William Ogburn, *You and Machines* (Chicago: University of Chicago Press: 1934), 6.

the recent upswing in business activity. Thanks to the way technical innovations had raised production capacity, Williams said, steelmakers and automobile companies could build up output without needing to expand their payrolls. Given such trends, "thousands of young men and women" growing up in the Machine Age were "destined never to become self-supporting."[30]

The WPA National Research Project: Organized Investigation of Displacement

While Washington personnel were now prepared to take technological unemployment seriously, the Roosevelt administration realized they still lacked data. Since the early 1930s, the Bureau of Labor Statistics had conducted some research on how the introduction of sound movies, dial telephones, and the teletype had affected jobs. Further investigations covered technological innovation and unemployment in agriculture and railroad work. But since bureau staff had to pursue many issues, from wages and prices to industrial health and safety, they could not concentrate exclusively on the topic of displacement. Their studies provided some valuable statistics on specific technical changes in certain professions but could not support any conclusions about more complex national economic patterns.

In the fall of 1935, WPA head Harry Hopkins established a new research program to improve understanding of the nature, causes, and effects of joblessness in contemporary life. The federal government had made establishing essential relief programs its first priority, Hopkins explained. With the recognition that unemployment was "likely to remain for some years," the time had come to start evaluating the factors behind disaster. A mandate to analyze job loss could lead in many possible directions. Significantly, the WPA elected to focus on the "role that technology played in current unemployment problems and long-term employment trends." In justifying that decision, assistant WPA administrator Corrington Gill observed that the "much-abused term, 'technological unemployment' entered our popular vocabulary with very little exact information" behind it. Certainly "no one can deny that in the last few years it has been voluminously written and talked about," but given the lack of any real knowledge about how machines affected labor, it was "not surprising that the average man is confused."[31]

The agency hoped that its National Research Project on Reemployment Opportunities and Recent Changes in Industrial Techniques might clear up such confusion. Hopkins' initiative made front-page news; in committing significant resources to this major initiative, Roosevelt demonstrated his readiness to treat technological unemployment as a genuine concern. WPA staff started with the hypothesis that the force of mechanization, which had brought "tremendous increases in the volume of goods and services," had also caused "greater economic insecurity of the individual worker." To support that premise, researchers pointed to statistics showing that manufacturing production had jumped 37 percent between 1920 and 1929, while manufacturing employment had fallen by 2 percent. Such a trend had continued despite the Depression, they noted; man-hour output in industry had risen another 25 percent between 1929 and 1935, even as joblessness became a national emergency.[32]

Such an approach met immediate opposition in some quarters. A March 1936 editorial in the periodical *Product Engineering* attacked the WPA for holding machinery "guilty until proven innocent." The McGraw-Hill journal charged the Roosevelt administration with failing to see that increasing productivity "invariably creates much employment outside of the factories." As mechanization allowed manufacturers to make automobiles more affordable, jobs multiplied in gasoline manufacture and sales, insurance, garage work, and road construction. Any criticism of production endangered the very course which had created the "present trend toward an ever higher plane of living for all." In short, the publication warned, "take away machinery and civilization will vanish."[33]

Defending his study, Gill insisted that the WPA fully recognized the ways in which mechanization could spur "tremendous multiplication of labor opportunities." However, a significant time lapse might exist between the disappearance of old jobs and the appearance of new ones. Earlier economists had only "vaguely understood" the complexities of such phenomena, but Gill promised that the WPA could establish the "relative potency of the forces" of labor displacement and labor absorption. He referred again to indications that recent rises in productivity and profits had not yielded corresponding employment gains, "disturbing" evidence that demanded careful investigation.[34]

Project director David Weintraub, who had been an economics instructor at the City College of New York before joining the National Bureau of Economic Research, planned to approach the matter from three angles. The project would start with a general statistical analysis to trace national trends in mechanization, productivity, and labor displacement. Second, researchers would study technical changes in specific industries, drawing on information and input from company executives, the Bureau of Labor Statistics, the Department of Agriculture, the National Bureau of Economic Research, and other appropriate sources. Finally, the WPA would explore what happened to displaced workers, how individual circumstances, local conditions, and the national economy all affected chances of reemployment. From the beginning, Weintraub stressed that the WPA would adopt a scientific perspective in "considering the economic conditions under which men and machinery are employed." The issue, he emphasized, could not and should not be oversimplified into an emotionalized battle of "men versus machines."[35]

Business leaders had a record of unpleasant experiences with Roosevelt's New Deal, and their distrust of the administration's motives extended to the WPA's study of technological unemployment. Some reportedly perceived the project as a means of hiding the president's real motives, a possible cover-up for attempts to revive the NRA or otherwise interfere with private enterprise. Frequently, WPA researchers who approached corporations with requests for information encountered a flat refusal. Firms that manufactured industrial machinery proved especially reluctant to cooperate. The machine-makers' trade association complained that economists often looked only at whether new equipment added or subtracted jobs in the workplace that adopted them, failing to factor in the way improved capacity created employment through the development of entirely new lines of business and additional demand for raw material. Such tactics presented a misleading picture, the industry protested, writing an unfair "bill of indictment against the machine."[36]

WPA staffer J. V. H. Whipple admitted that recent discussion of displacement had thrown business "distinctly on the defensive." The tide of popular feeling had made people "very resentful against any machinery . . . advertised as saving labor." Whipple agreed that machine-makers had a valid point regarding methodology. Statisticians needed to account for the way that improved production technology contributed to reemployment by lowering prices and stimulating market growth. Once experts devised a "straightforward" system to give the industry full credit for adding jobs, he assured his supervisor, the project would win "enthusiastic cooperation from the machine building fraternity." Accounting for mechanization, productivity, and labor change proved anything but straightforward. In internal memos, government economists puzzled over problems in classification. When a company came out with a 1937 model refrigerator, should that be designated as an old or new product? For that matter, did the entire electric refrigerator business count as a new industry, or just a derivation of icebox manufacture?[37]

While WPA staff accepted the idea that technological improvements created some jobs, most believed that spokesmen for the Machinery and Allied Products Institute stretched that point too far. Weintraub accused manufacturers of inflating employment numbers, making arbitrary distinctions between "mechanized" and "nonmechanized" business, and selecting only the most favorable periods for comparison. Such tricks produced a one-sided interpretation, he complained, and allowed the organization to make unrealistic assumptions about economic behavior. The Machinery and Allied Products Institute assumed that an inevitable rise in prosperity must ultimately absorb displaced workers, a generalization that "glosses over the problems inherent" in facing the changing economy. The fact that many people remained out of work for a relatively lengthy period, with accompanying financial hardship and personal stress, "does not receive even passing mention" in industry material, Weintraub grumbled.[38]

Observers waiting for Weintraub's own economic assessment found plenty of material for discussion in 1937, when the National Research Project's first major publication appeared. The report, entitled *Unemployment and Increasing Productivity*, suggested that mechanization had displaced a significant number of American workers during the 1920s and that, for all the prosperity during that decade, the economy had not automatically reabsorbed that labor. Researchers did not place a figure on technological unemployment, declaring that methodological difficulties and lack of data made it impossible even to guess. Drawing on data from the Census Bureau, Interstate Commerce Commission, Bureau of Labor Statistics, and other sources, however, Weintraub showed that between 1920 and 1929, production had risen 46 percent, with only a 16 percent employment growth. Those statistics made it "reasonable to conclude that in any given year a considerable proportion of the unemployed consisted" of workers displaced by new technology. The expanding service sector had created some work, but not necessarily enough. People needed certain skills and experience to win a position in a hotel, garage, or restaurant, which made it "extremely unlikely that all the workers displaced from basic industries obtained new jobs in service."[39]

Such statements reinforced the suspicion voiced by President Roosevelt and labor leaders that the accelerating pace of mechanization explained why joblessness had remained even as business began to recover from the Depression. Employers just no longer needed the same manpower, since technology had raised productivity so dramatically. To complicate matters, the number of Americans seeking jobs had increased by roughly 4 million between 1929 and 1936. Given those figures, production rates of 1929 were no longer sufficient to absorb all workers, the WPA calculated. In order to bring unemployment back down to the 1929 level, the United States would need to push up its production of goods and services to 20 percent beyond 1929's total.

The country did not appear ready to meet such a challenge, government economists warned. "Since our economic system has not evinced an ability to make the necessary adjustments fast enough, it may be expected that the dislocations occasioned by technological progress will continue to present serious problems of industrial, economic and social readjustment," the report continued. The likelihood of continuing displacement, Gill wrote, should not lead Americans to abandon the pursuit of scientific knowledge and invention. Over the long term, he stressed, any hope of improving living standards "depends upon technological progress." Instead, the government might establish new agencies to keep abreast of developments and recommend appropriate means to offset them. The idea of reducing working hours to accommodate declining labor needs seemed logical, for example. The country might create retraining programs and employment offices for displaced workers, or provide special unemployment insurance and relief. Without such steps to adjust labor conditions to the Machine Age, he cautioned, "dark prophecies of ever-increasing unemployment" might come true. Future generations might face permanent impossibilities in finding work.[40]

Those harsh words, which effectively established an official concept of permanent technological unemployment, attracted widespread comment. The

Cleveland Plain Dealer hailed Weintraub's report as a much-needed contribution to economic understanding, but the *Indianapolis News* grumbled that the "WPA has not won the confidence of the people as an authority on unemployment and the report may be considerably colored . . . by the environment in which it was prepared." Editors grudgingly acknowledged that the WPA's assessment did seem to fit "conclusions based on other surveys," that mechanization had cut into labor needs. In the more negative vein, the Elizabeth, New Jersey, *Journal* objected to the fact that the WPA still had not placed any exact figure on technological unemployment. Because of that defect, editors indicated, the government's work remained "a long ways from being a comprehensive, satisfying presentation."[41]

Critics at the *New York Times* challenged the WPA report, describing it as "misleading," while the *Christian Science Monitor* called it "unfortunate that the WPA should reinvoke that fear of the machine as a job destroyer." Both papers argued that, far from trailing business recovery in the 1930s, employment had actually surpassed it. They cited a recent study by the corporate-sponsored National Industrial Conference Board (NICB), which found employment running only 9 percent below 1929's figure, while production lagged 17 percent behind. Giving credence to NICB numbers over WPA data, the *Monitor* declared that "re-employment is proceeding at a pace which has laid the specter of a progressively jobless nation" to rest. Continued recovery might even lead to labor shortages.[42]

In response, Weintraub pointed out that the *Monitor* (like several other papers) had reported the WPA study incorrectly. According to their articles, the agency had claimed production would need to rise 20 percent over 1929 "to bring employment back to its 1929 level." In actuality, the report referred to returning *unemployment* to 1929 rates, a vital difference. Regarding the differences in numbers, Weintraub accused NICB economists of artificially inflating employment by counting entrepreneurs, family farm workers, and self-employed Americans. He complained that *Monitor* editors had unfairly condemned the WPA as being antimachine, when its analysis explicitly stated that future well-being would rest on continued technological innovation. Gill wrote the *New York Times* to protest its editors' similar readiness to accept NICB figures over WPA statistics. The NICB had been able to twist facts into predictions of a labor shortage only by starting from the "particularly doubtful" assumption that national production would soar 220 percent between 1932 and 1940. Gill ridiculed the NICB for projecting that construction employment could jump 126 percent and manufacturing jobs rise 39 percent, estimates not "justified by any known" evidence. Authors of the NICB study explicitly admitted that they had chosen a set of economic assumptions without regard to probability, opting instead for "setting the extreme limits of possibility." In overlooking that important caveat, Gill declared, the *Times* had become a booster for shoddy research.[43]

In a personal reply to Gill, the *Times* acknowledged that its "real difference" with the WPA work lay not in "differences of factual calculation" but in the editors' predilection for favoring the most optimistic economic suppositions. The newspaper believed the WPA must be wrong, that, by definition, mechanization could not create any meaningful trouble. Editors admitted they had started from an assumption that "neither technological improvements nor increase in working population create any permanent unemployment problem," the very proposition that the WPA study claimed to disprove. The *Times* chose to trust that "given a restoration of normal conditions there would be sufficient increase of production to absorb the unemployed." In other words, the complicated, multidimensional issue of displacement boiled down to a matter of faith in American progress.[44]

Harry Hopkins sought to drive home the argument that mechanization threatened the nation's chances of resolving its unemployment crisis, drawing attention to the WPA's research in speeches he delivered all across the country. Business boosters might trumpet the restoration of prosperity, he warned, but a rebound in production would still leave four to five million Americans victims of long-term if not permanent displacement. He declared, "When industry stopped in 1929, the research workers didn't stop, so that we now need less people to maintain 1929 production." Thanks to new machines and methods, he said, nine men in 1937 could perform as much work as had ten in 1929. President Roosevelt's 1937 State of the Union address echoed the idea that it might become ever harder for the economic system to provide jobs for all: "As a result of the natural increase in our population, each year at least 400,000 new workers are seeking work," he said, even as mechanization sharply reduced labor needs.[45]

Over the months, WPA researchers pursued the implications of such economic and demographic trends from many angles. Studies traced patterns of mechanization and employment across mining, agriculture, and construction. Staffers investigated developments in fifty-nine fields of manufacturing, including rubber, glass, leather, iron and steel, chemicals, motor vehicles, tobacco products, textiles, and the production of different foodstuffs, from canned and baked goods to candy and ice cream. As one example, Boris Stern found that cotton and woolen mills of 1936 could manufacture a given length of cloth with less than half the labor time required in 1910. Employers had introduced so many kinds of automatic machinery that "hardly a department . . . has not been affected by this sweep of modernization." Across the board, the WPA indicated, the labor required to yield a given unit of output had fallen by 10 percent between 1929 and 1939. Anyone who hoped that rising prosperity would create enough new consumer demand to absorb displaced workers should realize that in order to bring unemployment down to its 1929 level, 1939 national income would need to rise "more than 1/4 again as high as in 1929."[46]

WPA experts sought to learn exactly how much dislocation occurred, how fast, and how such harm balanced out against economic gains. To establish the length of time it took displaced laborers to secure new positions, staffers interviewed more than twenty-two thousand men and women. One study, produced in cooperation with the University of Pennsylvania's industrial research department, detailed the work history of 357 former weavers and loom fixers displaced when Philadelphia textile plants adopted more automated looms. For those who had devoted their entire lives to weaving, the prospect of switching jobs presented major psychological barriers. Many tended to "cling to their trade" despite the obvious fact that with automated looms, mills would never need as many operators as before. Some hunted for months for new weaving jobs and, if forced to pursue different work out of desperation, refused to accept the fact. Some who had not worked in textiles for over ten years still referred to themselves as weavers. Only the youngest managed to retrain, and even then, they generally had to settle for less skilled positions at a sacrifice of both money and prestige. A similar study of six hundred New Hampshire cigar-makers displaced by the introduction of rolling machines showed that 25 percent remained out of work for at least five years.[47]

WPA evidence confirmed that the nation's economy had started undergoing a fundamental shift away from traditional agriculture and industry. Researchers demonstrated that employment in farming, mining, manufacturing, construction, transport, communication, and utilities had grown only 3 percent during the 1920s, while service positions had nearly doubled. By 1929, service work accounted for 44 percent of total employment, a new high. Weintraub did not believe, however, that such expansion would naturally absorb people who lost jobs in older fields. One study by Isador Lubin showed that of 410 displaced workers who succeeded in finding new places within a year, only 53 went into service positions or found employment in fields derived from recent inventions such as automobiles, radio, or movies. Furthermore, the bottom rungs of service tended to mean low pay and low job security. Economic theory often treated old and new occupations as interchangeable, but in real life, human beings could not shift gears so easily.[48]

Hard experience showed, Weintraub wrote, that improvements in production could not translate immediately into lower prices and higher consumption. Any advantages of mechanization tended to trickle down slowly, and "until such benefits materialize, increasing productivity augments . . . unemployment." He included disclaimers carefully stipulating that still, "technological advance has been and is a necessary condition for the development of our nation's economy" over the longer term. Despite such words, many in the business community continued to perceive the WPA as preaching an antimachine sentiment. The agency wanted to win acceptance, since its research depended on getting information from industry. In 1938, staffers invited M. F. Behar, editor of *Instruments: The Magazine of Measurement and Control*, to review a draft of their report on how new instrument technology affected production, labor, and consumer goods. The project authors accepted some of the changes Behar recommended, and the final version of *Industrial Instruments and Changing Technology* did not sound bitterly antitechnology. The study described the use of relays, switches, and sensors as direct substitutes for manual labor and human inspection, but it also praised the role of instrumentation in facilitating development of automobiles, airplanes, radio, plastics, and other new industries. Nevertheless, that balance did not satisfy Behar, who proceeded to denounce the WPA as a bunch of "crackpots" who "unduly cry 'job-elimination!'"[49]

Over roughly five years, the National Research Project published more than sixty reports, representing the largest collection of research to date on trends in technological change, production, productivity, work, displacement, unemployment, and reemployment. The government distributed six thousand to ten thousand copies of each monograph. Columbia University and other educational institutions adopted some as texts for courses in economics, social science, and technical subjects.[50] The WPA approach had its gaps. The group concentrated on reviewing conditions in manufacturing and agriculture, saying relatively little about employment and technological changes in newer fields and the service sector. Nevertheless, the project made it ever harder for America to ignore the issue of technological unemployment. Weintraub presented evidence that appeared at least to validate the possibility that no matter how vital mechanization might be to long-term economic advance, it had factored into the nation's immediate crisis. Still more disturbing, WPA research seemed to support the fear that displacement might intensify down the road, a sign of imbalance in the modern world.

The value of WPA work lay in the attention researchers devoted to defining how mechanization affected labor prospects in specific types of work, such as agriculture. The latest model tractors had grown increasingly sophisticated, bringing new power and flexibility into crop production. To its admirers, such a machine represented "an ideal . . . robot farm worker," but unskilled farm hands complained they had been "tractored out" of jobs. The U.S. Department of Agriculture predicted that over the next decade, the number of tractors used on American farms would virtually double, displacing up to 400,000 more workers. Multiple technological changes accumulated; with motor vehicles, tractors, cultivators, and harvesters, a single farmer could cultivate 170 acres or more on his own. In dairy operations, the use of milking machines reportedly changed milking a herd of twenty-six cows from a three-man, 156-minute operation into a one-man, 43-minute job. According to agricultural engineer Henry Giese, total power resources available to American farmers had risen from seven million horsepower in 1850 to 47.5 million in 1924.[51]

Secretary of Agriculture Henry Wallace suggested that while old-fashioned methods of farming might sound romantic in retrospect, no one could deny how much technical and scientific research had aided production. Yet, even as he exulted about how much the reaper, the combine, the truck, the tractor, and the gang plow had multiplied crop yields, Wallace expressed reservations about their impact on rural populations. Since "we have now come to days of real soul-searching about all the things . . . hitherto called progress," he told the American Association for the Advancement of Science, it was "high time . . . to analyze these various labor-saving devices a little more critically." Farm mechanization would not truly improve life until the nation made proper "social adaptations."[52]

In 1936, the Department of Agriculture's Extension Service and the Agricultural Adjustment Administration issued a booklet entitled "Is Increased Efficiency in Farming Always a Good Thing?" The publication encouraged rural citizens to form discussion groups and start "thinking about these questions for themselves." It directed readers to reflect on the changes they had seen and asked, "Do you know of cases where the increased use of machinery in factories has thrown people out of jobs?" The text presented a fictional debate in which one farmer commented, "THE MORE MACHINES, THE FEWER JOBS." Given the prospect of cotton-picking machines that could handle as much work in one day as one hundred hand-pickers, the booklet's farmer declared, rural America had come "FACE TO FACE WITH THE UNEMPLOYMENT PROBLEM." Still, changes had indisputably improved the quality of agricultural products and slashed their cost, suggesting that the United States should "KEEP THE MACHINES" and concentrate on finding ways to "LESSEN THE HARDSHIPS CAUSED." The pamphlet promoted an activist government policy, suggesting that New Deal agricultural programs and the creation of Social Security could "cushion the bumps" in the "machine economy."[53]

Taking a personal survey of economic and social conditions across many states, journalist Lorena Hickok told Harry Hopkins about the toll that changing methods of farming had taken on rural people. On Maine potato farms, "as in many other places, improved machinery is taking the place of hand labor." Devices for digging potatoes could let one or two men complete the work of fifteen, Hickok wrote, and soon thousands of common laborers would find their chances of "earning any living from that source . . . practically nil." To supplement such informal observations, the WPA analyzed the details of how agricultural innovation affected production and employment. Since trends varied by region and by crop, the National Research Project produced separate studies of corn, wheat, potatoes, sugar beets, and cotton farming. Under John Hopkins, the project's head agricultural economist (on leave from Iowa State College), staff collected data from the U.S. Department of Agriculture and Census Bureau and also conducted their own field surveys of 4,300 farms.[54]

The study of changing technology in the cotton belt, conducted by government farm economist Paul Taylor, conveyed an explicit picture of how the "driving force of mechanization" had started "accelerating profound changes in the rural structure." Over recent decades, the spread of tractor farming had started "expelling families" in several states, Taylor repoi ted. Landlords who bought new equipment often combined several 160-acre parcels into one giant operating unit, dismissing the tenant families who formerly had worked the land. One cotton planter who had purchased new tractors and cultivators subsequently eliminated 130 out of 160 sharecropping families. Most displaced field workers could not even secure steady employment as tractor drivers, Taylor explained, and the rural economy could not absorb such an impact. Between 1930 and 1935, the farm population of one Texas panhandle county had fallen 24 percent, altering the entire foundation of community life. Especially in the South and the Great Plains, mechanization multiplied the strain in a population already experiencing a crisis due to drought, land erosion, and economic depression.[55]

Problems for the Southern farm society might explode, social scientists worried, once inventors succeeded in perfecting the cotton-picking machine. Experimental versions developed by 1936 used moistened rotating wire spindles to snatch cotton from open bolls. The potential for full-scale manufacture and adoption of such devices attracted significant attention in Depression-era discussions of technological unemployment. The Department of Agriculture used the cotton-picking machine to symbolize the whole issue, placing a full-page picture of one on the cover of its 1936 pamphlet under the question, "Is Increased Efficiency in Farming Always a Good Thing?" Government experts kept a close eye on the state of invention; in 1936, WPA staff conducted a special inspection of the Rust cotton-picker, one of the versions that appeared closest to practicality. Project economists estimated that, given the Rust machine's current speed and capacity, a final version might offer a cost advantage of $3.50 per bale—sufficient to trigger widespread job loss.

According to one article, the men behind the picking machine appreciated "the misery and tension that any further insecurity" in cotton regions would cause. Southerners John and Mack Rust reportedly felt reluctant to market their device "until they can discover some hopeful method of softening its bitter blow upon the back of the sharecropper." To get advice on the human situation, some developers turned to government experts. In a 1937 letter to Paul Taylor, Isador Lubin explained that International Harvester had promising plans for a picker but would need to spend several hundred thousand dollars to iron out the bugs and manufacture a few test machines. The company was "fully conscious" that a picking machine would "have a terrific effect upon the economic and social situation in the South." Its executives had trouble deciding whether to commit the extra funds, and Lubin commented, "I think they hesitate to push the picker for fear that they might be accused of disrupting the economic order." International Harvester accordingly asked federal economists for their opinion, and USDA analysts predicted that an efficient machine might displace 25 to 75 percent of the country's two million cotton-picking tenant families.[56]

In a sweeping report entitled *Changing Technology and Employment in Agriculture*, the WPA said that farm employment had dropped from 12.2 million to 10.6 million between 1909 and 1939, as the number of tractors grew from 10,000 to 1.3 million. During those decades, improved design made farm machines faster, lighter, and more powerful. The versatile, all-purpose tractor could both prepare fields and cultivate crops; with power-takeoff driveshafts, a farmer could attach a small combine, a mechanical corn-picker, or other implements. By the most conservative estimates, each tractor eliminated 150 man-hours of labor a year, a total of 195 million man-hours by the end of the 1930s. As farmers reduced the number of horses they raised, another 370 million man-hours were saved. With the additional 335 million man-hours eliminated in producing fodder, a total 900 million man-hours of farm labor had vanished. Manufacture, supply, and maintenance of tractors required 345 million man-hours; therefore, John Hopkins calculated, tractors had eliminated a net 555 million man-hours of work. Other innovations added up to a revolution in agricultural operations. Corn-picking machines eliminated work for small-town men who had previously earned six to eight weeks' wages by picking. In grain states, use of harvester-thresher combines had caused "a great army of migratory harvest hands" to disappear "almost entirely within a decade." The WPA noted that new machinery, along with improvements in seed, stock, insect control, and veterinary medicine, had dramatically improved farm efficiency and helped reduce food prices somewhat. Labor "almost invariably" paid the cost for such advances, and, with increasingly intensive mechanization, upcoming decades would see more "direct economic waste" in the form of technological unemployment.[57]

Manufacturers of farm machinery understandably resisted the idea that their business bore any blame for the social and economic trauma in rural America. Tractors were not labor-displacing but labor-serving, the industry insisted. Fowler McCormick, vice president of International Harvester, testified to Congress during the 1939–40 TNEC hearings that, far from endangering rural culture, mechanization strengthened the family farm. By relieving people of the most back-breaking chores, the exhausting side of agricultural labor, tractors would make country life more attractive and prevent the next generation from heading for the city. Mechanization promised to give farm families both greater profit and increased leisure. Under questioning, McCormick acknowledged that machines often displaced hired farm hands and migratory workers, but he argued that "if any form of labor has to be obviated, the floating or migratory form is possibly socially the best one." WPA economists might fret over displacement, but land owners did not like having to rely on outside help. Farm wives praised the tractor as "the most wonderful thing" ever, as emphasized in public relations material from Caterpillar. One woman wrote that while in previous harvest seasons she had been "tied down at home all the time cooking for a bunch of hired help," her husband's purchase of a tractor had halved their need for extra hands and had given her a chance to "enjoy life" at last.[58]

For Paul Taylor, the harm done to migratory and tenant labor could not be so readily downplayed. While McCormick had implied that surplus farm workers would magically vanish as machines appeared, North Dakota sociologist J. M. Gillette had shown that mechanization fostered "a self-perpetuating class of socio-economically submerged individuals." In California and the South, stress over displacement had contributed to bitter disputes between landowners and desperate tenants. WPA research indicated that mechanization had led to a giant economic gulf in rural towns, creating a psychological "stratification" that affected the whole community. Stores lost business as sales of food, work clothes, and shoes plummeted, while the loss of population also made schools and churches into "victims of mechanization." Back in the eighteenth century, Thomas Jefferson had defined pastoral virtues as the foundation of democracy. Over the decades since, the images and ideals of rural life had often stood as a shorthand for national character and a measure of American well-being. Discussion of displacement during the 1930s reflected a fear that mechanization had corrupted farming, transforming a wholesome calling into an unhealthy chase after power and efficiency. The *Des Moines Register* summed up the matter by asking, "What kind of an agriculture do we want?" Mechanization had spread beyond large wheat, corn, and cotton operations, bringing "factory methods of production" into fruit and vegetable growing as well. Displacement had grown into a nationwide threat, the WPA concluded, creating stress for rural and urban working people alike.[59]

In addition to the reports produced by the WPA's National Research Project, the Roosevelt administration got further information about the economics of mechanization from the National Resources Committee. In 1937, the NRC's special Subcommittee on Technology published an ambitious document entitled *Technological Trends and National Policy, Including the Social Implications of New Inventions*. Back in 1929, the *Recent Economic Changes* report had argued that workplace innovation promoted long-term economic gains and hence social progress. The 1933 *Recent Social Trends* report had classified displacement as merely one among many trends, from those in education and leisure to those in religion and residential life, all reshaping the future. The 1937 NRC study (coordinated by none other than William Ogburn) focused more narrowly on mechanization, analyzing in detail its consequences for labor. The chapter written by David Weintraub emphasized evidence of rising technological unemployment, stating flatly that the business expansion of 1920–29 had not been enough to compensate for both mechanization and an increase in the size of the workforce. Weintraub repeated the WPA's conclusions that in order to bring joblessness down to 1929's rates, output would have to rise to a level 20 percent above that in 1929. Meanwhile, technology did not stand still, and continued pursuit of mechanization would make it ever harder for job growth to catch up with job elimination.[60]

The NRC report echoed Ogburn's "cultural lag" theory on technological unemployment, as had the Social Trends Committee four years before. *Technological Trends* started from an explicit premise that "the greatest general cause of change in our modern civilization is invention." By treating mechanization as a force in itself, rushing ahead independent of any other factor, the committee suggested that any attempt to constrain technical development must prove futile. No matter how overwhelming the future appeared, Americans could not and should not challenge the natural advance of research and innovation. It became "obvious" that industry must go forward with development of the photoelectric eye, more acute and reliable than human vision and able to operate twenty-four hours a day. When inventors combined such devices with advanced calculating machines that "almost parallel powers of the human mind," the NRC declared, "no limit can be set to the work which might be taken over by machinery." Such developments would take a natural course, and humans would simply need to adjust.[61]

Columbia social scientist Bernhard Stern, who justified the Luddites' resistance to machines as understandably grounded in social, economic, and psychological marginalization, did not see any way for modern Americans to resist the pace or direction of change. It was plain, he declared in the NRC publication, that mechanization under the capitalist system "overwhelmingly" served the interests of a small ownership class to the disadvantage, "often permanent, of the masses." When workers saw how often employers adopted machines to weaken the power of unions and break strikes, they could "hardly be expected to be receptive" to the promise of progress. Yet labor's unhappiness, Stern argued, only made it more urgent for business and government to find ways of bypassing that popular discontent, of overcoming objections to new technological reality.[62]

The report considered it inevitable that in years ahead, science and engineering would revolutionize all economic sectors, from agriculture and mining to transportation, communication, and electric power. The final section, over 250 pages long, assessed prospects for further innovation and emphasized that, of the many effects of technical change, the one "most in people's minds is that of displacing labor." In regard to farming, USDA staffers S. H. McCrory and Roy Hendrickson wrote that the likely development of good cotton-picking machines would give many Southerners "reason to cringe before the possible consequences." McCrory and Hendrickson acknowledged the argument that forcing the nation's tenant farming population to switch to new lines of employment might be good over the long run, giving their children access to better education and other advantages. The immediate result, however, could be disastrous. Since most tenant farmers had "almost no experience with industrial discipline and complicated machinery," their hope of finding new employment would be "hemmed in by limited opportunity."[63]

By keeping an eye open for such changes, the NRC declared, government could try to minimize the harm. Its report assured readers that experts could easily anticipate the course of workplace revolution, since invention "*never comes instantaneously without signals.*" Ogburn calculated that it took thirty years on average to bring an idea from conception to commercial success, a fact that offered plenty of "opportunity to anticipate the impact upon society." To justify such a large-scale commitment to planning, the report invoked the cautionary analogy of Frankenstein. Mary Shelley's monster had turned on its inventors and destroyed them because "victims had failed to make adequate preparations," but the NRC promised that Americans could save themselves from similar misfortune. The authors of the *Technological Trends* study proposed some specific guidelines for setting a system to prepare for mechanization. The NRC's Science Committee could track ongoing research projects, accepting a special responsibility to pinpoint the "more socially significant" angles. The *Technological Trends* publication itself singled out a few discoveries demanding immediate review, such as recent improvements in the cotton-picker, air conditioning, the photoelectric cell, and plastics. Once scientists had identified potentially troubling areas, a committee representing the Departments of Commerce, Labor, and Agriculture could calculate how much labor displacement might result and pass their verdict on to a permanent national planning board.[64]

The *Technological Trends* report did an excellent job of proposing multiple paths of information flow to monitor technological advance. In the end it was not clear where such assessment might lead, given the group's reluctance to suggest that society might choose to redirect the path of mechanization. Considering the complexity of economic and social behavior, simply creating a general planning board would not guarantee that experts could devise any effective plan to avert displacement. Nevertheless, the NRC report did move beyond its predecessors in addressing technological unemployment head-on, and President Roosevelt endorsed its emphasis on expertise. Americans could learn to anticipate discovery, he agreed, and "make plans to meet new situations that will arise as these inventions come into widespread use." Even if industrial growth could eventually absorb many displaced workers, those individuals would "pay a very heavy price" in the meantime. The job of preparing for change had become too vital to leave to chance or to private initiative.[65]

By the late 1930s, the federal government had devoted a noteworthy effort to collecting statistics on mechanization and unemployment. Labor leaders and other concerned parties considered it a national priority to get experts to compile such information, both as a means of proving to skeptics that a serious problem existed and as the crucial first step toward deciding on a solution. After all, as William Green argued, nobody would expect a doctor to prescribe effective treatment for a patient without first properly diagnosing the disease. In a different analogy, William Ogburn declared that just as an engineer could not construct a giant dam without measurements, so a state could not hope to manage its affairs without understanding the conditions under which it operated. By the end of the 1930s, America's official capacity to examine the problem had improved significantly. In part, increased ability to gather data on machines and jobs reflected a more general awareness among officials of the power of numbers. During previous decades, the Bureau of Labor Sta-

tistics, Department of Commerce, and other bureaus had accepted a responsibility to collect and publish information about a limited set of social and economic questions. FDR's new alphabet soup of agencies brought an army of
statisticians and economists into Washington, turning entire offices into factories for figures. The WPA's National Research Project became one of the
most intensive investigations of a single economic topic undertaken to date.

The Roosevelt administration's effort to untangle the relationship between
engineering and employment trends yielded an unprecedented volume of
information. Analysts could not pursue all angles or resolve all methodological disputes, of course. Some observers grew impatient; the *New Republic*
complained in 1938 that Washington continued to avoid answering questions
about displacement. Back in 1936, at the creation of the National Research
Project, Hopkins had optimistically declared that "once we have diagnosed the
trouble" underlying massive joblessness, "we will be in a better position to
work out a permanent cure." By the end of the decade, Washington had far
more information than ever before about mechanization, but a permanent
cure for unemployment seemed as far off as ever. Administration officials had
started discussing ideas to restructure the educational system along lines that
could "constantly prepare and re-prepare our citizens" for the "new needs of
our economic and social order." Men and women who saw fellow workers displaced by new machines understandably and "instinctively" opposed change,
said Isador Lubin, but they also appreciated how much modern invention had
enhanced daily life through the production of automobiles, electric lights, and
entertainment. The promise of consumer abundance could restore people's
faith in technology and defuse the danger of job loss. If the average family
earned just an additional two dollars per day, Lubin predicted, movie attendance would triple and car sales would rise $119 million, creating new work
in the process. Such simple adjustments, giving workers a greater share of
Machine Age benefits, could allow modern people to reconcile mechanization
with genuine social well-being.[66]

Department of Agriculture economist Mordecai Ezekiel similarly looked
to expanded consumer activity to solve economic distress. Over recent years,
he complained, corporate price control and monopolies had diverted resources
away from public needs and into big profits. If the federal government stepped
in to supervise industry's production targets, demonstrating its willingness to
absorb any surplus of goods, it could defend the interest of workers. Under
such a rationalized system, manufacturers could avoid chaos while raising production enough to bring each family an annual income of $2500. Armed with
that purchasing power, consumers would increase food spending by 50 percent, housing 66 percent, and clothes and personal items almost 100 percent.

That rise in demand would then create six million new industrial jobs, three million professional and service jobs, and another three million in trade, transportation, and construction.[67]

True, employers would continue to introduce more and better machinery, eliminating old positions in the process. The federal government would need to provide vocational education for displaced workers, Ezekiel said. To advise people on where to find new jobs, he proposed creating a new Occupational Outlook Service in the Department of Labor which would predict future labor supply and demand in every field, just as the Agricultural Outlook Service projected crop supply and demand. Workers could not jump between places instantaneously, Ezekiel warned, and economic growth might yield only a limited number of jobs. Planners should supervise industry to ensure that introduction of new machines did not outrun the nation's capacity to absorb workers. If necessary, to maintain a balance between labor supply and demand, experts might consciously choose to hold mechanization in a slight check. Though the idea of slowing innovation would appall anyone who associated technology with progress, Ezekiel argued that it seemed far better than the risk of escalating unemployment. A course of invention which came only at the cost of "degrading a constantly increasing proportion to social and economic outcasts, is not true progress," he concluded.[68]

Trying to put any such plan into practice would have entailed an astounding battle between government and private interests, and no one knew better than Roosevelt the hard truth that no measure could make the problem vanish overnight. In his 1940 State of the Union address, the president seized the occasion to repeat his ongoing concern about technological unemployment. Though the bulk of his speech dealt with the international situation and dangers of war, Roosevelt's discussion of domestic affairs zeroed in on problems of Machine Age labor. Though national production had finally returned to its 1929 level, millions of men and women remained out of work, "a symptom of a number of difficulties in our economic system not yet adjusted." The country had to find ways of adjusting, the president declared; "we have not yet found a way to employ the surplus of our labor which the efficiency of our industrial progress has created." Roosevelt warned that mechanization had cut off opportunities for young people just entering the job market, forcing the United States to "face the task of finding jobs faster than invention can take them away."[69]

By singling out the dangers of workplace change for special attention in his most prominent speech of the year, Roosevelt painted technological unemployment as an ongoing threat to American well-being. Predictably, that assessment stirred controversy. Republican Thomas Dewey condemned the president for not realizing that since mechanization had always generated both new jobs and new luxuries in the past, it surely would do so in the future. H. W. Prentis, president of the National Association of Manufacturers, insisted that technological change never had been the source of unemployment. If antibusiness Democrats would only stop obstructing the natural progress of free enterprise, he promised, industry could start "putting inventions to work to create new industries and new jobs."[70]

Congress Debates Displacement and Proposals for a Technotax

The Depression-era discussion of technological unemployment in Washington stretched beyond the White House. Through the 1930s, as first Hoover and then Roosevelt faced rising popular concern about displacement, politicians on Capitol Hill also started to consider the economics of machine production. Democratic Senator Robert Wagner led the way in advising that Americans balance the immediate social costs of mechanization against unrealistic promises of a future consumer utopia. While pressing for passage of public works and federal employment measures in January, 1930, Wagner explicitly attributed job loss at least in part to technological change. The twentieth century, he cautioned, had reversed the rightful priority of men over machines. He denounced a system that gave business free rein to discharge workers at the least sign of slack, trusting that "a beautiful providence will feed, clothe, and house . . . and keep them fit until it is ready to re-employ them." Meanwhile, "it would never occur to a manufacturer to set his machinery out on the street during depression in the hope that the Red Cross would maintain it for him until the recovery."[71]

To compensate for how much mechanization had increased per capita productivity over recent years, some members of Congress advocated an across-the-board reduction of working hours. In 1932, Senator Hugo Black proposed a bill that would have required all firms engaging in interstate commerce to cut the workweek to thirty hours. For decades, advocates of mechanization had promised that it would lessen man's drudgery, but business owners and employers had allegedly failed to pass on the gains in efficiency to labor and consumers. Switching to a six-hour day would absorb six million displaced people, Black told Congress, thereby reducing relief needs and restoring purchasing power. Equally important, shortening the workweek would redefine habits of work and leisure, enhancing the life of future generations. The Roosevelt administration criticized Black's bill as unnecessarily rigid, and it ultimately became sidetracked. Nevertheless, the senator's efforts highlighted the fact that concern about technological unemployment had added a new element to America's longstanding debate over working hours.[72]

By the second half of the 1930s, other members of Congress also echoed the warnings from Roosevelt, Hopkins, and the WPA economists that modern life might involve chronic problems of technological unemployment. Wagner continued to lead that debate; 1936 and 1937 *New York Times* articles under his byline argued that current joblessness had deep roots in the growing mechanization of telephone service, railroads, cigar and cigarette manufacture, and automaking. Free enterprise could not be trusted to cure such ailments, he wrote. Optimists did workers no favor by suggesting that in about "fifteen years something new may turn up." If anything, future inventions would multiply in a geometric ratio, with accelerating rates of displacement. For Wagner, civilization imposed an obligation on citizens to defend their weaker fellows against adversity. If people had no other protection from the danger of displacement, then the federal government must pass unemployment insurance, housing construction, and pension programs.[73]

Pennsylvania Representative Charles Faddis declared that over the last decade or so, the American economy had become "supersaturated with machinery." Promoters of new technology had started to "resort to trick figures and theoretical answers to prove that machinery produces more jobs than it displaces," he wrote. For his part, Democrat Homer Bone repeated the familiar statistics indicating that business recovery had not restored employment. Americans could "no more disregard the ruthless implications of these figures than we can snub a cyclone," he declared. As long as businesses could move to produce more goods with fewer workers, "we will never be able to divorce ourselves from the ghastly anomaly of desperate poverty for millions in the midst of the greatest productivity." Bone described the logging machines in his home state of Washington, which supposedly allowed nine men to perform tasks formerly requiring twenty-one. To offset such a "terrifying problem," Bone also supported plans for reducing the workweek. Pointing to the use of combines in farming, Senator Joshua Lee suggested that inventors would soon find a way to attach a portable flour mill and a small baking machine behind the combine, so that one man could turn grain into bread in a single operation. "Labor-saving inventions have displaced many workers," he declared flatly, and Washington could not paper over the situation by throwing more money into relief.[74]

The question of whether mechanization even counted as progress if it brought pain for a certain number of workers troubled Texas Representative Hatton Sumners. Faced with stubbornly high levels of joblessness, Sumners no longer felt sure that a worship of scientific and technical advance represented the best guarantee of national progress. Ultimately, everyone would suffer if the United States directed too much energy toward improving machinery and too little to caring for the people displaced. Sumners laid part of the blame on the nation's patent system, laws formulated in a much earlier age by men who could not have anticipated the sheer scale of twentieth-century mechanization. To Sumners, a system that offered seventeen years' monopoly rights to any inventor "who can figure out a machine that will put somebody else out of a job" seemed little short of criminal.[75]

Democratic Representative Jennings Randolph, assistant majority whip, worried about the partisan politics of the technological unemployment debate. He feared that Republicans might use Democrats' expressions of concern to paint the incumbent administration as antiscience, staking out a monopoly on the idea of progress. The Democratic campaign must find ways to "prevent the opposition from using such arguments against us," Randolph told Roosevelt. The party did not have to drop the issue of technological unemployment, he wrote. To the contrary, it ought to highlight the notion that business had co-opted science and engineering into a "mad scramble for lower production costs." The Democrats' platform should incorporate planks to "favor the promotion of scientific research" as a source of new products, new industries, and hence new employment. Randolph proposed the idea of creating a federal Scientific Research Commission, which would establish "laboratory force beds" to cultivate investigations in physics, chemistry, and biology with the promise of economic applications.[76]

Other Congressmen expressed interest in another means of dealing with displacement, the technotax. Some union leaders, such as the AFL's William Green, favored taxing profits on machine-made goods to provide relief for affected workers. Cigar-makers expressed special interest in the technotax, which they praised as a "scientific method of carrying industry through the readjustment from handicraft to machine operation." During NRA code negotiations, the union called for giving each former cigar-maker a ten-dollar weekly stipend for the time he remained jobless, financing the scheme by adding a bit to the price of all cigars produced with new machines. In 1934, a California group called the American Technotax Society mounted a campaign against what it called industry's obsession with mechanization. In a "gigantic conspiracy against American labor," its pamphlets complained, "economists of big business with their journalistic hirelings have made a fetish of machine progress." To break through that conspiracy and confront problems of unemployment, government should begin taxing technology according to how many workers it replaced. A device that substituted for ten workers would pay ten units of relief funds.[77]

The technotax had a certain appeal during the Depression as a one-step solution to make those businesses that introduced mechanization bear some financial responsibility for supporting displaced employees. In Ohio, supporters formed the National Organization for Taxation of Labor Displacing Devices, which claimed to have the endorsement of officials from five national unions. In a 1935 *Kiwanis Magazine* piece, New Orleans judge William H. Byrnes, Jr., tried to sell the idea to the public. A technotax would not hit all machines indiscriminately; technology could be divided into two classes, he explained. Equipment that substituted for good human labor would be taxed, while those machines that performed jobs humans could not do would be exempt. An electric generator, for example, actually created employment by powering factories and streetcars. The technotax would not destroy progress, Byrnes declared, since the government would not want to set it at punitive levels. A company that adopted a device handling the work of five hundred men might be taxed three cents per day per job. That tax would cost only fifteen dollars a day but provide over five thousand dollars a year to help labor. All the businessmen of his acquaintance professed themselves willing to accept such a tax, Byrnes said, provided it would apply equally to their competitors and that the revenue would go toward restoring consumers' purchasing power.[78]

Technotax advocates attracted Congressional champions. California Democratic Representative John Hoeppel lectured Congress on this pet cause at least six times during the 1935 session. He recited tales of displacement attributed to the Owens glass machine, continuous-process steelmaking, steam shovels, recorded music, teletypes, poultry-cleaning machines, and more. The technotax, he promised, would stop machine owners from commandeering the lion's share of profits. It would finance federal public-works projects while helping craftsmen and small entrepreneurs survive against competition from highly mechanized big business. Hoeppel specialized in cute gimmicks to draw attention. Holding up a playing card, he announced that by ignoring the technotax, Congress had "discarded ace legislation which would have solved our economic problems long ago." He excelled in colorful language, declaring that the United States had "submitted too long to the rape of the machine." On different occasions, Hoeppel compared the machine to "a Frankenstein monster which has all but devoured us," an octopus "which has enslaved our entire Nation . . . to make millionaires on the one hand and paupers on the other," and an "economic tidal wave which has engulfed the American workers."[79]

As Hoeppel recognized, critics attacked the whole idea of a technotax as a fine on technical and social progress. The California representative mounted a preemptive strike, denouncing economists with "atrophied, or one-track minds" who believed "that a mechanized, mass-production industry is an absolute symbol of progress." The technotax would not cut off the pursuit of knowledge, he declared, but would redirect it into truly important lines such as medical research. Neither would a technotax block all invention, since it applied only to machines directly substituting for labor. Pure intellectual investigation would remain free of charge, as would the development of technologies such as the airplane or radio, which created jobs. When a fellow member of Congress asked whether such a plan would tax the tractor used on his farm, Hoeppel had to admit the details had not been worked out. In his opinion, the tax should "get the big boys first." The legislation he proposed in 1935 called only for a complete investigation into displacement, "with a view to imposing a graduated tax on mass production, machines, and equipment." Hoeppel further stipulated that all revenue from such a tax would finance public works projects to support displaced men. He claimed such a tax would prove practical, "easy to enact, enforce, and collect," but he apparently did not convince many colleagues.[80]

As long as concern about displacement remained, the idea of a technotax would continue to draw attention. In 1940, Wyoming Democratic Senator Joseph O'Mahoney proposed a "labor differential tax." All employers would receive tax credits based on the amount of wages paid, balanced by extra taxes on employers who enjoyed profits deemed "excessively large as compared with total wage payments." In recent years the profit of major industries, railroads, and utilities had risen 89.1 percent while employment remained stagnant. According to O'Mahoney, such evidence proved that without outside pressure, business leaders would simply disregard the national problem of displacement. A tax penalizing employers "who use less than the average of human labor" would reduce incentive to substitute machines for men. Far from undermining business prosperity, he argued, such steps would restore employment to optimal levels and thus renew consumer purchasing power. Wagner supported O'Mahoney, reminding listeners that the "seriousness of the technological unemployment problem is obvious." Nebraska Republican George Norris denounced the measure as "a tax upon human progress." All history rested on the advance of technology, Norris declared, the force which had brought humans from barbarism to civilization. A technotax would kill all prospect of future invention and thus reverse the course of world progress. O'Mahoney responded that he "would be the last person . . . to raise a stop sign on the road of technological improvement." The technotax would simply help "balance men and machines," he promised.[81]

Calls for a technotax, for federal laboratory "force beds," or for rewriting patent law never advanced very far. Opponents easily labeled such proposals attacks on free-enterprise progress, and the ideas remained too vague to sound practical. Nevertheless, the persistent interest in such measures reflected a heartfelt conviction that the American government must consider taking some type of tangible action. Especially after President Roosevelt himself began referring to the possibility of permanent technological unemployment, the issue came to play a central role in Depression-era politics. The WPA's National Research Project collected evidence that appeared to undercut easy assumptions that mechanization always meant progress. The mid-1930s economic rebound had revived production but had not restored jobs, researchers Hopkins, Gill, and Weintraub pointed out, suggesting that the continued advance of workplace technology had permanently eliminated some capacity for work. Critics complained that an antibusiness bias had led WPA investigators to exaggerate the modern risk of displacement, but for many segments of the American public, especially the labor community, the WPA reports validated preexisting concern about Machine Age trends.

"No Power on Earth Can Stop Improved Machinery"

Labor's Concern about Displacement

AS PRESIDENTS, CONGRESSMEN, and government experts discussed the possibility that technological unemployment had turned into a long-term condition of modern life, working-class Americans voiced more immediate anxieties. In previous decades, machines had seemed to threaten specific jobs, such as weaving or cigar-making, and workers in each trade addressed the matter separately. The Depression stretched fear of displacement beyond a few particular occupations, making it a broad-based concern for labor as a whole. Mechanization seemed likely to affect an ever-greater number of workers, as inventors devised more equipment to substitute for human skills, experience, and mental ability. Office employees, accountants, telegraph operators, and professional musicians felt newly vulnerable, adding to the unease in an already troubled economic climate. Workers expressed their discomfort both as individuals and through unions. The AFL, CIO, and other labor organizations mounted major campaigns during the 1930s to bring the problem of displacement to national attention. By definition, the issue placed unions in a tricky position. How could labor condemn the effects of technological change and demand corrective action without appearing to oppose progress?

Labor's Sense of the Problem

Through the 1930s, working men and women worried that they could literally see employers moving machines into the workplace to take over jobs. Coal miners feared for their future as they watched the industry more than triple the number of loading machines in use, from 523 in 1933 to 1,720 in 1940. One miner's wife protested to President Roosevelt that "the bosses made the coal loaders take their tools out of the mine and they are letting the machines do all the work." Men in one Kentucky local requested government help to protect their livelihood, declaring that loading technology had displaced "flocks of miners." Society was moving toward "modern 'barbarism'" rather than modern progress, the group told Secretary of Labor Frances Perkins, when it allowed inventions to destroy honest work. Across the country, Americans echoed the miners' dismay. New conveyor belts in New England iron and steel mills let a single worker perform the work of twenty-five, one machinist observed. Similar systems in flour mills reportedly helped fifteen men load cars faster than one hundred men had done before. CIO Textile Workers president Emil Rieve declared that his industry had installed knitting machines and looms equipped with self-governing speed controls and photoelectric eyes, which could automatically stop production when a run finished or problems arose. Someone visiting a modern textile mill might have difficulty just spotting a human operator amidst the huge rows of machinery, Rieve complained.[1]

In interviews with 150 unemployed Americans, the National Federation of Settlements found that at least twenty-six blamed their plight specifically on mechanization. Those men and women spoke of seeing their positions eliminated as employers introduced cigar-rolling machines, automatic paint-sprayers, printing machines, and other new devices. To jobless individuals, labor-saving machinery seemed like "the worker's worst enemy," said settlement officers. Some economists and businessmen might insist that mechanization created more jobs than it destroyed over the long run, but such reassurances rang hollow to those desperate for work. One steel manufacturer had displaced 1,500 employees at once when it closed an old mill, replacing it with a continuous-strip facility that did not provide work for even two hundred of those men. The company's former workers nicknamed continuous-strip machines the "Grim Reapers" and reportedly referred to the new mill as the "Big Morgue," the "place where all our jobs went dead."[2]

Some unions had tackled the issue of technological unemployment prior to the Great Depression. Glassblowers had protested for years against introduction of automatic bottlemaking machines. Cigar-makers had seen their lives upset; an experienced craftsman could roll 133 to 266 cigars per day, but a machine (tended by an unskilled woman at far lower pay) might turn out at least one thousand cigars daily. Large cigar companies adopted rolling machinery partly to insulate themselves against the strike threats and wage demands of the veteran cigar craftsmen and also as a means to force small competitors out of business. The resulting industry shake-up, combined with declining cigar consumption and the economic downturn, caused at least fifty-six thousand cigar-making jobs to vanish between 1921 and 1935. Though the spread of cigar-rolling and bottlemaking machines resulted in significant distress for some workers, such developments might be dismissed as isolated instances in the period before the market crash. As long as overall economic prospects remained favorable, fear of technological unemployment could be contained within certain occupational boundaries. Once depression set in, more and more Americans began to wonder whether the Machine Age had turned displacement into a universal working-class hazard. Union leaders argued it was no coincidence that economic collapse had come in a time characterized by an unprecedented pace of invention.[3]

Though new textile machinery had displaced spinners and weavers back in the Luddites' time, labor organizers contended, other occupations had remained unaltered. Even in the early years of the twentieth century, mechanization had seemed to come relatively slowly. By the 1930s, however, the accelerating advance of science and technology set off destabilizing changes in many areas of employment simultaneously, causing displacement faster than the economy could handle. Some modern devices had revolutionized entire industries almost overnight, said William Green, president of the AFL. Whereas a glassblower of 1918 had made forty light bulbs a day, new machines could turn out 73,000 bulbs each day. Manufacturers of the 1930s could produce 32,000 razor blades in the time it had taken to make 500 in 1913. When innovation could reshape production so radically in so brief a time, Green declared, labor had no chance to adjust.[4]

For Green, the changes in specific industries, such as the manufacture of light bulbs and razor blades, added up to a fundamental economic proposition. Figures from the Federal Reserve Board and the U.S. Census of Manufacturing, he said, proved that twentieth-century technology had altered the very nature of production. Across the board, manufacturers had started producing more and more goods with less and less labor. Between 1909 and 1919, industrial output had risen 35 percent, accompanied by a 38 percent growth in jobs. Between 1919 and 1929, as production grew another 42 percent, industrial employment fell by 7 percent. The material abundance of the 1920s had come at the expense of workers' opportunity, Green concluded. Responding with scorn, the *New York Times* accused labor leaders of misleading the public by presenting highly distorted information. "The error, or the trick, consists in taking an isolated instance of extraordinary saving effected by machine and suggesting that it is typical," editors wrote. Sophisticated equipment might be capable of producing ten thousand pins in five minutes, but it still took almost as many men to build a house in twentieth-century America as it had in seventeenth-century England. In short, the newspaper warned, "The sudden popularity among economists of 'technological unemployment' has let loose clouds of statistics which prove so much that they end by proving nothing at all."[5]

The AFL insisted that Americans had seen enough extreme examples of displacement to give rise to genuine alarm. In articles with titles such as "More Machinery, More Jobless," the union's *Weekly News Service* attacked the Hoover administration for ignoring the problem. When Commerce Department officials urged companies to take advantage of economic depression by replacing outdated facilities with increasingly mechanized ones, the AFL condemned the statement as "tragic." The big business interests that dominated government had turned values upside down, labor complained. At a time when the economy desperately needed firms to reduce hours and otherwise help displaced workers, Washington had reinforced the "cold, paganistic determination of employers to keep . . . substituting more and more iron men for human beings."[6]

With the change of administrations, labor interests found Washington more sympathetic and willing to talk about loss of jobs to mechanization. As Roosevelt, Hopkins, and other officials called attention to evidence that reemployment lagged behind the recovery in productivity, unions underlined the issue by asking, "Recovery for whom?" If anything, big companies had intensified their pursuit of new technology during the 1930s, continually making it harder for employment to catch up. The Textile Workers Union declared that even though their industry had almost completely rebounded from its Depression slump, employment in 1939 remained 5 percent below the 1929 level, since managers had introduced faster automatic looms, winders, and other machines during the intervening decade. Railroad unions calculated that while the amount of rail traffic in 1940 formerly would have required 1,610,369 workers, labor-saving devices let railways handle it with just 987,943 people.[7]

Through the last half of the 1930s, labor leaders argued that the sheer persistence of high unemployment showed the folly in facile assumptions that a strong free enterprise system would automatically reabsorb workers displaced by change. Finding enough work to go around in the Depression would have been difficult enough, considering that the labor force expanded each year, but mechanization had turned simple displacement into a full-scale job crunch. Philip Murray, chairman of the Steel Workers Organizing Committee, emphasized the irony: even though 1939 industrial production had grown past the previous all-time high of 1929, approximately ten or eleven million Americans still lacked jobs. Mechanization had severed the link between business welfare and employment, Murray suggested. Soaring production no longer guaranteed a restoration of jobs.[8]

Gender Dimensions of Displacement: Teletype and Office Machines

In addition to spreading general unease, fear of technological displacement heightened tension surrounding other labor questions, especially that of women's employment. Two decades earlier, during World War I, female workers in the arms industry and other nontraditional fields had made critical contributions on the domestic front. With the return of peace, women had been expected and, in some cases, pressured to return jobs to men. Female workers never completely vanished, of course. The 1930 census showed that over 25 percent of women past age sixteen were employed outside the home, accounting for 22 percent of the total American workforce. Although prevailing mores . did not encourage women to pursue full-time, fulfilling careers, the number of working women grew from 8.5 million in 1920 to ten million (almost two million of them married) in 1930.[9]

Economic collapse ignited new arguments over gender roles. The idea of women working while a large number of men remained jobless roused alarm about females stealing male opportunities. Of course, men would not have wanted most domestic jobs or other stereotypically female employment, but critics complained that employers had too often hired women (at lower salaries) for industrial work, clerical functions, and other positions to which men should enjoy first claim. As job openings became scarce, the notion of married women as breadwinners proved especially controversial. One national poll showed that 82 percent of Americans believed married women should not hold paid employment if their husbands could earn money.[10]

Legislation in 1932 required federal agencies to remove one member of a government-employed couple when cutting back on personnel. Though the rule did not specify gender, in practice it disproportionately affected women, given their lower pay and lack of status. Within a year, 1,600 individuals, mostly female clerical staff, had either been fired or forced to resign. Similar concerns at state and local levels endangered more women's jobs, and many school districts threatened to discharge teachers when they married. Fear of social disapproval reportedly also made some private employers reluctant to hire married women, especially for positions in contact with the public. Such apprehension may indeed have been justified; appalled by advertisements for women's pants, one *New York Times* reader warned that once men let "the formerly weaker sex wear the trousers," women might take over the workplace completely and reduce men to housework. One cartoonist of the 1930s showed men tending baby carriages in the park as one admonished a toddler, "I'll tell your mother how bad you were tonight when she comes home from work."[11]

Aiming to dismantle such prejudices, the Women's Bureau of the U.S. Department of Labor mounted a major effort during the 1930s to establish women's rights to employment and decent pay. Employers often justified offering women lower wages by saying that after all, females only worked to earn "pin money." Mary Anderson and other bureau staffers collected evidence that 80 to 90 percent of the married women who sought jobs did so because men in their family were unemployed, ill, separated from the household, or received wages insufficient to support everyone. Ten percent of working women, single and married, bore the entire burden of maintaining a family, while most others devoted at least half, if not all, their pay to keeping up a household. When women performed as well as men, they deserved fair wages, the Women's Bureau insisted. Research conducted during and after World War I showed that women in manufacturing could equal or surpass men's performance, especially in positions requiring physical dexterity or mental alertness.[12]

Nevertheless, the fear of men losing jobs to women led to talk about whether mechanization had fostered that reshuffling. Observers asked not only *how many* but also *which* employees new equipment displaced. The introduction of machines could shift the balance of power between different segments of the labor force, seemingly pitting skilled workers against unskilled counterparts or favoring younger men at the expense of older, experienced ones. Similarly, some worried that mechanization undermined men's position in the workforce. In the manufacturing sector, conveyor belts, cranes, and other machines that took over heavy lifting could negate men's advantage in physical strength. Special-purpose machinery with built-in routines could undercut the value of men's skill and experience. Automobile manufacturers had eliminated hand-stripers by hiring unskilled, lower-wage women to wield painting guns; in cigar making, women could handle rolling machines. As future innovation encouraged more employers to reduce the payroll by replacing a large number of men with a few female machine-tenders, men could find it even harder to compete, economist Carroll Daugherty advised. Workplace technological change might set the stage for a battle of the sexes.[13]

Talk about the gender aspects of technological unemployment spread to many occupations during the 1930s, including telegraph work. Communication had formerly required operators skilled in Morse code at both ends; the sender transmitted messages by manipulating an electrical circuit, while the receiver monitored the rapid flow of dots and dashes. Nineteenth-century telegraphers had prided themselves on their talent in working the key, and those with the speed and accuracy to handle wire service reports and stock market quotations had enjoyed special status. For years, inventors sought to develop a printing telegraph, but the high cost and technical shortcomings of early versions limited their appeal. By the 1930s, developers had made the machines practical, and more and more telegraph offices began replacing their old-style systems. Printing machines allowed users to bypass the Morse operator; using the common typewriter-style keyboard, virtually any typist could send messages. At the receiving end, messages came directly printed out in ordinary language, eliminating any need to wait for someone to decode them. With continued improvement, printing telegraphs could handle sixty or eighty words a minute, while, on average, human telegraphers only approached about twenty-five.[14]

While the printing telegraph alone would have presented a significant threat to operators' jobs, other inventions further transformed labor needs in this arena. The multiplex telegraph system could carry up to eight messages at once by using synchronized distributors to carry alternating impulses over the same line, literally multiplying its capacity. Some offices installed machines that encoded words on perforated tape; by replaying the tape, they could send the same message repeatedly, without needing skilled operators to type it out each time. New York newspapers used such technology to send articles and financial updates up and down the east coast and out to Chicago, where another machine duplicated the tape to send the message further. One operator in the 1930s could handle as many telegraph wires as had four men in previous decades, the AFL Commercial Telegraphers' Union reported. According to the Washington, D.C., local, the Western Union office in Richmond, Virginia, had recently reduced its work force 70 percent thanks to new technologies. Other cutbacks had come in the railroad system, which had begun using dial telephone systems rather than the telegraph to communicate vital information about train safety and routing. Old-style manual-block signaling, which had required station agents to supervise passage of trains, gave way to newer automatic-block systems, in which the train's progress itself sent signals to other traffic. Because of such mechanization, the U.S. Bureau of Labor Statistics concluded, the number of telephone and telegraph operators employed by railways had dropped from 79,346 in 1923 to 58,522 in 1932.[15]

By 1931, the bureau found, new telegraph technology had displaced another eight thousand operators from commercial telegraph offices, press organizations, and newspaper offices—up to 50 percent of those once employed there. Innovations not only eliminated jobs for telegraph workers but also altered gender balance, allowing a smaller number of lower-paid women to perform work once handled mostly by skilled men. In 1915, men accounted for 75 percent of telegraph operators, but by 1931, women made up 64 percent of the operating force. Men still handled most of the old Morse systems still in use, but young female typists worked on nearly 84 percent of the new multiplex printer systems, at roughly half the wages. One Chicago company reported, "Teletype was installed in October 1931. Four regular telegraphers laid off. Work done previously by telegraphers can be done just as well on teletype with much cheaper operators, labor cost being about one-half."[16]

The Bell System had connected more than twelve thousand teletype machines in the United States by 1935, offering business the advantages of "typing by wire." The device could handle multiple functions, promoters emphasized, giving employers who installed it a chance to streamline the office. One salesman, J. M. Tuggey, Jr., laid out charts of old-style office routines and then crossed out 50 to 75 percent of those steps, which he claimed the teletype could eliminate. The machine could print multiple copies of documents at once, using carbon paper held in place with sprockets and a feeding mechanism, allowing one employee to do the work of four in taking orders and preparing them for distribution. Tuggey suggested that the new efficiency would let managers reassign surplus clerks to other departments, but during those belt-tightening days workers feared being let go instead. While offices still needed a human typist to enter information into the teletype, further developments threatened to bypass even that labor. By the end of the 1930s, Western Union had started touting facsimile transmission, promising that the "automatic telegraph" would let users send their own messages, in their own handwriting.[17]

Though facsimile machines remained primarily outside the realm of practicality, bookkeeping devices and other business machines had already entered Depression-era offices, making white-collar workers worry about displacement. The twentieth century had brought increasing clerical employment for women. In 1870, only 100,000 Americans, overwhelmingly men, had engaged in office work, but the 1930 census showed almost two million women alone in clerical positions, making up 51.5 percent of all office employees. Advancing technology seemed to jeopardize that valuable avenue of opportunity. According to reports, "girl victims of the Machine Age" felt "quite bitter" about the new "robots in the office." One employee of Armour Foods complained to the state labor department that the company's new clerical system could displace about two thousand workers. The letter warned "anyone that has anything to do with typing, printing, labelling, etc." to watch out, because "this machine takes care of all that." Young people considering office careers should think twice, the Civic Education Service advised in 1938, since "bookkeeping machines are coming into increasing use, and this is tending to reduce employment."[18]

Other trends might compensate for some of that impact, the Civic Education Service added. Despite mechanization, both government and private enterprise could end up hiring additional bookkeepers and clerks as operations grew increasingly complex. New Social Security legislation and sales taxes demanded that business keep track of extra information. F. W. Nichol, vice president of International Business Machines, maintained that the very speed and accuracy of office machines would produce jobs. Executives who installed the latest accounting devices would want to compile more data than ever, creating a need for human recordkeepers and managers. The resulting insights would help business save money and experiment with improvements, with the resulting success generating even more employment. By using machines to track production, inventory, and sales, one manufacturer reportedly had gained so much efficiency that within a year, shopfloor employment had risen from three hundred to eight hundred workers and office jobs from five to eighty. Innovation might occasionally create temporary displacement, Nichol admitted, but offices also needed people to run and supervise the new machines. In 1939, IBM's own two-week classes taught 3,645 workers to handle the technologies that had altered or ended their old jobs. Furthermore, Nichol added, the manufacture and sale of equipment generated substantial work; just keeping machines supplied with paper amounted to a business worth $2 million a year.[19]

IBM claimed that overall, 80 percent of accounting machine installations caused no net change in jobs, while employment rose in 70 percent of the other cases. In its own investigation of office jobs in seven cities, the Women's Bureau found the effect of mechanization far more tricky to analyze. In some instances, machines had not displaced employees but had allowed a business to expand without adding the people it would otherwise have needed. One St. Louis insurance company had used four clerks to perform statistical work for 20,000 policyholders, but after it merged with another firm and adopted new office machines, four clerks could handle 65,000 cases. The effect of mechanization varied, depending on how managers integrated new technology with existing procedures. One department that centralized its clerical routine after adopting dictating equipment found that eight women could equal the former output of twenty-five, but another division in the same firm that did not reorganize after mechanization experienced no change in labor needs. To make matters more complicated, machines adopted in one place might hit workers elsewhere. After one department adopted expanded punch card technology, it no longer kept so much information in conventional files, leading the filing office to cut three women. Changes in one company could even undermine employment in a completely different business; a firm adopting machines that addressed letters, printed bills, or typed form letters might cancel its outside contracts with printing services or letter bureaus.[20]

On balance, the Women's Bureau survey suggested, new technology tended to create a net loss in clerical labor. "Machines have tended decidedly to curb the rapid rise in number of employees with the increased office functions of modern business," researcher Ethel Erickson wrote. In other words, mechanization enabled employers to do more with fewer workers, so that even when companies enjoyed sizeable growth, that new business might not generate a proportionate expansion of job opportunities. As with the teletype, office mechanization highlighted gender dimensions of the displacement issue, shifting the balance of work between men and women. After installing machines, one Philadelphia firm replaced its male clerks with women, whom it paid at least 35 to 50 percent less. A St. Louis enterprise had saved $4,000 a year in salaries by switching from seven male and female clerks to three bookkeeping machines run by women.[21]

Through the 1930s, the Women's Bureau devoted substantive effort to studying how new office machines and other technologies reshaped patterns of women's employment. While William Green and other national union leaders spoke and wrote extensively about the consequences of technological change, organized labor concentrated on episodes in steelmaking, coal mining, auto manufacture, and other primarily male workplaces. Professional economists, news reporters, and even the NRA's National Research Project similarly relegated women's concerns about mechanization to the background. Those parties left it to leaders of the Women's Bureau to emphasize that new workplace equipment affected workers of both sexes. The impressive power of modern machines, plus the speed with which so many companies had begun installing them, would determine the future of women's jobs, good or bad. Mary Pidgeon worried that, on balance, mechanization subtracted opportunity. The telephone industry's adoption of dialing equipment had started closing off an important avenue of employment for girls, she noted. "In the manufacture of sewing machine needles, one girl can now inspect as many as nine could before; as bean snippers in canning factories, machines have made it possible for twelve women to do the work formerly done by two hundred; in textile mills, a machine ... made it possible for one woman to do the work formerly done by seventeen drawers-in of the warp."[22]

To document displacement, the bureau compiled data on the results of technological change in 115 Midwestern and eastern factories from 1921 to 1931. Ninety-four percent of employers who adopted new or improved machinery reported resulting reductions in labor cost, and almost half of them had gained labor savings of at least 50 percent. The number of people engaged in the relevant work processes fell from 6,401 to 3,604, a job loss of 43.7 percent. That Women's Bureau study, which preceded the WPA's more general research, offered still more dramatic evidence of machines directly substituting for hand labor, in a plant where 624 workers had been cut to 184, a 70.5 percent drop in employment. During the 1920s, within a climate of general business expansion, companies gradually making the transition to new machinery had often been able to reassign displaced workers in alternate (though sometimes less desirable) positions. The Depression threatened to reduce such possibilities of absorption and to raise the psychological toll of mechanization. Whenever machines brought displacement, employees not directly affected began to fret that their jobs too might fall to future change. Highlighting the ironies behind technological advance, one knitting mill superintendent commented, "Before long we won't need workers, the machines are so perfect; but unfortunately machines don't wear stockings."[23]

To draw public attention to how mechanization affected working women, the Women's Bureau produced a three-reel movie in 1931, *Behind the Scenes in the Machine Age*. Thanks to the wonders of technological advance, the film indicated, the "infant industries" of early America had grown into gigantic business for a "mighty nation." Opening shots contrasted a colonial printer's hand press, which turned out one page at a time, to the sheer speed of a modern automatic press, which churned out a continuous stream of papers. Yet, "along with the wonders of the machine age," invention had turned technological unemployment into a major source of economic inefficiency. In the twentieth century, "human beings as well as materials must be saved from the scrap heap." Accompanying cartoon-style graphics interpreted those words literally, showing a wastebasket piled high with little figures of women and men.[24]

To illustrate the toll of mechanization, the film presented case after case of how technological change reduced the number of workers needed. Visual images contrasted old production processes with new machine-centered methods. One scene showed that in previous years, the task of packing cereal employed four women, one to open a bag and place it in the box, a second to scoop in cereal, a third to weigh the filled box, and a fourth to close it. The next scene showed a machine preparing the box, filling it with a regulated measure of cereal, then sealing it. A caption explained, "Formerly twelve girls packed 17,000 boxes of cereal a day. With machines it takes only five girls." Other factories in the food industry had similarly begun to mechanize packaging operations. "Formerly one girl wrapped 960 lollipops a day. Now two girls and the machine wrap 60,000," the film observed. "By hand, three girls wrapped nine boxes of crackers a minute. The machine with two girls wraps fifty-five." Subsequent scenes showed how mechanization had multiplied per capita efficiency and thus potentially reduced labor needs in the manufacture of cigars, bandages, books, automobile cushions, and other products.[25]

Despite the power of technology to displace humans, the movie continued, the solution did not lie in banishing new inventions. In order to keep step with the Machine Age economy, workplaces had to pursue the latest innovations, leaving people no choice but to accept and adjust. The Women's Bureau promised that through careful management, business could avert most of the distress associated with mechanization. The Bell System, it suggested, had done a model job in smoothing the transition from manual to dial switchboards by making the changeover gradual and by reassigning as many operators as possible. With expert planning, adaptations should benefit everyone; companies could adopt new devices while retaining their experienced employees, workers would continue to receive a paycheck, and the national economy would avoid the burden of joblessness.

That upbeat conclusion radically simplified the complex problem of technological unemployment. Indeed, the Women's Bureau's own research had indicated that mechanization could affect labor even across department or company lines. Moreover, the Bell System case remained controversial, as critics had accused the phone company of covering up real problems with self-congratulatory publicity. If the film exaggerated the ease of finding an ideal accommodation between humans and machines, it nonetheless served an important purpose. Movies offered a chance to reach a wide range of audiences, since they could be distributed and shown relatively easily. The bureau made *Behind the Scenes in the Machine Age* readily available to organizers of labor union meetings, educational institutions, women's clubs, and other public forums. With scene after scene showing women at work in industry and offices, the footage visually reinforced the idea of female employment as an integral part of the modern economy. That message underlined the bureau's argument that in any serious discussion about the course of America's new Machine Age, women's interests deserved consideration.

The American Federation of Musicians' Fight

Despite such hopes that labor might yet come to terms with mechanization, the issue left workers of the early 1930s with a sense of powerlessness. Most knew only too well that the attempts of unions to minimize displacement over previous decades had usually failed. In the first several years of the new century, the National Window Glass Workers had passed rules that banned members from working with machines and barred machine operators from joining the union. That futile strategy had backfired on the union, leading to its demise. As glass manufacturers introduced more and more machinery, craftsmen tried to hang onto work by making pay concessions. But even with reduced labor costs, old-style operations simply could not compete with the efficiency of mechanized facilities, and the distraught and divided union disbanded in 1928. Similarly, stonecutters had tried to demand that employers not use new stone-planing machines more than eight hours per day and that companies hire at least four traditional stoneworkers for each planer adopted. Through repeated strikes, labor won temporary concessions, but no more. Restrictions proved impossible to enforce, and national leaders abandoned the effort by 1908. Some locals continued their own fight, refusing to perform the fine hand-finishing needed for machine-cut stone and trying to block shipments of machined stone from entering their cities. Employers had ways to counter such tactics; when metalworkers tried to limit the number of nail-making machines one man could operate, companies simply switched to nonunion workers.[26]

When attempts to limit technological change collapsed, unions had sometimes tried imposing an obligation on employers to protect labor. Garment workers scored a victory in the 1920s when the Chicago Trade Board and Board of Arbitration made its approval of new tacking and basting equipment conditional on having companies reassign displaced workers to equivalent jobs. In other instances, when industry introduced mechanization, unions insisted that their members at least had a right to claim the job of operating new machines. That strategy created trouble for labor, though, since it raised complicated questions about retraining and seemed too much like a surrender. In the 1880s, the International Typographical Union adopted a policy that supported the adoption of Linotypes, providing that displaced compositors could handle the machines. One dissatisfied Ohio local rejected that compromise and mobilized an anti-Linotype strike, only to see their national union leaders cooperate with employers by sending in substitute workers from surrounding areas. Such cases suggested that while unions might sometimes win temporary accommodations to carry a few members through the transition to modernization, those isolated triumphs could not ensure a broader base of workers any long-term protection. History seemed to show that once companies set their heart on adopting the latest devices, labor had virtually lost the battle already. "Without an exception, any organization, since the beginning of the factory system, that has attempted to restrict the use of the improved methods of production has met with defeat," the head of the cigarmakers' union concluded in 1923. "No power on earth can stop . . . improved machinery."[27]

Nevertheless, with the risk of displacement hanging over their heads, some workers felt desperate enough to try fighting. During the Depression, the American Federation of Musicians (AFM) mounted an elaborate campaign that it hoped might stem the tide of technological change in the film industry. Through the early 1920s, before inventors had developed workable systems of sound reproduction for movies, local theaters had employed live musicians to provide accompaniment for silent pictures. Small houses featured only a pianist or violinist, but glamorous "movie palaces" engaged full orchestras. As long as Americans continued to exhibit a seemingly endless enthusiasm for moviegoing, such employment for musicians appeared secure. But in 1925, Warner Brothers became the first major studio to add orchestral music to a regular release, playing prerecorded disks on a turntable synchronized to projectors for its film *Don Juan*. Significantly, a desire to replace live musicians may in part have motivated Warner's experiment with Vitaphone technology. Unlike other big studios, Warner did not operate its own theater chains and so had to convince local owners to screen their productions. Theater managers would be eager to show sound movies, Harry Warner hoped, since they could save the expense of hiring musicians. Most other Hollywood executives approached the new technology with trepidation, but after *The Jazz Singer* drew more than $1.5 million in 1927, one studio after another hastily began converting to sound. By 1931, all large Hollywood producers had abandoned the silents.[28]

The switch from silent to sound movies affected employment across the industry. Demand for electrical technicians and sound engineers suddenly soared, since making a "talkie" required studios to keep technical experts on set as recording supervisors and emergency troubleshooters. Across the country, every theater owner who decided to show sound movies needed technicians to install and maintain the unfamiliar equipment. Projectionists also benefitted, since the sheer complexity of Vitaphone projectors often compelled theaters to employ not one but four people to handle the films. Projectionists demanded and won substantial pay raises, citing the extra effort involved in monitoring amplification and coping with the inevitable technical difficulties. Between 1929 and 1931, eight thousand new positions opened up for licensed sound projectionists, whom the Bureau of Labor Statistics described as being "unquestionably in the most favorable position of all the trades employed in the amusement industry." Such trends left the International Alliance of Theatrical Stage Hands and Moving Picture Machine Operators in a difficult position. Even as the popularity of sound film gave projectionists negotiating power, it sent employment for stagehands, the other half of the union, into a nosedive. Talking pictures also competed with live theater, and reduced audiences due to economic decline multiplied the damage. After 1930, vaudeville houses and acting companies began closing in all but the largest cities, displacing stagehands, performers, and musicians.[29]

Filmmakers' switch to sound opened up opportunities for some creative artists while closing them for others. Silent movies had not relied on extensive or sophisticated wordplay, and the resulting inexperience showed up in early talking pictures as stilted and unconvincing dialogue. Soon, as studios focused on the need for good writing, bidding wars pushed salaries for the best scriptwriters as high as $1,000 a week. After the industry started making musicals to showcase the versatility and excitement of sound technology, composers, lyricists, and choreographers flocked to California for the chance to land well-paid jobs and reach a national audience.[30]

For Hollywood actors, suddenly required to speak instead of mime, conversion to sound proved a make-or-break experience. Some profited from the change; stage actor Conrad Nagel, who had been earning $1,500 per week making silent films, found that his well-trained voice could command $5,000 a week in talkies. The new technology fostered success for Bing Crosby, Paul Muni, Jean Harlow, and the Marx Brothers, all of whom made distinctive voices and sounds part of their appeal. Other personalities whose looks had assured popularity in silent pictures failed to make the transition. German actor Emil Jannings, who had earned $5,000 per week in silents, proved insufficiently adept in English and returned to Europe. Vilma Banky's Hungarian accent severely limited her talking roles, while John Gilbert's pleasant baritone simply did not match the macho image audiences expected. Some stars needed assistance to master film speech; Douglas Fairbanks Sr.'s voice initially came across too strongly for microphones, and May McAvoy's too weak. By 1928, Hollywood boasted seven "colleges of voice culture," where vocal coaches and elocution experts offered to correct actors' pronunciation and adjust speaking tones. The stress that the advent of talkies placed on performers also contributed to employment for one more community: the period reportedly brought booming trade for Hollywood psychics.[31]

Overall, the technological revolution in film hit musicians hardest. The Bureau of Labor Statistics estimated that the installation of sound equipment in 11,828 theaters between 1929 and 1931 had displaced 9,885 musicians, half the total formerly working in movie houses. Examining in detail the situation in Washington, D.C., researchers found that the city's fifty theaters had all shifted from silent to sound movies between 1927 and 1930. Owners who either would not or could not quickly switch to sound saw attendance plummet. That trend forced some small theaters to close, creating unemployment for projectionists, ushers, and musicians. The introduction of sound had raised wages and created four new positions for projectionists in Washington, but it had displaced 221 musicians. A year later, twenty of those "technological casualties" had left the city, while thirty had found alternate work teaching music or performing in restaurants and hotels. Twenty-one had abandoned music for other jobs, while twenty-two remained unemployed. The plight of such men made news in New York City when a violinist arrested for smoking in the subway pleaded for leniency by explaining that he had lost his job to the talkies. After the man showed the court his bankbook, with a balance of thirty-seven cents, the judge waived the one-dollar fine.[32]

Talking pictures even hurt American musicians' employment overseas. Through the 1920s, dance bands from the United States had enjoyed tremendous popularity in other countries. By one estimate, American entertainers in Paris provided 80 percent of the music for Montmartre nightclubs and restaurants. One dance hall manager said that he only kept French musicians around to play while the Americans rested, the time when customers stopped dancing and sat down to order champagne. But as sound movies gained an international audience, employment for musicians tightened everywhere. In 1930, a group of French cinemas installed sound equipment and fired five hundred musicians in one day. To help the displaced entertainers, France's Ministry of Labor tightened regulations, insisting that American musicians obtain special work permits and requiring that cabarets hire at least as many French as foreign performers. With strict enforcement of the work rules, one American band that had not applied for permits in a timely fashion had to break its contract at an elite Paris restaurant and leave the country. Reportedly, French police even barged into establishments to interrupt performances, threatening business owners with a $4,000 fine and posting guards to ensure that the Americans ceased playing.[33]

Such developments unfolded rapidly and greatly unnerved America's musical community. Professional musicians had felt relatively confident during previous economic downturns, trusting that their artistic skill would always be in demand. Technological change ripped away that sense of security. Chicago Federation of Musicians president James Petrillo complained that "like a bolt out of the blue, comes a Frankenstein of mechanical achievement with its arrogant assumption of human proclivities." Just as the popularity of refrigerators had made the iceman superfluous, just as automatic equipment had displaced painters and streetcleaners, Petrillo said, musicians had now fallen victim to technological unemployment. In "one fell swoop," he wrote, the movie industry's switch to sound pictures "tears from our grasp the fruits of a lifetime of struggle and labor."[34]

Musicians knew that it would prove hard to confront Hollywood directly.

At the end of the 1920s, AFM conventions voted down any idea of trying to mount strikes against the studios making talking films. Such action seemed unlikely to prove effective, AFM president Joseph Weber observed, since studios needed only two hundred and fifty performers at most for recording film scores. If the Americans boycotted movie work, the industry could easily bypass them by importing foreign musicians. The AFM also worried that a strike might arouse popular antagonism, if movie-lovers felt that the union had set out to ruin their amusement. Artists could hope, Petrillo suggested, that in the end, the public would continue to support live music. Once audiences had satisfied their initial curiosity about talking pictures, he predicted, they would become exasperated with the poor sound clarity and lack of musical artistry. Popular discontent would compel studios to keep making silents and force theaters to rehire their orchestras. "Art cannot be permanently mechanized," Petrillo concluded. Rather than just waiting for the public to grow tired of the talkies, the AFM attempted to stimulate a backlash against Hollywood's obsession with sound. The union engaged a public relations firm and, starting in October 1929, initiated a radio and print campaign promoting live music. Eventually, AFM advertisements ran in more than nine hundred publications around the country.[35]

The ad series did not emphasize the misery facing displaced musicians, since Weber believed that such a tactic would prove pointless. "We are too experienced not to know that if men lose employment by reason of the introduction of a machine, no appeal to the public sympathy suffices." Instead, the campaign centered around a message that recorded music could never compare to the value and quality of live performance. Rather than taking a harsh, accusatory line, the AFM adopted a humorous slant, poking fun at recorded sound by labelling it "canned music." Illustrators added cartoons depicting talking pictures in the shape of a tin-can robot, a visual image that offered illustrators great scope for imagination. One ad showed a robot dressed up as a nurse, trying to induce a girl to swallow a spoonful of "canned music." The robot advises the poor young theater-goer, "Take it, dear, it's genuine music," to which she replies, "It's only more of that old canned sound, and I'm tired of it." Sound quality in early talkies could indeed prove unsatisfactory; a recording might become completely distorted or fall out of synchronization with the picture. To highlight the flaws of "canned music," one ad showed the robot in mythological form as the god Pan. As he plays his pipes, birds scold, rabbits run away, and squirrels cover their ears. Greedy studios and theater owners hoped audiences would ignore the obvious inferiority of talking pictures, the AFM asserted. One cartoon showed the robot as a bride, marching down the aisle with Hollywood's "Canned Music Promoter," who could "love anything that would reduce his overhead expenses." Providing that "theatre patrons can just be persuaded to accept less than their money's worth," the ad declared, "this happy couple can prolong their cacophonous honeymoon."[36]

The AFM attributed popularity of sound films to sheer novelty, which it suggested had already started to wear off. To encourage expressions of discontent, the group invited Americans to join its new Music Defense League. Union members and music stores distributed applications, and the AFM ads began including slips readers could send in to "vote for living music." Weber claimed that within sixty days, 1.78 million people had signed up with the Music Defense League, and by 1932, that number was over three million. One Maine resident allegedly wrote on the membership form that the "last time I was in a theatre I felt as if I were in some low down cheap circus." A supporter from New Jersey called the "foggy music" of the talkies "absolutely disgusting." Moviegoing had begun to feel "as rotten as to have to sit through a temperance lecture."[37]

The union did not organize mass consumer boycotts of sound films and did not call for the government to ban talking pictures. Nevertheless, its approach generated controversy. Some radio stations had rejected AFM material, Weber complained, under pressure from movie interests. Union leaders decided to abandon radio and concentrate on print, but newspapers in at least five cities also refused their ads. Editors for the *Milwaukee Journal* explained that decision by citing a policy against "advertising which contains destructive criticism of any other business." The AFM responded that it merely exercised free speech; studio executives, theater owners, and any others who took issue with the union had the right to produce their own ads defending the talkies.[38]

Between late 1929 and 1933, the AFM spent at least $1.2 million on the campaign against talking films, an effort financed in part by a 2 percent levy on the wages of members who still had work. In 1930, the union claimed some triumph, reporting that several large theaters had reengaged their orchestras. But by 1931, some union members suggested abandoning the fight. Conversion to sound had gone too far, they said, to permit any widespread reinstatement of live music. Conductor Walter Damrosch called the AFM ads "as futile as the efforts the hand weavers once made to stop the development of the machine age." Over the long run, he suggested, movie sound might actually increase employment for musicians by nurturing public appreciation for the art. Weber rejected that "gratuitous advice," implying that Damrosch, who had a generous contract with the National Broadcast Company, showed a "woeful lack of understanding" for rank-and-file musicians. Weber insisted that the AFM campaign had successfully "created a psychology against canned music," which ensured that talkies would never take over completely. "We realized that a fight against technological unemployment would have availed us nothing, hence we waged a cultural fight."[39]

THE KING'S FIDDLERS

Old King Cole was a merry old soul,
And a merry old soul was he.
He called for his pipe and he called
for his bowl,
And he called for his fiddlers three.

MY, my, what a rumpus when the robot appeared! It seems that the Prime Minister, in a fit of economy, had installed canned music and fired the King's rollicking fiddlers. The jolly old monarch was wroth.

But King Cole could remedy the trouble. He had only to order the robot to the attic, send for his beloved fiddlers, and have the Prime Minister publicly spanked.

Theatre patrons can't get action so swifty. But they can insist on having their money's worth in the theatre by joining the Music Defense League. Sign and mail the coupon.

AMERICAN FEDERATION OF MUSICIANS,   AF-6
1440 Broadway, New York, N. Y.

Gentlemen: Without further obligation on my part, please enroll my name in the Music Defense League as one who is opposed to the elimination of Living Music from the Theatre.

*Name*_____

*Address*_____

*City*_____*State*_____

THE AMERICAN FEDERATION OF MUSICIANS

(Comprising 140,000 professional musicians in the United States and Canada)
JOSEPH N. WEBER, *President*, 1440 Broadway, New York, N. Y.

Fig. 3. The advertising campaign mounted in the early 1930s by the American Federation of Musicians used cartoons and humor as a way of encouraging readers to express their dislike for the new talking pictures. The hapless robot became an object of mockery, representing all the flaws of "canned music." Recorded sound simply could never match the pleasant or soul-stirring qualities of live music, the drawings emphasized, and theater owners who installed sound movies effectively cheated audiences out of good entertainment. *American Federationist* 38, no. 6 (1931): 663.

Increasingly, AFM members began to question the ad campaign's effectiveness and protest the 2 percent tax. At the 1932 convention, one local denounced Weber for spending "hundreds of thousands of dollars per year on this useless propaganda when thousands of our members are in actual want." The union began phasing out its advertising in favor of a new campaign, organizing "Living Music Days" to show that live music had a social value that talking pictures could never match. By the end of 1932, one hundred locals had presented free concerts, dance parties, and parades. Such programs gave "business leaders a new understanding of the civic importance of musical culture," the AFM claimed, and led communities to hire musicians for more performances. But however welcome any work might be in the Depression, one-time engagements could not make up for the full-time theater jobs musicians had lost. By the mid-1930s, the AFM effectively conceded victory to the talkies. New York Local 802 glumly declared that "our industry is a sick one; technological improvements have indeed desimated [sic] our ranks."[40]

The AFM ran up against too many factors favoring the dominance of sound pictures. The technical conversion from silents to talkies entailed substantial capital investment, by some estimates costing each studio at least $250,000 and each theater between $5,000 and $15,000. Such a commitment made it unlikely the industry would decide to scrap its new equipment and rehire hundreds of musicians. Such a retreat would have amounted to the studios' admission of failure, and the talkies had not failed. Each year, Hollywood studios, directors, and actors grew more comfortable with sound technology and more accomplished in its use, while engineering advances in recording and playback devices improved quality. Audiences continued to fill theaters; even if some segment of the population disliked the talkies, the Music Defense League never approached the clout that would have been required to convince studios to return to silent films.

As if musicians had not been sufficiently shaken, social scientists warned that future technological change could further disrupt their employment. Television broadcasting would soon become practical, predicted the 1937 report *Technological Trends and National Policy*; the British Broadcasting Corporation had already started preparing studios and transmission apparatus. The *New York Times* offered a reassurance that television "shows no signs today of adding to social suffering or unemployment." Men and women with radio and film experience would find ample opportunities in the new industry, the *Times* writer reasoned, which would create a great need for cameramen, lighting engineers, scriptwriters, scene designers, and make-up specialists. Yet to demoralized musicians in the midst of the Depression, innovation had come to mean trouble.[41]

What Could a Union Do?

For all its frustration, the AFM had more options in dealing with mechanization than did many other unions. Musicians could claim a unique skill that had a certain cultural status, at least allowing them to try the strategy of appealing to the public's interest. Every theater-goer had noticed the replacement of live music by recorded sound, whereas other cases of mechanization had remained behind the scenes. Customers might not even realize that automakers had substituted welding machines for hand welding or that their breakfast cereal had been packed by automated equipment. In general, labor had little chance of getting consumers to weigh in against technical changes in production. Business interests ran their own publicity campaigns contending that mechanization improved their finished product. The telephone company suggested that by installing the dial system, it gave callers a more perfect service than human operators could ever provide. Cigar manufacturers performed an end run around the question of displacement by asking smokers whether they really preferred a cigar rolled in an "old filthy shop" by a man who "spit on the ends" to a cigar manufactured by machine. Food companies also played up the sanitary angle, advertising their products as "untouched by human hands." In all such cases, business leaders' argument that new production technology benefitted consumers provided a powerful counterweapon to labor's talk about how it hurt workers.[42]

Throughout the 1930s, labor leaders concentrated on drawing attention to the problem of displacement, making numerous speeches and radio addresses, writing popular articles, and repeatedly urging politicians to address the subject. More than that, union men attempted to turn technological unemployment into a rallying cry for organization. William Green explicitly cited the issue when launching new AFL recruiting drives, promising that unionization would help protect workers by reclaiming some of the power mechanization had stolen from them. Confronting the question "Is the world becoming robotized?" AFL vice president Matthew Woll responded, "Not so long as labor is organized."[43]

Some officials argued that labor relations would have to evolve in order to stay relevant in the face of workplace reengineering. Philip Murray called for the creation of a "scientific unionism" to help workers cope with the twentieth century's development of scientific management. Older wage agreements, dating from the years when industry had relied more on human power, no longer fit the modern context, he argued. Labor negotiators needed to start demanding contracts written along lines that compensated for the way machines had replaced muscle in work. Though the details of Murray's "scientific unionism" remained vague, fear of displacement did influence the course of labor relations in the 1930s. Talk of technological unemployment played a role in the creation of the Congress of Industrial Organizations (CIO). The AFL's craft-centered organization might benefit skilled workers, John L. Lewis suggested, but it had become "obsolete" precisely because mechanization had started eliminating that labor. Lewis turned his powerful voice and evocative imagery to emphasize that unions must be organized around new principles for defending the workers "driven out by arms of steel."[44]

Even as Lewis and other leaders called for action, it remained unclear what strategy unions could best adopt. The failure of the AFM's approach showed the problems inherent in trying to slow, stop, or reverse the introduction of new technology. Should labor acquiesce to new machinery and instead try to ease the inevitable transition by fighting for the support and retraining of displaced workers? Did unions have any power to affect workers' future in the Machine Age? Aside from the AFM, most unions did not attempt any serious campaigns in the 1930s to reverse the course of mechanization. In fact, the majority of leaders went out of their way to stress the opposite: they wanted business, government, and the public to acknowledge the existence of serious problems for workers, but they did not wish to block innovation. According to Green, labor understood the argument that economic well-being rested on the continued improvement of production methods. Union officials ended up walking a fine line, decrying machine-driven displacement while insisting that they did not hate technology. John L. Lewis drew a careful distinction, declaring, "Labor does not oppose labor-saving machinery, but it insists upon the saving of labor and not its destruction by the machine."[45]

Even if unions had wanted to do so, one AFL report explained, modern forces made it "idle to oppose technical progress." A sense of determinism pervaded labor statements on technology. "Further installation of industrial mechanical equipment is inevitable," Green declared in one speech, and "no persons with vision and understanding would attempt to stop it." The AFL welcomed the latest inventions, he insisted, though it lamented "the fact that technical progress has been permitted to usher in human want." Though business interests accused them of resisting progress, other labor leaders proclaimed, their groups actively encouraged change. The Steel Workers Organizing Committee ran an ongoing educational campaign in favor of technological advances, Philip Murray said. Such endorsemeOnts came with conditions; Fannia Cohn, executive secretary of the International Ladies' Garment Workers Union, approved of mechanization providing that industry started giving ordinary Americans a fairer share of the benefits in the form of greater wages and more leisure time.[46]

Speaking for the United Auto Workers, R. J. Thomas defined the issue as *control* of machines rather than the spread of technology itself. Society had permitted the "rulers of our financial and industrial systems" to monopolize the gains of mechanization while making workers bear all the cost. Thomas emphasized that workers could not let misery drive them to extreme reactions. "Organized labor has learned through centuries of experience," he continued, that "to check or destroy new machinery is plain social folly." Depression-era union leaders seemed to take special pains to avoid comments that might allow critics to paint them as Luddites, inciting workers to riot against mechanization. In the textile industry, the traditional focus of machine-breaking in past centuries, Emil Rieve insisted that "in no instance does the union say to management, 'This installation, this invention, you can't have. We are practical enough to realize we can't just stop invention and . . . probably wouldn't want to."[47]

I. M. Ornburn, president of the Cigarmakers' International Union, dissented from that chorus, becoming one of the few labor leaders willing to challenge the idea of inventions' inevitability in the 1930s. The mere fact that engineers had devised new equipment, he suggested, did not mean it automatically deserved to be incorporated into production. Considering how often company executives had held back discoveries themselves to protect patent rights or old capital investments, Ornburn declared, it seemed hypocritical for them to criticize workers who questioned the need to embrace mechanization. Especially during a period of severe depression, the state could well choose to "slow down our technological speed," regulating change to defend broad social interests. Ornburn did not advocate violence, vandalizing facilities, or banning research, but in merely talking about "putting a brake on technological progress," he already went further than other union officials. Ornburn's relatively extreme statements undoubtedly stemmed from the fact that cigarmakers had been fighting a losing battle against displacement for two decades. "It is not a sabotage of progress to temper it by what society can bear," he insisted. Doctors might be able to learn a great deal from vivisecting human beings, but few people would defend paying such a price for knowledge.[48]

Inside unions, people agonized and argued over the question of what price might be acceptable to pay for progress. In 1935, the AFL journal printed a piece declaring that, contrary to the predictions of optimistic economists, mechanization had not lowered consumer prices and thus generated work. The authors accused advocates of inflating the statistics on how many jobs automatic tabulators and other machines really created. They called for abolishing steam shovels, tractors, textile machines, and other devices "contrived for no purpose except to cut out labor and make more profit for a few manufacturers." At the top of the article, the *American Federationist* ran a special note justifying why the journal had decided to publish it. Editors liked the authors' "vivid description of the problem resulting from labor-saving machinery," though they disagreed with the argument for outlawing certain innovations. The fact that business owners had hogged the gains did not alter the case that mechanization still carried the potential to create cheaper consumer goods, higher wages, and shorter working hours. If only savings from increased efficiency were "more equitably distributed," the journal commented, production reengineering could yet provide "more comfortable living for all," plus "that leisure which results in a higher civilization."[49]

The AFL journal's stance underlined the fact that when push came to shove, most mainstream labor organizations accepted the premise that despite displacement problems, Americans could still associate mechanization with progress. When William Green said that "technical progress means more things at lower prices and consequently more physical comforts and greater ease of living for greater numbers," he could easily have been speaking for big business. Green, like corporate leaders, promoted mechanization as "the means to higher material civilization." Though he went on to call it "a sad commentary that individual wage earners have paid the social costs of technological progress," Green had acquiesced to the definition of consumerism as the primary measure of well-being. The promise of material abundance justified mechanization, and once workers won a more equitable distribution of profits, they would let industry follow its desired path of technological advance without objection. Implying that employers had been greedy might have called for a burst of inflammatory rhetoric, but again, mainstream labor carefully avoided any appearance of radicalism. R. J. Thomas argued that ordinary Americans should receive a greater share of the benefits from modern productivity, not as a step toward socialism but as a way to promote further mechanization. Far from aiming to topple capitalism, union men argued that giving men and women higher wages and increased leisure hours would restore purchasing power and jump-start consumption, thereby paving the way for still more extensive and rapid workplace change.[50]

Once labor leaders had effectively ruled out the concept of trying to limit mechanization, the question became one of how to minimize the damage, how society could help workers adjust to the inevitable. Adapting to the Machine Age, some unions suggested, required a fundamental rethinking about the nature of employment. Back in the late nineteenth century, American unions had fought for reducing the workweek, primarily on grounds of health and safety. The battle for shorter hours had raged ever since, and the alarm during the Depression over mechanization brought new life to the campaign. Union leaders argued that employers would invariably cut back their payrolls once machines allowed them to meet production targets with less labor. But if the standard workweek shrank in proportion to technological advance, business would need to compensate by dividing the work among more men. Given the seriousness of unemployment, Green suggested, the United States had to start following a basic "principle of relating the number of man-hours of work available to the number of persons who have to earn a living."[51]

Whereas unions of the 1920s had generally called for setting the workweek at forty hours, unions of the Depression decade put a more urgent spin on the matter. AFL and CIO conventions in the late 1930s adopted resolutions demanding a thirty-hour week as "the only answer to the machine age." American railroad workers, who had switched from a ten- to an eight-hour workday in 1917, now began pushing for a six-hour day. Union president A. F. Whitney estimated that mechanization and modernization had raised railroad operating efficiency by 70 percent between 1920 and 1935. Common sense demanded a shortening of hours "to absorb the great numbers . . . now denied employment in favor of the machine." Moving to a six-hour day would supply jobs for at least 350,000 men currently unemployed, the Brotherhood of Railroad Trainmen estimated.[52]

Union leaders insisted that their campaign would not hamper progress by imposing artificial limits on work. Shorter hours should evolve naturally, they suggested, out of the progress of technology itself. Just as the Western world had already moved away from the fourteen-hour days common in nineteenth-century factories, twentieth-century society now enjoyed an opportunity to continue adjusting working conditions. If anything, the AFL argued, it had been business that had upset economic life by resisting the historical trend toward a shorter week. If companies had followed labor's recommendations for cutting hours back in the 1920s, they could have prevented mass unemployment. Managers had been "unwilling to give up their doctrine that long hours offered . . . a better chance to get more out of their employees," unwilling to address the consequences of mechanization. Over recent years, Green charged, labor standards had been "moving backward while technical development went rapidly forward." Foundry and machine-shop workers continued to average over fifty hours per week, despite the fact that productivity had risen 50 percent between 1919 and 1929. The United States could not reverse the trend toward displacement, the AFL said, as long as business kept trying to apply "an 1870 labor principle to 1932 machine production."[53]

Of course, if employers simply slashed hours, people currently employed would lose income. Therefore, unions proposed that even as firms cut hours, they should still pay employees for a forty-hour week. Managers would gain more than they lost from such a step, labor insisted. Studies had shown that reducing hours actually raised productivity, since the move helped eliminate worker fatigue, time-wasting, absenteeism, and turnover. Furthermore, economic activity would pick up once companies began spreading work among more individuals. Business leaders should have been giving workers a greater share of the gains from mechanization all along, labor contended. Commerce Department figures indicated that between 1920 and 1929, industrialists' share of the profits had grown 72 percent and stock dividends had risen 256 percent, but employees' real wages had increased only 13 percent. Unions added that since the onset of the Depression, workers' pay had fallen another 10 percent on average. Economic trouble invariably resulted when ordinary Americans lost purchasing power, given the fundamental principle in Machine Age economics that mass production could not be separated from mass consumption. Merely to keep textile and garment industries functioning at 1929 levels, the AFL calculated, each American woman would need to purchase four new dresses, three pairs of shoes, and a dozen pairs of hose per year. To keep the flow of goods moving, business should have an interest in supporting family income.[54]

Labor organizations justified the call for shorter hours as a necessary and sensible adaptation in the Machine Age economy, adjusting conditions to correspond to the way technology had transformed the nature of production. Even if labor could get government and business to go along with that argument, such broad structural changes would not materialize overnight. Unions had to come up with short-term measures that would offer members more immediate protection. Some concentrated on the hope of retraining displaced workers, following a principle that in the modern world, job security came through flexibility. Men whose old skills had been devalued by mechanization might land positions operating the new equipment. During the 1930s, the printers' union ran classes to help members learn how to use the latest types of presses. When one coal company announced plans to switch from horse-drawn to motorized delivery, the teamsters' local bought a truck and hired a chauffeur and mechanic to show its men how to drive and repair vehicles. On some occasions, companies placed enough value on workers' skills to join in retraining efforts. In 1934, the New York transit system ran free classes on company time to teach streetcar drivers to handle buses, so that the city would not lose their experience in navigating crowded areas. When the Madison Avenue route switched over to buses, 180 former trolleymen made the transition successfully, and New York started retraining another four hundred to prepare for converting other lines.[55]

Of course, some social scientists warned that as the efficiency of mechanization reduced total labor needs, even retrained men might not find jobs, and in such cases, unions would need to take a more assertive stance. The American Communications Association, part of the CIO, had a 1940 agreement with Mackay Radio and Telegraph which stated flatly, "No employee of the Company shall be dismissed by the Company during the life of this agreement because of mechanization or technological changes." ACA head Daniel Driesen admitted that the restriction placed certain burdens on the employer but argued that management would ultimately benefit by winning worker loyalty. Such a clause offering men a blanket protection was unusual, but, according to one estimate, two hundred out of four hundred major labor contracts negotiated between 1933 and 1939 in manufacturing, transport, communication, and the extractive industries required that employers who installed new machines should at least make an effort to shift affected workers to alternate positions. One contract between the Textile Workers Union and the American Viscose Company specified that displaced individuals be put on a list for preferential rehiring when openings appeared in any company facility. The ACA established an arrangement with the Radio Corporation of America stipulating that, in order to give workers time for retraining, managers must give six months' advance notice before installing new equipment and, during that period, hold vacancies in other departments for those men and women before hiring anyone from outside. Such agreements might serve the interests of management as well; if the possibility of reemployment could satisfy enough people, an employer might head off any danger of radical protest.[56]

As part of life in the twentieth century, union leaders indicated, Americans should start holding companies accountable for how their mechanization affected labor. The existence of "machine-made unemployment is not 'an act of God,'" textile union representative Max Danish declared, but "part and parcel of evolution of industry," which businessmen who chose to install new devices needed to acknowledge. Phil Murray asserted that just as the courts had compelled factory and mine owners to assume some responsibility for workers injured on the job, employers now had to bear some obligation for workers displaced from the job. The United States could no longer afford to let business just discard surplus laborers, forcing society at large to bear the cost of supporting them and their families.[57]

In a few instances, labor succeeded in making industries accept their alleged responsibility for the consequences of workplace innovation. In 1932, the International Ladies' Garment Workers Union challenged the clothing industry over its adoption of six hundred electric pressing machines. In arbitration, George Alger, chairman of New York's cloak and suit industry, ruled that employers should pay $8 per week into a union fund supporting unemployed garment workers. Such money might not "offer a panacea for technological unemployment," said union official Max Danish, but the settlement did show that "industry owes the workers made jobless through the progress of machinery a measure of relief" to help "tide them over." Alger also granted machine operators a $12 weekly pay raise, accepting the union's contention that since new pressing machines had almost doubled production rates, workers were entitled to share the reward. Editors at the *New York Times* criticized Alger's ruling as an irrational strike against progress, comparing it to the "ancient law of deodand by which any inanimate object . . . accidentally the cause of a man's death is forfeited to the crown."[58]

Railroad workers fought their industry over the transition from steam to diesel locomotives, which eliminated any need to have a fireman on board stoking the engine. In theory, the engineer could run a train alone, the rail unions maintained, but as trains became longer and faster than ever, entrusting their operation to a single person entailed grave risk. Condemning union attempts to create "make-work" jobs, railroad executives protested that keeping an extra man on the locomotives would impose wasteful cost and inefficiency. Labor mounted a public and political campaign against one-man trains, backed up with the threat of strikes, and eventually won agreements with some major railways stipulating cases in which a fireman or "helper" would join the engineer.[59]

Such isolated victories felt sweet. Retraining programs and reemployment clauses undoubtedly helped a certain number of people through difficult times, proving that some workers could indeed survive technological change. Such arrangements, however, did not constitute a thorough solution to the problem of displacement. Labor did not have infinite finances to support retraining, and workers who acquired new skills might still have trouble finding positions. Few unions could count on negotiating protection for workers or receiving favorable arbitration rulings. The railroad unions might be able to put up a good fight for jobs, but smaller and weaker labor organizations had less chance to win.

Labor officials such as Green, Murray, Lewis, and most major union leaders really preferred not to wage outright war against the mechanization of production. They professed their willingness to support workplace innovation, providing that ordinary people enjoyed a larger share of the benefits through wage hikes, increased leisure, and material abundance. Yet many Depression-era men and women expressed a deeper pessimism and anger. A 1930 survey by the University of Michigan's Bureau of Industrial Relations showed that a large number of individuals doubted whether advances in production method would ever raise wages or lower consumer costs. Any savings would only go into the pockets of owners "already making excessive profits," said one respondent, and therefore he saw no reason to let industry "reduce its labor cost further" by introducing machines. Decades of dispute over wages, hours, union recognition, and other contentious issues had generated lasting antagonism, which fed suspicions about mechanization. Workers had come to "resent both the new processes and the management which has installed them."[60]

When workers feared the immediate dangers of displacement more than they trusted promises of future progress, they directly expressed their opposition to change. Whereas union heads went to great lengths to avoid any suggestion of outlawing innovation, some rank-and-file members insisted that the United States could and should stop business from introducing machines at workers' expense. Railroad workers proved especially militant, demanding "drastic legislation . . . to regulate and even to prohibit" employers from the "irresponsible" adoption of machines that had as "their sole purpose the elimination of jobs." In a letter to the *Railroad Trainman,* one man wrote, "If the introduction of new machines into our already over-crowded mechanical civilization cuts men off from their means of living, then they must wait; the genius who invents them must be content to waste his sweetness on the desert air for a few years more; and the capitalist who sees . . . only an opportunity to double up his profits must be taught self-control." The best efforts of Roosevelt's New Deal to restore economic health would fail unless society took control of technology, he warned, since "machinery can displace man-power faster than all the government employment agencies could absorb it." Union members needed to demand an answer to the fundamental question "Are the 'rights of machinery' more important than human rights?"[61]

Representatives of local branches challenged national union leaders over how to address the issue of technological unemployment. At the 1934 United Mine Workers convention, delegates stood up one after another to describe how mechanization threatened their jobs. One Indiana man reported that his mine had cut back from 375 to fewer than 185 workers while maintaining the same level of output, thanks to new coal loaders. An Illinois delegate demanded that government use relief and work-project funds to "buy up those loading machines and have them scrapped and let us men go back to work and make our living with the shovel and pick." That suggestion threw the meeting into bitter dispute. Referring to the way glassblowers had destroyed their union in an unsuccessful attempt to block bottlemaking machines, some speakers argued that any effort to fight mechanization would prove hopeless. One delegate worried that his fellows had forgotten the evils of nineteenth-century mining, with its child labor and catastrophic accidents. Rather than even "think of going back to the dark ages," he declared, unions should "adopt the machines and use them for our benefit and not for ... Andy Mellon." Another man declared, "You may call me a young modern fool, but I stand up here in favor of modern machinery." The crowd indeed responded with derision, rejecting any notion of coming to terms with change.[62]

One Indiana local proposed a resolution calling for "gradual removal of the machinery," a step that prompted John L. Lewis to speak. Miners had no chance of forcing the powerful coal industry to accept a ban on machines, he cautioned. Trying to impose an ultimatum would only backfire, since unions could not afford to create the impression that they opposed progress. Moreover, an effort to put miners to work by eliminating coal-loaders would be as absurd as trying to create jobs for carriage-makers by forcing people to give up automobiles. "You can't turn back the clock and scrap all ... modern invention," Lewis declared. "What we must do is to see to it that we also share in the benefits wrought by the machines, in the form of reduced working time and increased compensation." In the end, Lewis convinced the hostile audience to accept a substitute resolution calling for "subjecting machines, science and modern development of industry to human-kind." Instead of protesting against new equipment, the UMW pledged to fight to reduce working hours and give members a "greater participation in the blessings ... of progress."[63]

Lewis had headed off resistance, but controlling the debate had proven difficult, and miners' unhappiness continued to run high. At the 1936 UMW convention, West Virginia and Illinois locals introduced at least ten separate resolutions calling for a technotax. Some delegates hoped to slow the rush to mechanization, suggesting that if society imposed a fee on loading devices, "coal operators may not be inclined to buy these machines to put our brethren out of employment." Others viewed any thought of blocking mechanization as unrealistic, since "the human body cannot compete with the machines of this age." Those delegates supported the technotax as a means of at least minimizing the pain, a way to provide some basic income for displaced miners and their families. Most of the resolutions remained vague about how a technotax would work; the one version that specified details called for a tax of twenty-five cents a ton on all machine-loaded coal. At the 1940 convention, delegates from Kentucky, New Mexico, Ohio, Pennsylvania, Tennessee, and West Virginia introduced another twenty-eight technotax motions, one of which raised the proposed levy to seventy-five cents per ton. Despite the popularity of the technotax among members, most officers hesitated to embrace such a concept, lest it allow the business community to stigmatize labor as antiprogress.[64]

Workers could avoid clashing with national union leadership over how to deal with mechanization by developing their own forms of resistance. University of Michigan researchers reported that employees in one auto parts factory had flatly refused to help install a new machine that would have substituted for three men. Other workers admitted that if they had ideas for how to improve production, they would not tell supervisors. "The company doesn't need any help in improving efficiency," one explained. "The engineers discover too many ways to cut down on labor costs now. If a production worker finds a better way to do a job, he should keep it to himself." One man in the glass industry said he had done just that, kept secret suggestions that would have made fellow workers redundant, since he did not trust his employer to help them find new jobs.[65]

Some locals defied the national office outright by taking their own stands opposing labor-saving technology or imposing conditions on its use. In Washington, D.C., painters working on the new Social Security headquarters resisted pressure to use a new type of paint which could be applied in just two coats rather than four. Some electrical workers passed make-work rules requiring that contractors keep a qualified electrician on all jobs involving temporary lighting, "though he has nothing more to do than turn off the lights when the other men are through work." Obviously, such restrictions might backfire, raising production costs to the point of discouraging business and thereby reducing work. Employers could also transfer business to nonunionized locations or cities with more accommodating locals. After New York unions insisted that printing companies employ five operators per press, some companies moved to Chicago, where labor settled for three.[66]

Aggressive locals might get into trouble with union superiors, who tried restraining or disciplining men for adopting rules that threatened to undermine the national line on technology. Nevertheless, on a few occasions, workers' distress exploded into outright defiance. In 1933, Lorena Hickok reported to WPA leader Harry Hopkins that when an Iowa builder started to excavate the site for a new post office building in Sioux City, using "as much machinery and . . . [as] little labor as he could get away with—the unemployed went down there and threatened to wreck the machines, staging one of the nicest demonstrations you ever saw." But often, those upset about change confined themselves to verbal protest or symbolic gestures, such as holding a ceremonial burial of a miniature machine model. Dealing with the issue just left many frustrated. Trying to negotiate with employers had its limits. Attempts to block or slow down the introduction of new devices usually failed. The AFM campaign did not put a dent in the popularity of talking pictures, and telephone operators' complaints could not stop the spread of dial equipment. Especially as long as their union leaders endorsed the inevitability of change, workers feared that they had lost control over their fate. Mechanization could render all their skill and experience meaningless, changing familiar work processes into something strange.[67]

What factors governed the strategic choices of labor during the Depression? Why did most leaders remain so cautious, insisting that, despite the seriousness of technological unemployment, they believed in the power of mechanization? The question turns not on the strategies labor considered but on the extreme tactics it omitted. Union officials frequently referred to Britain's history of Luddite revolts, reminding the public that resentment of machines had once spawned vandalism and even murder. Yet all but the most paranoid listeners would have been hard-pressed to interpret such references as direct threats. For all the despair and fury American workers expressed over actual and potential job loss, mass machine-wrecking remained rare, if not unknown, in the 1930s.

While there is no reason to assume violence would have availed labor anything in the struggle over displacement, it is worth considering why radical action failed to develop. After all, science fiction stories and other narratives written in the 1930s (and in decades since) often entertained the possibility that hatred of the machine might drive workers to rebellion. In reality, several factors limited any eagerness to take an aggressive stance. Out of a desire to hang onto work at any cost, labor might bargain with companies for preferential rehiring. Under such circumstances, union locals might end up promoting mechanization. When Philadelphia hosiery-makers began losing business to nonunionized Southern states, the American Federation of Hosiery Workers supported measures to protect their jobs by updating local plants with faster and larger machines.[68]

An awareness that attempts to block mechanization had backfired on unions in earlier decades also conditioned labor's response during the Depression. History left no obvious course of action. Attempts to assert workers' rights against displacement had failed too often. It seemed safest for labor to campaign for shorter hours and higher wages while implementing retraining programs and other measures to help at least a small number of workers. Furthermore, in the 1930s, American unions were emerging from a period of decided weakness, still wrestling with basic questions of securing recognition and rebuilding membership.[69] Resources and energy which otherwise might have gone toward more direct fights against mechanization may have already been monopolized by events such as the AFL-CIO clash.

The character of union leadership during this time undoubtedly shaped the reaction to mechanization. Even as unemployment rose, the AFL's William

Green clung to a conservative vision of peaceful, mutually beneficial labor-management relations. In his nonconfrontational ideal, workers would help make production more efficient and would be repaid with higher wages, shorter hours, and fair treatment. Though John L. Lewis possessed a charisma and flamboyance which sharply contrasted to Green's personality, the CIO leader also shied away from extreme tendencies. Back in the 1920s, Lewis had maneuvered to marginalize radical forces within the United Mine Workers, heading off a nascent interest in nationalizing mining and blocking mechanization. Like Green, Lewis wanted to limit the conflict with business. Introduction of machines would make the entire mining industry stronger over the long run, he reasoned, and enable unions to extract higher wages and shorter hours.[70]

Throughout the 1930s, business leaders repeatedly accused labor of taking an extreme, antiprogress stance. In truth, that was precisely what union representatives most wanted to avoid. Although the AFL and CIO sought to persuade business, politicians, and the public to acknowledge workers' fears of technological displacement, they scrupulously avoided any basic challenge to the notion of progress under an existing capitalist structure. Machine-breaking obviously would have destroyed union attempts to appear reasonable and might even have fed sympathy for the businessmen targeted. The International Ladies' Garment Workers Union and some other unions had managed to gain small footholds against displacement, and machine-wrecking might have reversed such success. Finally, if unions turned too destructive, companies could feel justified in using even greater violence to restrain vandals. In such a light, the avoidance of Luddite tactics could be understood as a pragmatic acknowledgement of reality. While locals and individual workers sometimes rebelled against their leaders' readiness to accommodate mechanization, national offices had tactics to maintain control. By reducing dues for unemployed members and helping them sign up for relief, unions kept idle men tied to the organization and could thus deter them from any thoughts of revenge. Finally, after Roosevelt's election, the NRA and other New Deal programs provided support for the unemployed and encouraged labor's hope that the tide of national political events might be swinging in their favor. Given all these constraints on radicalism, the absence of Depression-era outbreaks of machine-smashing starts to look less surprising.[71]

As long as labor representatives went along with the assumption that technological change followed a predetermined path as part of the evolution of modern society, it made no sense to try slowing, stopping, or redirecting that course. More than that, Green and other mainstream union leaders hoped that, while industrialists and investors had grabbed an unfair share of the profit from mechanization so far, labor could still win a more satisfactory redistribution in times ahead. Machine-breaking would have destroyed any ambition toward this goal. Even while members agonized over the job losses they saw around them, their leaders clung to a more optimistic vision of Machine Age progress through mass production and mass consumption, preferring to believe that mechanization might yet bring a consumer and worker paradise. For all its caution, labor made a real difference in the Depression-era discussion. By insisting over and over again that recent years had brought serious problems, union leaders helped establish the issue of mechanization as one commanding national attention. The media and politicians responded to powerful images of workers and even entire towns being thrown onto relief as a result of technological advance. Throughout the decade, talk about men losing jobs to machines remained widespread, bringing the idea of technological unemployment into popular currency.

4

"Machinery Don't Eat"

Displacement as a Theme
in Depression Culture

WHILE UNION INTERESTS produced literally hundreds of speeches, articles, pamphlets, and advertisements decrying the harm mechanization had done to workers, technological unemployment amounted to more than just another labor issue in the 1930s. A significant number of Americans whose jobs were not immediately endangered nonetheless expressed deep reservations about the new Machine Age economy. With the Depression, more and more people seemed ready to challenge or at least rethink the assumption that improving the technologies of production always meant progress. That urge brought talk of displacement into common parlance, and criticism of new workplace equipment developed into a staple of popular culture. Through the 1930s, American books, movies, humor, and artwork provided space for expressions of alarm, writing a type of collective textbook documenting uncertainty.

Just looking around in their everyday lives, ordinary Americans believed they could see a dangerous trend toward elimination of jobs. "I had a cousin who worked in D.'s checking bills. They got in adding machines and three girls got laid off," one person commented. Another observed, "It used to take four days to load a boat down at the river. Now a machine loads it in nine hours, with only one man working." A chef from California protested to the Labor Department that, on Nevada road construction projects, the new Le Torneau scraper allowed one laborer to move twenty-four yards of earth at once, single-handedly accomplishing a task that previously would have employed seventy-two men and 288 horses. Common sense, the man suggested, should tell anyone that such a sharp reduction in labor needs would create social and economic havoc. One New Yorker reported that he found proof of the displacement problem just in his daily commute. "When I left my apartment this morning, I pressed a button . . . self-service elevator, one man out of job. I then went to the subway station and found one man making change and nine turn-stiles: nine ticket-choppers out of jobs. I then boarded a ten-car express manned by two guards: eight more men out of jobs," he wrote. "Who knows where it will end?"[1]

Where it would end, to some in the Depression, unfortunately seemed to be permanent technological unemployment. Once business had installed "machinery to do the heavy work," they simply "do not need men to work any more," one Texas man complained. Adoption of increasingly powerful work-place equipment would be "fine if we could find something else to do, but there is nothing." A Tennessee man put the issue more bluntly in a letter to Eleanor Roosevelt, asking the First Lady, "Will you please warn the people of what's going to happen in America if these property owners don't quit mak-ing industrial slaves out of their laborers . . . or else installing machinery and laying the common laborer off of his job to starve to death?"[2]

During her cross-country tour of 1933–34, Lorena Hickok observed how much people in different states blamed the force of mechanization for dis-rupting their community. Residents of Minnesota's Iron Range didn't expect unemployment problems to end any time soon, she reported, because they had seen how easily mine owners had installed labor-saving devices. "Electric shov-els that can be operated by one man, instead of six or eight needed on a steam shovel. Electric cranes that, in moving track about, can enable three or four men to do the work that used to require 40 or 50." Elsewhere in the Midwest, locals expressed outrage at a public works contractor who brought in $75,000 worth of heavy machinery and finished a road job using just eight men, while a comparable site nearby employed forty men using shovels. The aim of relief should be to provide work, not to save it, they complained, and the govern-ment should stipulate specific manpower targets for all public works projects to prevent further abuse. All across the nation, Hickok told Harry Hopkins, she had "heard a lot about that business of machine versus hand labor" from ordinary Americans who didn't like what they saw.[3]

Religious organizations and other groups joined individuals in expressing unease over the pace of modernization. In 1930, an Episcopal Church confer-ence on industrial relations called on employers and workers to cooperate in addressing the urgent problems caused by mechanization of production. Dis-placement had recently taken a tragic toll on people in the printing, mining, textile, and railroad industries. The substitution of machines for men raised disturbing questions of social justice, the Episcopalians suggested. The Pres-byterian Board of National Missions, through its Committee on Social and Industrial Relations, issued a special Labor Day message in 1932 acknowledging that it seemed impossible to halt the momentum of innovation. Nevertheless, the Presbyterians insisted, workplace equipment must never "be thought of as possessing more importance than the man who operates it."[4]

Throughout the 1930s, mass-circulation magazines and newspapers reported on such expressions of dismay (and in the process undoubtedly fed worries as well). The *New York Times Magazine, Current History, Survey, Survey Graphic, New Republic, Living Age, North American Review, Literary Digest, Review of Reviews, Commonweal, Rotarian,* and a host of other periodicals published articles bemoaning the human consequences of mechanization.[5] Some pieces offered a general overview of recent changes in the nature of work, while others described specific examples, such as the continuous-sheet steel mill or the cotton-picking machine. The Roosevelt administration's statements on the issue of displacement drew media attention, as did reports produced by the WPA National Research Project. While the tone of popular press analyses varied widely, most reporters started from a premise that Americans should consider the possibility that companies' rush to mechanize had contributed substantially to present-day joblessness.

In 1939, *Harper's* published a piece that mounted an unusually direct challenge to the idea that Americans ought to resign themselves to workplace mechanization. All too often, Peter Van Dresser complained, people tended to "talk and think as if Scientific Technology were a kind of willful genie whose gifts we must gratefully accept while we accommodate ourselves as best we can to his bad habits." Engineering did not represent some inevitable force, he reminded readers angrily, and machines did not invent themselves. Modern technology was a "child of the wishes" of men who adopted it, executives and managers of big corporations. Left to the control of power-hungry employers, mechanization had crippled agriculture, fostered monopoly, and forced government to adopt a no-holds-barred "policy of oil-imperialism, raw-material imperialism, foreign-market imperialism." American workers, intellectuals, and professionals must resist that destructive trend, Van Dresser wrote. Engineers should concentrate on developing a wider range of equipment, devices that would serve small business as well as large. The nation could not return to economic and social health so long as people "accepted utterly without criticism the blueprints for America's technological future (and present, for that matter) formulated by the industrial empire-builders."[6]

While Van Dresser addressed mechanization as an abstract philosophical matter, other articles in the popular press put a more personal spin on the subject. In 1933, *Scribner's Magazine* presented a piece whose emotional immediacy transformed technological unemployment from a problem in economic theory into a drama of human struggle. Once the glassmaking industry adopted machines for making bottles, Ruth Crawford wrote, her life collapsed.

Her father had been a skilled glassblower earning good money, until the "Jersey Devil" displaced him. The family had to abandon its nice house, following her father as he travelled from town to town, hoping at least to find a job operating the new machines. When the force of invention left a honest craftsman despondent and uprooted families, Crawford concluded, it became the "great evil" of modern existence. Her father had tried "to make peace with a machine that was a better blower than he," but there "was no peace to be had."[7]

As the subject of labor displacement began to arouse concern at the highest levels of government, even *Good Housekeeping* weighed in on the controversy. The magazine attempted to reassure readers, to downplay any idea that the installation of new workplace machinery had caused trouble. An article entitled, "Will the Goblins Finally Get Us? No!" quoted Henry Ford as saying that "people are going to be all right." William Wickenden, president of the Case School of Applied Science, promised that the invention of still more advanced production equipment would usher in an age of material plenty and spiritual revitalization. The article urged ordinary Americans to place their trust in such men of authority, to join them in welcoming mechanization as a force for good. "Who would not be enthusiastic over the promise of a New Man, given leisure and security by the changing, advancing machine?"[8]

For those who wanted more detail, publishers turned out a number of books on technological unemployment during the Depression decade. One popular author, Stuart Chase, had established a reputation back in the 1920s for writing about how economic developments affected American workers and consumers. Some of his books written before the crash had drawn attention to the power of production technology, cautioning that the accelerating trend toward factory and office mechanization might endanger future job prospects. His 1931 work *Men and Machines* explored that question in depth, weighing the many pros and cons of modern life. The twentieth century had brought an unprecedented speed to technological change, he wrote, creating much excitement but also making employment more tenuous. Chase acknowledged that the spread of innovation created some positions; after all, it took human labor to build, sell, and repair machines. Other Americans, though, felt the continual threat of displacement hanging over their heads, like a modern sword of Damocles. Introduction of turnstiles had cut the number of platform workers in the New York subway from 1500 to 470, while on modern steamships, three supervisors watching gauges could perform the work formerly done by 120 stokers. "Machinery did not inaugurate the phenomenon of unemployment, but promoted it from a minor irritation to one of the chief plagues of mankind," Chase wrote. A "new job can no more be created as fast as the machine tips a man out of an old one."[9]

Inventions themselves had not created unemployment, Chase suggested; the difficulty came when government, economists, and business leaders refused to recognize that human well-being must take precedence. A properly designed society would exert control over the adoption of workplace technology. "Haste on the part of any machine—however praiseworthy from the standpoint of efficiency—to supplant workers faster than you can relocate them or adjust their hours of labour, shall be punished by a prompt withdrawal of lubricating oil." Unfortunately, Chase indicated, the business community had only made problems worse in recent years by putting the savings gained through efficiency into profit rather than using them to lower consumer prices. Purchasing power had not kept up with the system's increased ability to turn out more goods. Chase asked, "Is it not all tragically ridiculous? Men are to tramp the streets by the thousands because machines can provide more than enough to go round." Evidence suggested, he told readers, that "from now on, the better able we are to produce, the worse we shall be off."[10]

After publishing *Men and Machines*, Chase continued to pour time and energy into talking about displacement. In numerous speeches and articles, Chase called for reducing the workweek to reflect the way machinery had reduced labor needs, for arranging a system of national economic planning to ensure that workers and consumers received a fairer share of the benefits from modernization. Such proposals were neither new nor unique, but Chase's access to the media helped bring them before a popular audience. His dramatic yet clear language, combined with the directness with which he addressed the issue, made him a visible and influential figure. In a 1934 radio address, Chase warned that the Depression itself intensified the risk that the future would bring still more serious problems of technological unemployment. As the economic crisis squeezed business, it only increased managers' obsession with mechanization, their eagerness for finding ways to raise productivity. Moreover, as workers would soon discover, thousands of unemployed scientists and engineers with "a lot of extra time on their hands" had kept busy by inventing and improving labor-saving devices.[11]

Technocracy

A particular burst of press coverage came with the brief rise of the Technocratic Movement. In 1932, Columbia University's Department of Industrial Engineering announced it would conduct an "energy survey," analyzing patterns of employment and energy use in American industry and agriculture over the past century. Within months, director Howard Scott reported that the project had gathered statistics showing that the nation had passed through a crucial juncture in economic history. Before 1900, technological change had occurred slowly and production levels remained relatively constant, but the three decades since had brought enormous and frequent oscillations in output. The nineteenth-century economy had been like "a slow-moving ox-cart which suffered little damage in collision," Scott said, while the twentieth century resembled "a high-speed racing car hurtling down a highway." Its sheer velocity increased the danger that the system might spin out of control and multiplied the cost of a crash. The Machine Age had turned ordinary economic life into a flirtation with disaster. Scott and his fellows argued that the force of mechanization had already left the United States far closer to a full industrial breakdown than most people recognized. Between 1900 and 1929, total steel production had risen fivefold, from eleven million metric tons to fifty-eight million. During that same period, innovation and machine power had cut the man-hours required to produce one ton of steel from seventy to thirteen. Such revolutionary development had created a chaotic situation for labor and proved beyond doubt, Scott declared, that "the fundamental cause of the depression is not political, it is technological."[12]

Of course, by 1932, Technocrats were neither the first nor the only observers who wondered about the relationship between mechanization and job loss. Stuart Chase had already written *Men and Machines.* Labor leaders such as William Green repeatedly spoke about the problems that had resulted from the advent of continuous-strip steel mills, office machines, and other new devices. Scott claimed that Technocracy had moved beyond mere anecdotes to quantifiable and therefore irrefutable proof. The energy survey's data indicated that levels of industrial employment had peaked in 1918 and declined thereafter, although production had continued to move upward until 1929. To Scott, those figures represented undeniable evidence that employment had become "an inverse factor in the rate of production." That new economic reality, he declared, made it pointless to hope that future inventions could create enough work to absorb all displaced laborers. Even if popular demand for television and other novel consumer items gave birth to entirely new industries, the business community would still be trying to eliminate as much labor as possible. Engineers and inventors would keep improving production machinery, and so employment could never catch up with ingenuity.[13]

To Scott, such trends seemed to lock the modern world into a self-destructive course of advancing technology and declining employment. The United States could only escape that vicious cycle, he decreed, by revamping its entire mode of economic assessment. Technocracy sneered at mainstream economists and social scientists for clinging to ideas as backward "as physical theory in the time of the Greeks." Principles of price and profit might have made sense in an age when manufacturing depended on human power and human skill, but in the twentieth century, they became as ridiculous as "attempting to measure height with gallons of water and multiplying potatoes by automobiles." Such outdated concepts "must be thrown away," so that scientific and engineering experts who understood the Machine Age could create a new economic system, one "derive[d] from the nature of the machine itself." Such Technocrats focused on patterns of energy use, defining their units of measurement in terms of calories and foot-pounds. Economic sanity would only be restored, they argued, once the United States had learned how "to maintain a thermodynamically balanced load" across the entire world of production.[14]

Scott never spelled out the precise details of any plan for recovery, but that fault did not necessarily seem fatal at first. Though his economic theory might sound like gibberish, his predictions of technological unemployment came across clearly enough. Furthermore, by claiming to have extensive statistics documenting dangerous trends in mechanization, Scott lent his warnings an air of scientific authority. Throughout late 1932 and into 1933, Technocracy became the subject of widening discussion. Scott's pronouncements on the danger of epidemic displacement drew coverage in the *New York Times, Survey Graphic,* and other media outlets. As Scott gained public attention, however, his combination of dramatic rhetoric and pretensions to expertise soon drew fire. By early 1933, bashing the Technocrats became a popular pastime. Criticism of Technocracy made a good news story in itself, creating something of a circus atmosphere around the issue of machines and jobs. One New York politician denounced Technocracy as "the great Columbia rackety-rax," a "multitude of miscalculations" by a "flock of dithery young scientists" who had come up with "a scientific snake dance right under the hard-boiled brow of Nicholas Murray Butler," Columbia University's president. The business community also joined in attacking Scott's motives and credibility. Virgil Jordan, president of the National Industrial Conference Board, condemned the Technocrats as a bunch of "intellectual terrorists who proclaim the technocrack of doom." Their alarmism had cast "a paralyzing spell ... over responsible sections of the community," he moaned, "because our faith in ourselves had been shaken by depression." The United States had gone "technocrazy."[15]

While such ridicule made for cute reading, more sober criticism of Scott's methods and conclusions came from some of the professionals he scorned. Meredith Givens, secretary of the Social Science Research Council, challenged the idea that technological unemployment had become rampant. A less hysterical look at the evidence suggested that mechanization might account for 10 to 15 percent of total joblessness at most, Givens announced, and the remaining cases resulted from the problem of "idle machines rather than the busy machines." Wilford King, a New York University economist, echoed the sentiment that very little job loss "can legitimately be ascribed to technological improvements." Extrapolating from 1930 census data, King came up with even more reassuring numbers than had Givens, suggesting that out of 2,686,145 unemployed, fewer than 97,000 had lost jobs to machines. The very effort that King, Givens, Jordan, and other economists, politicians, businessmen, and scientists poured into discrediting Technocracy underlined how much they feared that talk of mass labor displacement due to automation had become credible. Yale economist Irving Fisher condemned Technocracy as "one of the most dangerous sophistries now feeding" on people's desperation. He accused Scott of instigating "a deliberate campaign of fear" at the very moment when leaders needed to restore public confidence. For Fisher, Scott's work proved the adage about a little knowledge as a dangerous thing, since his interpretations of economics sounded just plausible enough to catch on. The Technocrats might churn out "many charts as to foot-pounds, " Fisher warned, but nobody should take them seriously "until they include a few other economic factors besides energy."[16]

Defenders argued that, while it might be easy to poke fun at the Technocrats, these doomsayers had nevertheless put their finger on a genuine economic problem. Stuart Chase assured the public that available statistics indeed suggested a disastrous trend toward increased mechanization and decreased labor needs per unit of output. But Scott's rhetorical flourishes and his exaggerated claims of technical qualifications made him vulnerable, and by early 1933, the tide had swung against the Technocrats. The group's early statements had drawn attention partly through the project's link with Columbia, but that January, President Butler explicitly disavowed any notion that his institution endorsed Technocracy. Columbia had merely provided working space, he explained, out of a well-intentioned desire to support needy researchers.[17]

Even as the media frenzy over Technocracy subsided, even as it appeared that critics had fully discredited Scott, the attention drawn to mechanization did not dissipate. If anything, subsequent years brought some vindication for the Technocrats, as respectable observers and institutions came to agree that displacement represented a serious problem. During Technocracy's heyday, in 1932, Scott had warned that even a burst of prosperity that instantly restored industry to its peak 1929 production levels still would not alleviate unemployment. Mechanization had spread so far in just the three intervening years, he said, that in 1932 employers could achieve 1929's level of output while using a mere 56 percent of the displaced workers.[18] While naysayers scoffed at Scott's analysis, Franklin Delano Roosevelt expressed similar concerns just three years later in a major press conference. The president cited less apocalyptic figures, suggesting that a return to 1929 production would absorb 80 percent of the unemployed, but he shared the Technocrats' fundamental conviction that ongoing innovation would make it ever harder to compensate for displacement. The publication of the 1937 study *Unemployment and Increasing Productivity* and other government reports appeared to indicate that such predictions had come true. The nation's manufacturing sector had rebounded from its postcrash implosion, yet unemployment remained stubbornly high.

The issue of technological unemployment did not arise with the Technocrats, and it did not die with them either. Stuart Chase, William Green, and many other Americans had begun expressing alarm over displacement months before the Technocrats arrived on the scene. The impression of Technocracy as a historical curiosity, a one-time movement inspired by the odd character Howard Scott, belies the truth that the Technocrats succeeded in attracting attention and gaining some credibility precisely because concern about mechanization had already become familiar.

The very public rise of Technocracy did play an important role in energizing the debate over workplace mechanization. Scott's controversial pronouncements grabbed headlines, and coverage of Technocrats' predictions undoubtedly helped acquaint many Americans with the whole question of technological unemployment. Popular treatments such as Scott's *Introduction to Technocracy* and Frank Arkwright's *The ABC of Technocracy* allowed people to follow the discussion without becoming embroiled in complex technical disputes over economic methodology and conservation of energy. Even in the depths of Depression misery, there had never been much likelihood that Americans would act on Scott's bold yet vague schemes. His quirky personality proved detrimental, providing ammunition for those who wanted to dismiss any link between machines and job loss. Throughout the latter half of the 1930s, businessmen, scientists, and other critics often responded to talk of displacement by ridiculing anyone who raised the issue, calling them heirs to the misguided Technocrats. But despite the rapid debunking of Technocracy, the issues Scott and others had explored stayed very much alive. Scott's fundamental belief, that mechanization presented real problems for workers and for the American economy as a whole, found support from less outlandish sources and became part of Depression-era culture.

Technological Unemployment in Depression Literature

Authors of the 1930s brought the discussion of workplace change not only into nonfiction books and articles but also into just about every genre of writing. References to displacement appeared in works of classic literature, in detective novels and science fiction, in poetry and children's books. Writers voiced the deep tensions behind popular awareness of new technologies, the misery of men and women who felt that the Machine Age had ripped away all their control over life. Concern about mechanization predictably turned up in the Depression's "social novels," or "novels of protest," works by some of the country's most famous writers, which explored how "progress" had come to haunt ordinary Americans.[19]

While WPA statisticians compiled formal reports analyzing how tractors, combines, and corn pickers had reshaped rural life, John Steinbeck translated such images into the story of "a simple agrarian folk" who had been "caught in something larger than themselves." *The Grapes of Wrath* described the lives of people "who had not farmed with machines or known the power and danger of machines in private hands" until "suddenly the machines pushed them out." Land owners disclaimed responsibility for driving off tenants, explaining that bank managers had forced them to adopt the latest, most efficient machinery. To displaced families, the tractor resembled a tank, in that "people are driven, intimidated, hurt by both." One helpless character in the novel comments, "If this tractor were ours it would be good," but in the hands of a land owner, that same tractor "turns us off the land." The Joads became "a people in flight, refugees from . . . the thunder of tractors and shrinking ownership." Mechanization seemed like an inescapable force; even after fleeing to California, the Joads find workers there afraid that introduction of the "cotton-pickin' machine" would soon "put han' pickin' out." Though critics charged Steinbeck with oversimplifying and emotionalizing complex agricultural economics, *The Grapes of Wrath* made the issue of farm labor displacement into a powerful human tragedy.[20]

For Upton Sinclair, it was the latest industrial technology which best suggested the problems facing labor. His 1938 novel *Little Steel* contained one scene in which an observer describes the sheer power of continuous-strip mills. "You slide a hot ingot in at one end and set some levers, and it goes at twenty-five miles an hour, and out at the other end comes rolled steel of any thickness, any size, any shape you want." Such machines would "put tens of thousands of the most highly paid rollermen out on the streets," the expert warns. Such a prospect did not faze steel magnate Walter Judson Quayle, whom Sinclair characterized as a man so deeply in love with the idea of innovation that he lost all sense of how changes in methods affected people. Like Steinbeck, Sinclair saw the rapid onrush of technology as *the* defining force of the twentieth century, something with an increasing power to destroy working-class families.[21]

Mechanization tended to exacerbate economic disparities between classes,

Theodore Dreiser wrote in his 1931 book, *Tragic America*. In the first half of 1930, workers' wages had fallen by $700 million, he said, while shareholders' dividends rose by $350 million. The "unscrupulous and selfish" managerial elite kept pursuing innovation out of a desire for "underhand gain," heedless of the resulting overproduction and job loss. The "developing machine age speeds, reduces, and discards men," Dreiser wrote. He had seen older workers especially, "like the worn-out machine, thrown on the scrap-heap to rust away." When shipping companies adopted conveyor systems to move enormous cargoes, dock workers joined telephone operators and cigarmakers on the bread line. Owners of the Botany Worsted Mill in Passaic, New Jersey, had introduced 360 new automatic looms, despite the likelihood that such change would add to a local unemployment rate already hovering around 33 percent. For Dreiser, even more than for Steinbeck or Sinclair, joblessness represented the damnable outgrowth of a system stacked against workers. The "capitalist failure" had turned the promise of production into catastrophe. "Twenty years ago, labourers dreamed of a halcyon machine age, with six or seven hours' work a day. Yet to-day, with machinery at almost the perfection point, they are beggars, receivers of charity, while 40,000 millionaires bestride the land."[22]

Like Dreiser, Sherwood Anderson told readers that he had personally witnessed the harm done by mechanization. In his 1931 essay *Perhaps Women*, Anderson wrote, "Have I not myself seen how every year machinery becomes more and more efficient? Does not efficiency in machinery mean less men employed? If men are not employed, how are they to receive wages?" Men in previous centuries had earned a living through their physical and psychological strength, Anderson explained, but engineering had "taken from us the work of our hands." Soon the only jobs left would be those for machine operators, work which destroyed men by reducing them to mere attendants of a far more powerful device. The "age has moved too fast," Anderson concluded, and "modern men have lost maleness because we have not really dared face the machine." Though modernization might render man meaningless, it could never duplicate a woman's fundamental ability of child-bearing. The female spirit would remain unbroken, Anderson noted, giving the world one last hope of resisting dehumanization.[23]

The works by Anderson, Dreiser, Sinclair, and Steinbeck reflected a common concern that the recent pace of mechanization violated basic principles of human decency. Men and women who simply wanted to earn a living had ended up in dire straits because of forces beyond their control. By giving the development of technology precedence over the well-being of ordinary people, civilization had headed down the wrong track. Such pronouncements

made for a weighty message, but readers who preferred lighter forms of literature would still find the issue creeping into their books. Some authors, in
fact, simply assumed that audiences would understand passing references to
the concept of machine-based job loss. In the mystery *Death on the Aisle*,
Frances and Richard Lockridge threw in a brief, bitterly humorous mention.
A female character, bemoaning her fiancé's choice of police work as a profession, asks, "Oh, Bill—why don't you sell ribbons?" The fellow immediately
shoots back, "Didn't you know about ribbon clerks? Technological unemployment—dreadful thing." The very casualness of that line underscored the
reality: Depression-era economics had brought the phrase "technological
unemployment" into common parlance.[24]

In *Death on the Aisle*, as in *Grapes of Wrath* and *Little Steel*, authors incorporated real-world technological facts into their literary settings. Steinbeck's
Joads reacted to the same advances in farm mechanization that concerned
WPA researchers; Sinclair's description of continuous-sheet production
mills came straight from the pages of *Steel*. Fictional concern about job loss
gained plausibility because it was so closely rooted in genuine fears. For other
Depression-era authors, true horror came not from those machines already in
operation but from the potential development of infinitely more powerful
devices. Science fiction writers displayed a special fascination with the idea
of a future built around the work of robots. As most commonly envisioned,
such devices would resemble a person in form yet operate far more efficiently.
Though the actual construction of independent, productive humanoid robots
remained a dream in the 1930s, writers used the possibility to reflect on the issue
of technological unemployment. In exploring the most dramatic scenarios for
how mechanization might transform life in centuries to come, writers revealed
their feelings about economic and social conditions of the present.

The whole notion of technological unemployment created a division
within the 1930s science fiction community. Some devotees denounced the
slightest suggestion of linking mechanization to job loss, considering that any
such idea betrayed a shameful lack of faith in the wonders of modern science
and invention. The genre of science fiction had been founded on and should
remain defined by the tenet that technical knowledge opened up wonderful
opportunities for improving life and expanding man's horizons, they argued.
"Science fiction is based upon the progress of science," sci-fi pioneer Hugo
Gernsback explained in 1931. "If you admit that machines and science are all
wrong and that they are destroying humanity, then there should be no such
thing as science fiction."[25]

Gernsback did his best to exclude the issue of displacement from the field
he had helped establish. As editor of *Wonder Stories*, he said, he had rejected

a number of submissions that represented "out-and-out propaganda against the Machine Age." Authors of such stories set out to portray machines as "Frankenstein monsters," Gernsback charged, attempting to "inflame an unreasoning public against scientific progress, against useful machines." His own optimistic philosophy maintained that even if workplace mechanization might temporarily displace some labor, it could never present any deeper problem. History proved that over the long run, all machines actually "created employment where none existed before," Gernsback insisted, and so the world must continue its evolution toward a superior "machine civilization."[26]

Though Gernsback fiercely defended technological change as the keystone of modern progress, other members of the science fiction community seized eagerly on the theme of machines eliminating labor. The Depression-era treatment of machines as job-destroyers carried over some precedents from earlier science fiction, most notably Czechoslovakian playwright Karel Capek's 1921 classic R.U.R. Capek had envisioned a society in which the perfection of robots made human beings appear hopelessly inefficient. In the play, the makers of "Rossum's Universal Robots" advertise that their "intelligent labor machines" could handle the work of two and a half men for less than a penny an hour. By substituting machines for men, manufacturers could churn out so many goods that everyone could just help themselves to an unlimited supply. "Yes, people will be out of work, but by then there'll be no work left." As Capek's later scenes showed, such utopian assumptions missed the mark. Far from elevating people to a wealthy, leisured aristocracy, in the play the elimination of labor render man completely superfluous. When robots take over all valuable functions, humans lose their sense of purpose, their strength of character, even their capacity to reproduce. Man's desperate battle to reclaim the world comes too late, since robots made invincible warriors. Having proven their superiority, the machines then see no reason to keep a weaker species alive. Capek's play established the concept that in trying to replace workers with machines, even out of the most praiseworthy intentions, society might end up committing suicide.[27]

The concept of technological advance setting the stage for a final man-versus-machine conflict, which humans were destined to lose, emerged as a powerful motif of science fiction. In Capek's version, man starts the ultimate battle as a futile effort at self-defense. In other writers' scenarios, robots initiate attacks on humans when they became tired of having such inefficient, parasitic creatures around. Even if the spread of robots did not lead to actual war, science fiction suggested, the problem of technological unemployment might still annihilate civilization. John Campbell's 1934 story "Twilight" featured a mysterious visitor from seven million years into Earth's future, "the twilight of the race." In Campbell's world, as mechanization advanced, human beings gradually lost all curiosity, rationality, and other mental abilities. The final victims of technological unemployment could no longer even understand the machines around them, and so man quietly disappears. With no one around to order them to stop, the hundreds of "perfect, deathless" robots carry on forever, repairing each other and maintaining all the long-abandoned cities.[28]

Science fiction writers of the 1930s transformed the old Puritan work ethic into a Machine Age morality play. Honest labor represented the foundation of humans' self-respect and motivation, and if the future advance of technology robbed men of those values, life would disintegrate.[29] Such science fiction operated according to an extreme social Darwinism; without either the need or the opportunity to maintain their physical and mental health through work, humans would deteriorate biologically and move toward extinction. Each labor-saving improvement helped ensure the final triumph of technology; it seemed only logical that machines would be more fit than human beings and have a better chance of surviving in a new Machine Age. Technophiles such as Gernsback might let their imaginations loose on a future full of amazing new gadgets, gleaming cities, and the glories of space travel, but writers such as Campbell projected an alternative message, one that tapped into contemporary fears about unemployment.

The doomsday version of science fiction writing suggested that engineering "progress" represented a slippery slope toward an irreversible loss of control. By contrast, the genre of utopian fiction conveyed a lesson that as bad as displacement had gotten, society could still correct the problem to construct a future in which mechanization brought genuine well-being. Like earlier utopian writers such as Edward Bellamy, Granville Hicks used a futurist setting as commentary on the social present. In *The First to Awaken*, Hicks told the story of a twentieth-century man, George Swain, who awakens from a hundred years of cryogenic preservation to discover that the United States had successfully sorted out all its economic and social difficulties, including technological unemployment. Hicks built an entire fictional history of the future, which postulated that economic chaos and job loss had persisted in the United States through the 1940s, leading to election of a fascist president and then civil war. Finally recognizing the fatal flaws of capitalism, Americans had created instead an ideal system of democratic socialism. "Civilization depended on the machine" in the perfect world Hicks described, where a systematized process of invention had raised production to its highest level. Photoelectric cells, conveyor belts, calculators, and other manufacturing technologies yielded a material abundance which cooperative communities distributed for mutual benefit. Machines had taken over all dangerous, stressful, and unrewarding tasks, freeing people to concentrate on meaningful functions such as teaching and scientific research. Society had completely redefined the meaning of employment. Nobody had to work more than four hours a day, leaving them plenty of time and energy for exercise, culture, and other leisure activities. The happy citizens of 2040 could not understand how their ancestors had put up with a long workweek and an unjust distribution of wealth, why it had taken so long for them to revolt against a "cruel, crazy" capitalism in which mechanization only spread misery.[30]

A more lighthearted look at the relationship between mechanization and work turned up in an unexpected place, children's literature. In Virginia Lee Burton's 1939 story *Mike Mulligan and His Steam Shovel*, the hero brags that his beautiful steam shovel, Mary Anne, "could dig as much in a day as a hundred men could dig in a week" (a boast that repeated almost word for word labor's concern that technology had revolutionized the entire basis of employment). Burton's illustrations show steam shovels like Mary Anne digging canals, cutting mountain passes, and building highways, airports, and skyscrapers. The book also noted that construction stimulated jobs; Mary Anne and Mike could "keep as many as thirty-seven trucks busy taking away the dirt." Ironically, as the tale develops, Mary Anne and Mike themselves fall victim to displacement. Though Mary Anne remained in excellent condition after years of service, powerful new gasoline, electric, and diesel shovels "took all the jobs away from the steam shovels."[31]

Burton created an impression of technical change as inevitable, with drawings that show electric and diesel machines advancing over a hill while steam shovels are unceremoniously dumped into junkyards. The story emphasized the emotional costs of displacement, telling children that Mike and Mary Anne felt "VERY SAD." One illustration shows Mary Anne shedding machine tears, while Mike sits idle in front of a sign reading "No Steam Shovels Wanted." To find work, Mike and Mary Anne head for outlying areas, just as farm mechanization had driven Steinbeck's Joad family to the road. In a small place called Popperville, Mike and Mary Anne accept a challenge to dig the cellar for the new town hall in a single day, though skeptical observers thought the task would take one hundred men at least a week. Illustrations show Mike and Mary Anne furiously digging, so carried away that they neglect to leave themselves a way out of the hole. In the true fashion of children's literature, a little boy proposes the perfect solution: leave Mary Anne in the basement to serve as the new furnace and hire Mike as janitor. That requisite happy ending conveyed an interesting moral. Although even a temporary job loss caused pain, Mike and Mary Anne ultimately solved their own displacement problem. Rather than engage in a useless protest against change, the story suggested, Americans should appreciate that the system still left plenty of opportunity open for resourceful and talented individuals.[32]

Children of the 1930s might encounter the issue of machines and jobs in schoolbooks as well as in their leisure reading. Harold Rugg, professor of education at Columbia Teachers' College, created an influential series of social studies texts for older elementary and junior high classes which aimed to link students' sense of American history to an awareness of contemporary issues. In *A History of American Civilization: Economic and Social,* published in 1930, Rugg declared that the United States had not experienced any real problems of technological unemployment between 1865 and 1914. Advances in production had supported a rising standard of living, while the relatively slow pace of invention had left displaced workers a decent chance to find alternate jobs in a growing economy. After the World War, he wrote, displacement suddenly emerged as one of America's central problems. In lines that echoed Ogburn's cultural lag theory, Rugg warned that "thousands of research experts are working steadily, perfecting old machines or inventing new ones," too fast for society to handle. His text described the automation of New York's subway system, which had reduced a train's operating staff from eleven men to two and had eliminated over one thousand turnstile attendants and cashiers. A steam shovel run by a single operator could reportedly mine the same amount of iron ore as five hundred hand laborers. "For the first time in the world's history, the industrial experts tell us, our machine civilization has become so efficient that millions . . . cannot find work."[33]

Rugg did not hate technology; in fact, he had earned two degrees in civil engineering before moving into psychology, sociology, and education. But, in his mind, schools could not shy away from controversial subjects such as joblessness. To help youngsters visualize the power of mechanization, his 1931 textbook *An Introduction to Problems of American Culture* included a photograph showing the long rows of machinery in A. O. Smith's auto parts plant. Production formerly requiring two thousand workers could now take place with two hundred, Rugg wrote, and the employer wanted to develop a "manless" factory. In his eagerness to get students to appreciate the seriousness of displacement, however, Rugg covered up legitimate grounds for dispute. He bluntly asserted that the present unemployment crisis had originated from "clever machines . . . taking work away from men," despite all evidence that many complex economic factors had intersected in creating and prolonging the Depression. Although many experts remained reluctant even to estimate technological unemployment, Rugg did not hesitate to declare that between 1920 and 1927, more than six hundred thousand employees had lost jobs in industry because of productivity gains. Such oversimplifications made Rugg's work controversial, and critics accused him of trying to poison children's minds with anti-American, anticapitalist ideology. Nevertheless, by the late

1930s, more than five thousand schools had adopted his books, exposing almost half the nation's youngsters to them at some point in their education. With the authority inherently carried by classroom texts, Rugg's material encouraged readers to entertain the possibility that mechanization posed grave problems for America's present and future.[34]

Depression-era discussion of technological unemployment could take many forms, even rhyme. A verse printed in a 1937 issue of *Survey Graphic* described how cotton-picking machinery could potentially displace farm workers:

> And beneath blue Southern skies
> Many watch with anxious eyes
> And idle hands, distraught, afraid
> Before the thing that men have made
> To take their place, their ancient toil,
> Their lifetime work on Southern soil; . . . .
> And now—along the tawny rows
> The great devouring monster goes,
> To do the work a swifter way,
> Accomplishing within a day
> More than many countless hands,
> But oh, the cry along the lands:
> 'It does our work! If we are through,
> What shall we do? What shall we do?'[35]

Songs popular in coal mining regions expressed workers' frustration over the fact that between 1924 and 1936, the proportion of coal loaded with underground machines had soared from 0.4 percent to 55.6 percent. Surveying the mining areas of Kentucky, folklorist George Korson found a lyric that ran in part:

> Here is to Old Joy a wonderful machine,
> That loads more coal than any we've seen.
> Ten men cut off with nothing to do,
> Their places needed for another Joy crew . . .

A song originating in West Virginia asked listeners:

> Tell me, what will a coal miner do?
> When he goes down in the mine,
> Joy loaders he will find . . .
> Miners' poor pocketbooks are growing lean,
> They can't make a dollar at all,
> Here is where we place the fault:
> Place it all on that coal loading machine.[36]

The American Federation of Musicians reacted with equal dismay to the way theaters had jumped from silent to sound movies, eliminating the need for musical accompaniment. In 1930, the union journal printed a piece of doggerel bemoaning the pervasiveness of technological unemployment in modern life. The rhyme, written in the voice of a person walking around town, showed different people repeatedly expressing fear of job elimination.

> A motorman, I came upon. "I know just what you mean;
> It's called an iron man," he said, "a powerful machine."
> "You simply cannot tire it out, 'twill work both day and night.
> "Pretty soon we'll lose our jobs, the end of work's in sight". . .
> The next one was a carpenter. He sadly shook his head.
> "They're building houses by the yard. Robots set them up instead
> Of us who used to do it all. We'll soon be down and out.
> This thing's a dire calamity, and should be put to rout."
> And then I saw a factory, with people hanging 'round.
> "What's wrong?" I asked. The answer was, "A new machine they've found
> That does the work of ten of us. Whatever shall we do?
> They call the Robot 'Progress.' To us it's just 'Hoodoo!'"[37]

As part of its Music Defense League campaign, the AFM put together advertisements that included poetry ridiculing the quality of talking pictures. One cartoon headed "The Robot on the Run!" shows a harp-carrying robot being chased away by a crowd carrying signs reading "We want real art!" The accompanying limerick ran:

> Oh! I went to the canned goods fair;
> All the prunes and the tunes were there,
> And the tin-canny laugh
> Of a cheap phonograph
> Made me want to get right up and swear.
> The canned orchestra gurgled and squawked,
> All the voices gummed up when they talked;
> And the only thing good
> In that whole neighborhood
> Was the door, out of which we all walked.

Another AFM ad presented a long fable in rhyme about a monarch who wanted to gain complete control over the world through the mechanization of work. The grand Pooh-Bah got electricians to fill his cities "with great Robots that moved at his touch, (That laborers would suffer did not trouble him much.)" The king's technical tour de force had, however, overlooked the link between earnings and consumption. "The cashiers sat nodding, cosmetics on tin, The doormen stood waiting for folks to come in," but "the people he'd figured their money they'd spend, Were all them workers who had come to their end!"[38]

The AFM's limericks and cartoons represented ways that Depression-era humorists managed to find a light side to talk of technological unemployment. Clarence Day wrote a piece for *Harper's Monthly* recommending that people start training animals to take over jobs. "Think of great future factories where interested wolverines work, their eyes shining with happy excitement as they gallop about, pulling levers." Technological unemployment would not exist in such a system, Day suggested, since any superfluous animal employees could be turned into exotic gourmet dishes. The jokes created by unemployed workers themselves tended to exhibit a bitterer streak; one comment ran, "The only machine that increases the number of names on the payroll is New York's political machine." A 1930 humor column in a magazine produced by the League for Industrial Democracy conveyed the sense of hopelessness lurking behind the laughs. "When a machine throws a man out on his neck," the *Unemployed* observed, "fancy economists tell him not to worry, that he is 'technologically unemployed.' But let him try to tell that to his landlord when the rent comes due, and he finds that he is also 'technologically' evicted."[39]

The most prominent humorist of the day, Will Rogers, used his shows and newspaper column to raise some trenchant questions about the Machine Age meaning of progress. At the end of 1930, he wrote, "Well, the old year is leaving us flat, plenty flat. But in reality it's been our most beneficial year. It's took some of the conceit out of us. We was a mighty cocky nation, we originated mass production, and mass produced everybody out of a job with our boasted labor-saving machinery." Engineers had succeeded in their goal of devising production methods that eliminated the worker, "the very thing we are now appropriating money to get a job for." In the rush to replace humans with more efficient technology, the country's business leaders had lost sight of the bigger picture, Rogers declared. Corporate managers "forgot that machinery don't eat, don't rent houses, or buy clothes."[40]

Talk of Displacement Enters Radio and Film Culture

During the 1930s, radio broadcasts brought discussion of technological unemployment into the home. In 1930 alone, Senator Robert Wagner, Assistant Commerce Secretary Julius Klein, and a number of labor leaders took to the airwaves to deliver speeches and comment on the issue of displacement. Some stations set up two speakers to debate the pros and cons of increasing mechanization. Other programs offered a round-table format, in which several participants might exchange ideas about how new production technologies affected economic well-being. Among the era's more well-known commentators, "radio priest" Charles Coughlin frequently referred to the question of machines and jobs. Like many other observers, Coughlin considered the world war a dividing point in economic and social life. As the military effort put producers under pressure to maximize output, he said, applications of science and invention had reached a new speed. Though labor-saving efficiency helped win the war, it spelled disaster in peacetime, when more than four million veterans returned home and began seeking jobs. "Unemployment on a huge scale was an absolute certainty," he told audiences, when one person in 1918 could complete as much work as two and a half people in previous decades. During the 1920s, Coughlin continued, technological innovations kept helping employers get more production out of fewer workers. Comparing the years 1923–33 to the years 1913–23, Coughlin noted that factory production had soared 42 percent, even as manufacturing dropped 500,000 employees. Coal mine yields had risen 23 percent with 100,000 fewer workers, and railroad business had grown 7 percent with a loss of 250,000 positions.[41]

The trend toward mechanization had become unavoidable, Coughlin warned. After all, "the scientist is not going to vanish. The engineer of tomorrow does not plan to put his brains into cold storage." Five, ten, or twenty years down the road, businessmen would incorporate an even greater array of labor-saving techniques. Yet, despite the prospect of spreading displacement, Coughlin declared, Americans must not resort to desperate measures such as destroying machines. Offering a "bounty to every Dillinger and desperado for removing scientists from our universities" would not solve the problem. Instead of opposing change, working-class Americans needed to find some way to stop industrialists from monopolizing the profits of engineering advance. The National Union for Social Justice, the organization of Coughlin's listeners, vowed to convert an old-fashioned "economics of scarcity" into a new "economics of plenty," in which mechanized factories and farms turned out enough goods to satisfy everyone. Like labor leaders such as William Green and political figures such as Mordecai Ezekiel and Isador Lubin, Coughlin endorsed the promise of machine-made wealth at the same time that he denounced the incidence of machine-made unemployment. He remained vague about exactly how the country should distribute the benefits of change to ordinary people, and once Coughlin began devoting more energy to attacking Roosevelt, his broadcasts contained fewer references to mechanization. But for a number of months, the issue provided a nice line of argument for Coughlin, a seemingly plausible explanation for economic distress which audiences could readily follow.[42]

During the Depression, motion pictures provided another vehicle through which the American public encountered talk of technological unemployment. Pare Lorentz's 1936 film, *The Plow That Broke the Plains*, provided powerful commentary on agricultural mechanization as a source of stress for both workers and the environment. The documentary, made for the U.S. Resettlement Administration, suggested that farming had been redefined by a succession of technical revolutions. The opening image shows a plow pulled by a single horse, while the next shots show first a bigger plow pulled by a dozen horses, then a small tractor, until Lorentz had filled the screen with dozens of larger tractors advancing across a field. For years, the narrative explains, Americans had "reaped the golden harvest" with machinery that "turned under millions of new acres." All that stress on the land, combined with a relentless drought, had created a crisis that left both men and machines idle. Using stark photographs of a plow buried in dust, Lorentz emphasized the collapse of family farming.[43]

Two years later, when Lorentz filmed *The River* for the Department of Agriculture's Farm Security Administration, he again suggested that twentieth-century forces had pushed the land, the economy, and people beyond their natural limits. In both *The River* and *The Plow That Broke the Plains*, Lorentz had emphasized the link between social and environmental difficulties. For his subsequent project, he hoped to write and direct a movie that would center more specifically on the human implications of workplace mechanization. Lorentz wanted to investigate the puzzle of how a country that had so much powerful technology ended up with so many people out of jobs. Lorentz first drafted his ideas in the form of a radio script entitled *Ecce Homo! Behold the Man*. This story centered around a conversation between four unemployed men who met at a Kansas gas station, each travelling to a different part of the country in search of work. A New England mechanic who had been displaced when his mill relocated headed south, but an Alabama farmer heading north warned him that prospects didn't look much better down there. "Now they're talking about chopping cotton with machines," leaving "nothing but relief for the little man." Nevertheless, *Ecce Homo* emphasized, the same technology that had left so many desperate could still pull the United States out of depression. In a long concluding speech, a laid-off Detroit assembly line worker praised the engineers who had literally dug through quicksand to construct the wondrous Boulder Dam. "They can move mountains and they can shove rivers around! There's men and machines and there's sun and land and room for a man to turn around in. And there's a man-sized job to be done!"[44]

Lorentz, who had ended *The River* with a paean praising the Tennessee

Valley Authority, similarly wanted *Ecce Homo* to highlight the notion that the federal government might use regional planning to restore economic, social, and environmental health. After reading the first part of an *Ecce Homo* draft, President Roosevelt asked Lorentz, "How are you going to end it?" to which the filmmaker allegedly replied, "Sir, how are *you* going to end it?" Lorentz never completed filming, but a 1938 radio broadcast of *Ecce Homo* had at least two effects: the Ford Motor Company cancelled its advertising on CBS, and Lorentz began exchanging views on labor displacement with John Steinbeck. In a 1939 letter, Steinbeck urged Lorentz to cover problems of farm mechanization in the planned film version of *Ecce Homo*. Development of cotton-picking devices would prove disastrous for labor, Steinbeck wrote, and he wanted "to pin a badge of shame on the greedy sons of bitches who are causing this condition."[45]

Lorentz's interest in the impact of mechanization helped inspire Willard Van Dyke, who had worked as a cameraman on *The River* and then launched an independent documentary-making career. Van Dyke hoped to change the world through his movies in the 1930s, he later explained, helping to right social injustices by "calling attention to them." To focus attention on the problem of displacement, Van Dyke made a film in 1940 entitled *Valleytown: A Study of Machines and Men*. Rather than analyzing complex and abstract economic statistics, Van Dyke told an emotional, purportedly true story about how economic change had hit one community. Before the Depression, the movie's narrator explains, "Valleytown" had flourished. Growth of the steel industry had provided three thousand jobs, and locals had regarded the steel mill as "money in the bank." With scenes of bustling streets and Christmas shopping, Van Dyke showed how the steel mill's payroll financed consumer activity, which in turn generated business and retail employment. In those optimistic times, Valleytown residents could associate technology with progress. Mechanization helped make products more affordable, and though it eliminated some jobs, the change came slowly enough for people to adjust.[46]

Once depression hit, Van Dyke indicated, ordinary Americans could no longer trust that modern science and engineering guaranteed both good jobs and a rising standard of living. As the steel industry closed old plants to make way for more modern continuous-strip mills, Valleytown's streets and stores emptied. The situation "wasn't the fault of the machines; you can't blame them for the Depression," the film's narrator explains. "In fact, machines had often created jobs—but now there was depression." Displaced steelworkers, who had seen the sheer power of modern equipment firsthand, asked, "What good are the machines if they throw us out of work?" Van Dyke skillfully used visual shots to give audiences a sense of the enormity of technological change. He contrasted shots of old-style steelmaking, showing men physically yanking superheated steel sheets out of the furnace, with scenes of newer mills, where a much smaller group of men pulled levers and watched gauges as conveyors carried the steel along automatically. Van Dyke filmed Valleytown men watching helplessly as their old plant was gutted, the smokestacks torn down. Employment could never catch up with change, the movie implied. New strip-mill facilities allowed companies to produce more steel with less labor, so that "for every thirty men who worked before, now there is one." Van Dyke showed one former steelworker musing, "I hear that they tell you we're living in a wonderful age, the age of machines and gadgets and things, an age to wonder at. All right, I'm wondering." How could his family afford milk for the baby? Van Dyke later referred to *Valleytown* as the best of all his films, the one "that came closest to what I wanted to say."[47]

While *Valleytown* concentrated on putting the feeling of displacement on film, more casual movie references showed how much the question of machines and jobs had become part of popular culture in the 1930s. The last scene of MGM's 1933 film *Dinner at Eight* made technological unemployment the subject of a final exchange between world-wise actress Carlotta Vance (played by Marie Dressler) and social-climbing vamp Kitty Packard (Jean Harlow):

> Packard (proudly): "I was reading a book the other day."
> Vance (stops in surprise): "Reading a book?"
> Packard: "Yes, it's all about civilization or something, a nutty kind of book. Do you know that the guy says that machinery is going to take the place of every profession?"
> Vance (looking Packard up and down): "Oh, my dear, that's something you need never worry about."[48]

The decade's most famous take on modern industry, Charlie Chaplin's 1936 film *Modern Times*, did not address technological unemployment directly. Evidence suggests, however, that Chaplin believed that increasing mechanization made displacement an urgent matter. Referring to *Modern Times*, one historian has called it ironic that "a medium of artistic communication closely allied to the machine did not produce film masterpieces praising technology but ones . . . suspicious and critical of the machine."[49] Yet in the mid-1930s, the movie industry had just fought through its own hectic, traumatic technological transition. Considering the amount of controversy that had arisen over how talking pictures affected musicians, actors, and technicians, even Hollywood could understand why ordinary Americans found mechanization suspicious.

Visual Images and Symbols of Technological Displacement

While writers and filmmakers provided some powerful accounts of Depression-era concern about technological unemployment, book and magazine illustrators offered equally strong glimpses into the sense of crisis. Translating an abstract concept such as displacement into visual terms that would be instantly accessible to laymen was no simple matter. Artists of the 1930s succeeded remarkably well in inventing a pictorial language for the debate over mechanization, turning images of machines and robots into a type of shorthand symbolism. By definition, showing the concept of *displacement* presented a challenge: how could a drawing indicate a negative, the *absence* of labor from modern production? Many books and articles ran photographs of new machinery, combined with captions commenting on that particular device's employment ramifications. To accompany Harold Rugg's description of the way the A. O. Smith factory had incorporated extensive mechanization into building auto frames, his social studies textbook used a photo showing long rows of riveting machines, assembling machines, and finishing machines. Lines underneath read, "A scene in the factory where 200 men are taking the place of 2000." Some captions specifically instructed the viewer to observe how few human beings appeared in pictures of the modern workplace.[50]

A related strategy for illustrating technological unemployment required two contrasting photographs, one showing workers engaged in old-style production and a second picturing mechanized operations with fewer people involved. A number of periodicals adopted such a technique of visual comparison, as did the Women's Bureau in its film *Behind the Scenes in the Machine Age.* Such representations usually started from the machine itself as a presence in modern industry and then worked to convey a sense of the worker's absence. Occasionally illustrators took the opposite approach, beginning with images of displaced workers and then trying to link the fact of their unemployment to mechanization. In a 1937 photograph for the Farm Security Administration, Dorothea Lange showed six men standing in a row, dressed in overalls, hats, and jackets. The photo itself contains no visual clues to identify the men as unemployed or to link their joblessness to mechanization, but the title reveals them to be "Former Texas Tenant Farmers Displaced by Power Farming."[51]

For all its realism and versatility, the camera proved a limited instrument for capturing a visual sense of displacement. Publications could communicate the idea more easily through drawings and cartoons, in which artists took symbolic liberties to create a vivid impression of how the course of change overwhelmed man. One drawing for a 1930 issue of the *Magazine of Wall Street* showed banks of gears looming behind a worker who had sunk to his knees, not in worship but in an attitude of desperation. Artists still faced the same problem that complicated photographers' task, how to picture workers' *absence* from Machine Age production. In a clever solution, one artist for *Iron Age* drew one worker using a motorized trolley to move a heavy load and then outlined faint renderings of additional men pushing that burden by hand. Those ghost laborers, drawn transparently to indicate their elimination from present-day industry, effectively helped the reader visualize how motor power had substituted for manpower.[52]

Frequently, artists seeking to illustrate technological unemployment chose to avoid even vaguely realistic drawings of conveyor belts, riveting machines, or dial telephone equipment. Instead, many turned to the image of a humanoid robot to symbolize all modern devices. In its advertising campaign, the American Federation of Musicians constantly referred to talking pictures with the derogatory term "canned music" and incorporated cartoons of a robot resembling a stack of tin cans. Of course, machines for recording and replaying sound in no way resembled a robot, but a straightforward photograph of wires, spools, and switches would have offered far less creative scope. AFM cartoonists used the robot to poke fun at Hollywood's foolishness in thinking that recorded sound could ever substitute for real music; the AFM robot does not appear as a menacing Frankenstein monster but as the butt of jokes. In a parody of Roman history, one ad showed the robot dressed in a toga to represent Nero, fiddling away while temples labeled "Musical Culture" burn. Another ad updated the fable of Old King Cole, telling readers that the "Prime Minister, in a fit of economy, had installed canned music and fired the King's rollicking fiddlers." The cartoon shows a furious monarch sending the despicable robot off to the attic and ordering that his prime minister be "publicly spanked" (see fig. 3).[53]

The robot served as a wonderfully flexible signifier, one that could be revised and reinterpreted to fit different emotional overtones. Like the AFM's artists, cartoonist Fred Cooper adapted the humanoid image to depict economic and social issues. His drawing for the title page of William Ogburn's 1934 pamphlet *You and Machines* featured a workman gazing up at two towering robots, clearly thinking about mechanization as a new force in his life. Unlike the AFM cartoons, Cooper's drawings conveyed a sense of technology as ruthless. His illustrations showed robots trying to pull a hammer out of a worker's hand, punching a man in the head, and literally pushing humans away from workbenches and toward a sign marked "poverty." Such cartoons carried an instant message, one which could capture workers' feelings of frustration and resentment. Referring to immigration laws that restricted the entry

Fig. 4. Many artists of the 1930s chose to represent technological unemployment in a literal form, showing workers being swept off the payroll sheet or dumped into huge wastebaskets. The robot, shown here in a more dark and menacing form, represented the agent of transformation. Harold Rugg, *An Introduction to Problems of American Culture* (Boston: Ginn, 1930), 7; reprinted from the *Locomotive Engineers' Journal.*

of foreign labor, one drawing in Ogburn's pamphlet showed an American wrestling the robot and asking the bystander, a figure of Uncle Sam, "Hey, Uncle, how about this fellow?"[54]

An image showing a robot hitting a worker or grabbing a hammer away from him does lack a certain subtlety. Yet in spite of that brutality, the very cartoonishness of Cooper's robots removed some of the sting. Even when they were shown shoving people around, it was hard to consider them too terrifying. Cooper drew his robots the same size as humans, and their resemblance to tin cans makes them appear more comical than diabolical (see fig. 2). Like the AFM cartoonists, Cooper could use the robot to lighten up the serious issue of job loss. To accompany Ogburn's description of an automatic pancake-maker, Cooper drew a robot pouring batter with one mechanical hand and flipping three cakes with the other, while a loudspeaker mouth blares, "Come and Get 'Em!" The corniness of the image did not negate the message about displacement, but it did offer readers an excuse to smile.[55]

Other artists also chose to use the image of a robot pushing workers around to depict the problem of technological unemployment. The *Locomotive Engineers' Journal* used a cartoon (copied later in Rugg's social studies textbooks) that showed an enormous robot literally taking a broom to sweep tiny figures of people right off the payroll sheet. Many illustrators gave the robot a more ominous visage than either Cooper or the AFM cartoonists did. One frequently reproduced *New York Times* drawing portrayed a huge black robot looming over a fleeing crowd. Another cartoon tried to sum up technological unemployment in one panel; above the caption "Sign of the Time," a robot paints a sign reading "No Help Wanted" (fig. 1). The robot symbol allowed illustrators to embed detailed messages about complex economic and social issues in a single drawing. One oft-reprinted cartoon from 1940 drew on WPA research suggesting that because of the introduction of mechanical improvements, business did not need to bring workers back even after recovery began. The artist showed a robot staring at newspaper headlines reading, "Industrial production passes 1929 figure—Employment drops 1,000,000 under 1929." The caption, "Is the Robot Beginning to Think?" underlined the force of such trends.[56]

The Depression decade identified the humanoid robot as *the* symbol of technological change. In fact, when the U.S. Department of Labor put together an exhibit for the 1936 Texas Centennial Exposition, it dispatched a speech-making robot to address the crowds. "It is true that I, the machine, have taken the place of hand workers in some employments," the robot's prerecorded voice told audiences, but in virtually every instance, it continued, innovation

"IS THE ROBOT BEGINNING TO THINK?"

Fig. 5. Some cartoons of the 1930s could encapsulate complex messages about the relationship between men and machines in a single image. This panel referred to government economists' conclusion that even though business production had begun to recover from the depression, employment had not responded. Mechanization eliminated many jobs for good, some observers feared; industry would never need as many workers as before to turn out a given level of output. Talk about "permanent technological unemployment" might give even a powerful robot pause for thought. Illustration by Bishop in the *St. Louis Star-Times*, reprinted in the *New York Times*, February 25, 1940.

ultimately created more employment than it destroyed. Among all examples in the popular culture of the 1930s, it would be hard to top the sheer irony of government officials sending a machine to speak about the problem of displacement.[57]

Through cartoons and photographs, radio shows and movies, books and articles, Americans of the Depression encountered more and more talk about the substitution of machines for workers. Some authors described the actual machines that had entered factories and farms, while science fiction writers and illustrators drew more on the symbolic role of technology. Sometimes the subject encouraged light-hearted humor, while on other occasions it conveyed emotional bitterness. In virtually every case, popular culture references reflected a message that Americans had better start thinking seriously about whether workplace reengineering always contributed to well-being. That very suggestion, that people could and should challenge the path of mechanization, raised alarm in the nation's business community.

5

"The Machine Has Been Libeled"

The Business Community's Defense

AT THE MOST ELEMENTARY LEVEL, talk of technological unemployment violated some of the American business community's favorite assumptions about the social and economic advantages of mechanization. Business leaders responded aggressively, contending that aside from anecdotal evidence, no one had any proof that the modernization of production left a significant number of people without work. The National Association of Manufacturers (NAM), the Chamber of Commerce, and the Machinery and Allied Products Institute mounted major public relations campaigns insisting that mechanization would always bring lower prices, rising consumer demand, and thus more jobs—at least over the long run. Enthusiasts frequently pointed to Milwaukee's A. O. Smith factory, with its extensive use of labor-saving devices, as the ultimate in progress.

Naturally, not all those involved in business saw exactly eye to eye. Some small business owners actually echoed labor in expressing dismay over the way large operations rushed to install new technologies. Small business owners' concern stemmed from their fear of being left at a competitive disadvantage rather than any inherent sympathy with unions. The voices of small entrepreneurs, however, became swamped in the outpouring of promechanization sentiment from representatives of large companies and organizations such as the NAM. Big business interests maintained that even if some jobs vanished when new machines appeared, workplace improvements paved the way for a rising standard of living. Advertisements suggested that thanks to the genius of scientists, engineers, and businessmen, Americans remained fundamentally wealthy even during the Depression. Corporate research and development laboratories stood ready to invent new items for manufacture, which would excite consumers and provide opportunity for workers. Such rhetoric made talk of displacement seem both ungrateful and unpatriotic, a stubborn refusal to acknowledge that America's past, present, and future rested on the technical wonders created under free enterprise.

Business Leaders Downplay Displacement

While union representatives, social scientists, government officials, and ordinary Americans pointed to what they regarded as evidence of serious technological unemployment, the 1930s big business establishment refused to accept any such idea. "Blaming machines for breadlines is one of today's great fallacies," *Nation's Business* maintained. Benjamin Anderson Jr., an economist with Chase National Bank, said that no matter how bad things might seem, "rapid technological improvement is a dynamic and energizing factor rather than a factor slowing down business and reducing employment." Though temporary job loss might occasionally occur, the economic system would soon rebalance itself and move forward. While Americans naturally became upset to see people without work, Westinghouse president F. A. Merrick insisted, displacement amounted to nothing more than a "superficial and incidental" annoyance "in the readjustment period of progress."[1]

Subscribing to the most optimistic economic assumptions, business leaders argued that mechanization ought to ensure all Americans of a rising standard of living. Improvements in production would set off a "benevolent circle" of effects; when manufacturers gained efficiency and saved money by introducing machines, they could reduce the price of goods, thereby stimulating consumer activity and creating jobs. For proof, some cited the way commercial bakers had adopted large mixing machines, conveyor belts, and sophisticated new ovens. As the industry improved the quality and slashed the price of store-bought bread, it generated an escalation of demand which kept two hundred thousand people employed. John Van Deventer, industrial consultant and editor of *Iron Age*, hoped that even skeptics could follow such simple logic. "Average annual income, divided by average cost of things," he explained, "equals the quantity of things that can be bought."[2]

Real-life economic behavior hardly ever lived up to such rosy promises, labor representatives protested. Although in theory employers who added machines might want to pass the savings on to customers and thus increase business, critics complained that they rarely did so. Some statistics suggested that even at the height of 1920s prosperity, the benefits from mechanization had gone more toward raising profits than reducing consumer prices. Economic experts pointed out another set of flaws in the presumption that advances in production must guarantee prosperity. Consumption did not represent an infinitely elastic economic factor; no matter how cheap a loaf of bread became, people would eat only so much. Moreover, even if mechanization under certain circumstances might create a "benevolent circle," it remained conceivable that the onset of the Depression had driven the United States into a more negative feedback cycle.

Advocates insisted that even in depression, fundamental rules of economic progress made it impossible for displacement to ever rise to an emergency level. Only those who exaggerated the depths of distress, they indicated, could conclude that mechanization had thrown the country into crisis. Charles Kettering, General Motors vice president and research director, accused critics of underestimating America's continued economic strength and its enormous potential. The nation's financial status was "not so dark as some would have us believe," Kettering insisted. Rather than feeling constrained by "dark and gloomy" talk about technological unemployment, businessmen and inventors needed to seize opportunities for further innovation. Westinghouse vice president J. S. Tritle similarly maintained that despite four decades of mechanization, modern industry still created plenty of employment opportunity. Government figures showed that in 1889, sixty-nine out of every thousand Americans had held manufacturing jobs, while in 1929, those same businesses employed more than seventy-two out of a thousand people. Of course, any statistical analysis that took 1929 or 1930 as the point for comparison could not shed light on the question of whether the Depression had brought a surge in displacement. But for Tritle, Kettering, and like-minded fellows, such evidence proved sufficient to suggest that technological advances did not cause any major hardship.[3]

While Stuart Chase, William Green, and many others claimed to see evidence that mechanization had cost a significant number of Americans their jobs, the pro-mechanization community declared that under closer examination, such a story did not hold water. Benjamin Anderson Jr. calculated that only 0.437 percent of all workers in 1930 could possibly have been eliminated by the introduction of new production equipment. He observed that Commerce Department reports attributed just 102,170 out of 2.4 million cases of unemployment to "industrial policy," a category covering work-force reduction, dismissal of older workers, and substitution of cheaper labor as well as technological change. Even counting 11,403 general layoffs and throwing in 100,000 cases as margin for error, Anderson figured that machines had eliminated 213,573 men and women at most, less than one half of one percent of a total workforce of 48,832,589. Justin Macklin, assistant commissioner of the U.S. Patent Office, similarly contested the cases of displacement cited by critics. He had once heard a prominent senator insist that the development of refrigeration had forced thousands of icemakers and ice sellers out of work.

Macklin responded that the manufacture and sale of refrigerators provided employment for almost twice as many men as had ever been involved in producing ice. Moreover, the number of ice dealers in the United States actually rose from eight thousand to over nineteen thousand during the 1920s, since refrigerators had made thousands of lower-income families "ice-conscious" and prosperity allowed them to purchase newer and larger iceboxes.[4]

Even toward decade's end, as the WPA's National Research Project offered detailed studies suggesting that mechanization was making it ever harder to bring unemployment back down, the business community generally refused to consider it any real trouble. In 1940, steel manufacturer Charles Hook maintained there was still "no evidence for concluding that technological improvements cause permanent unemployment." Some "temporary dislocations" might be "inevitable in a progressive society," he argued, but prosperity would prevail as long as government stopped placing artificial restrictions on business strategies and the free flow of capital. Other business leaders blamed recent economic difficulties on the way high tax rates dampened incentive. In their view, talk about machines and displacement simply shifted attention away from the real enemy, the Washington political establishment and its tendency to interfere with free enterprise.[5]

To the extent that even temporary displacement existed, banker Henry Bruere spoke for many in calling that hardship simply the "price of progress." The business community's faith in mechanization reflected its extended love affair with the sheer power of science and technology. For workers, introduction of new technology threatened disaster, but for employers, the prospect symbolized greatness. "We have done it," one steel company president rejoiced, "rolled strip steel from its molten form in two hundred foot coils," thanks to continuous-strip machines. "I can't watch our experimental laboratory roll molten steel for more than a few minutes. It almost makes a fellow go crazy thinking about the millions of dollars worth of equipment it will make obsolete and the thousands of jobs it will eliminate. It's terrific."[6]

In their enthusiasm for innovation, those men often talked as if mechanization came free. In truth, companies incurred sizable costs and risks in adopting the latest devices. Simply maintaining unfamiliar machines could prove expensive and problematic, while a breakdown might bring operations to a standstill. In *Men and Machines*, Stuart Chase had commented that in addition to causing unemployment, mechanization created a "technological tenuousness" in modern life. A failure of power systems or transportation could disable an entire city, he warned, halting industry and blocking the flow of food supplies. The multiplication of complexity and the "sheer piling up of technical services" made twentieth-century people vulnerable. Science fiction writers loved to toy with the theme of Machine Age catastrophe, but business leaders spoke as if technology must always live up to the most utopian predictions. Employers focused on the notion that mechanization could free them from depending on human workers, who all too often proved lazy, uncooperative, or demanding. In promoting the dial system, telephone company executives emphasized the inherent superiority of machines over human beings. Even the best switchboard operators remained too slow, they argued; the rush of city life and business demanded an engineering solution. Human beings might fail, but the Bell System promised that dial equipment could provide customers with swift, perfect service.[7]

In installing mechanization, business actually exchanged one form of dependence for another. Even as the phone company looked to undercut the role of switchboard operators, adoption of the dial system meant that it had to rely more than ever on engineers and technicians. Continuous-strip technology might eliminate hundreds or thousands of floor workers, but a steel mill president must keep men who understood how to run the new machines. Business managers of the 1930s did not perceive that need for technical knowledge as a liability. They trusted engineers as fellow professionals, ready to advance the cause of greater production through mechanization. Over the preceding decades, the rise of in-house research and development laboratories had made scientists and engineers part of the modern corporation. In fact, the lines of distinction had blurred, as men with technical backgrounds rose to executive ranks; Gerald Swope's presidency of General Electric represented a classic case.[8] Such developments set the stage for business leaders to accept engineers as sympathetic allies in the push for advancing production technology. If anything, the Depression-era controversy over displacement drew the two groups closer, through a common faith in mechanization as progress.

Advocates argued that new inventions would always emerge from the workshops of talented scientists and engineers to create wealth and opportunity, but, in reality, the process of invention rarely ran smoothly. Technical difficulties and practical complications had repeatedly delayed the process of getting a good cotton-picking machine to market, for example. Nevertheless, businessmen maintained that, inevitably, the course of mechanization must go forward. Just as development of textile machinery and steam engines had driven the original Industrial Revolution, so twentieth-century inventors would proceed to create devices of still greater power. The substitution of machines for inefficient, troublesome, or expensive human workers represented part of the evolution of civilization. By depicting mechanization as an inevitable force of history, corporate leaders set out to downplay fears of job displacement. The Depression must not distract industry from its natural path of progress. Some executives had already become too dilatory about investing in new equipment, *Rand McNally Bankers Monthly* complained. One study from 1930 revealed that 48 percent of the machine tools in metalworking shops and 73 percent in railroad shops were more than ten years old. The present economic climate presented a golden opportunity, the magazine advised, since managers could get purchases delivered quickly and installed at less cost, just in time to boost profits during a slow period. Since New Deal policies threatened to force wages up, it made sense for companies to mechanize as a way to increase output without expanding the payroll.[9]

Manufacturers of workplace equipment often made explicit promises to employers about how much labor their new machines could save. A 1931 advertisement for the Acme record-keeping system criticized the inefficiency of old office methods, which might require as many as twenty people. "Now we have Acme. *Seven* girls do the work." Employers should not hesitate to buy such labor-saving equipment or feel guilty about its effects, ads stressed. Any displacement that resulted did not count, since, over the long run, mechanization would improve a firm's financial position and thus increase job security for its remaining workers. Moreover, by buying new equipment, managers helped keep more than twenty million people at work in the machine tool industry. The National Committee on Industrial Rehabilitation, a group promoting the sale of machinery, promised that putting money into modernization brought a "geometrical" expansion in employment. For each dollar spent on replacing outdated facilities, chairman James Harbord declared, another "three dollars will be spent in consumer goods and materials industries and more people will be put to work." Regardless of talk about technological unemployment, *Rand McNally* concluded, companies performed a good deed for labor by introducing new machines as fast as possible.[10]

A. O. Smith: The Quest for Total Mechanization

The ideal of mechanization appeared most clearly in the nation's automobile industry. By the 1930s, the car had become established as a Machine Age icon, encompassing all the hopes or fears about what mechanized mass production and mass consumption meant. For those who worried about the disappearance of work, Detroit exemplified the dangerous trend of modern companies substituting machines for men. For those who considered mechanization a social and economic boon, the case of automobiles seemed to prove that new production technologies ultimately created both new jobs and a higher standard of living. Proponents calculated that the spread of automobile ownership had provided employment over recent decades for 3,732,000

Americans, a figure that included 795,000 positions making vehicles and parts, 1,400,000 chauffeurs and professional drivers, 455,000 dealers and salesmen, and 20,000 machine tool builders. Henry Ford estimated that car manufacture had generated 2.5 million additional jobs in road construction, service, and repair and oil, rubber, and gasoline production. As families increasingly used vehicles for recreational trips, business also boomed at vacation spots, hotels, camps, restaurants, and amusement areas. Although such indirect economic stimulation proved difficult to tally, a 1935 *Automotive Industries* article declared that automobiles generated one out of every ten dollars earned by American workers. *Nation's Business* asserted in 1940 that eleven million men and women earned a living in ways related to the automobile.[11]

Advocates of mechanization frequently repeated such claims, and yet, critics accused Detroit of having contributed to technological unemployment. *Survey Graphic* reported in 1935 that just over the last five years, one major automaker had installed machines that enabled nineteen men to make 250 engine blocks in the time it had formerly taken 250 men to produce one hundred. According to the U.S. Bureau of Labor Statistics, a single man using a spot-welding machine could perform the work of eight hand-riveters, while the latest enameling machines needed only 30 percent as much labor as the old hand-dipping process. Molding equipment could produce 900 pistons per man per day, whereas an experienced molder and his assistant had averaged just two hundred each. Devices that incorporated a photoelectric eye could duplicate not only the actions of human hands but also the value of human judgment and senses. In 1937, Ford introduced a machine that could perform eleven different inspection tests on each valve push rod at a pace of about forty rods per minute. The device checked for hardness by dropping a hammer on the rod and using an electric eye to see that it rebounded to the proper level, and it screened rods for internal defects by using microphones and amplifiers to gauge their tone when struck.[12]

Detroit's monomania with mechanization had gone too far, the NRA's Research and Planning Division concluded. After holding hearings on the status of automobile workers, the agency charged that manufacturers had acted in a socially irresponsible fashion. In the midst of rampant unemployment, automakers had a duty as the nation's leading industry to uphold the value of human labor. In response, the Automobile Manufacturers Association accused Washington of letting the American Federation of Labor dictate its conclusions. The industry did not try to downplay its pursuit of technology; chairman Alfred Sloan Jr. boasted that General Motors spent tens of millions of dollars on new equipment every year. Automobile companies declared that it was their quest for efficiency that had built the industry into a mainstay of the United States economy. By using machines to turn out a high volume of cars at a reasonable cost, the industry could pass savings on to consumers and thus keep business rolling. Between 1910 and 1925, the number of auto manufacturing jobs had risen from 51,294 to 197,728, despite the fact that production of a 1925 car required only one-sixth the man-hours of 1910. Such a mutually reinforcing economic relationship, the *New York Times* editorialized, made any talk about permanent technological unemployment ridiculous.[13]

Automakers promised that, far from causing any permanent harm to workers, their commitment to mechanization would allow the industry to maintain its position of economic strength. If the Ford Motor Company rejected the cost-saving advantages of manufacturing technology, officials calculated that the cost of a V-8 would soar to $17,850, a figure well beyond most consumers' reach. Just making a hub cap shell by hand would cost $2.50, whereas Ford had brought the price down to twelve cents by investing in $44,000 worth of automatic stamping equipment and dies. Such cases proved that "without machinery there would be no automobile industry," executive W. J. Cameron declared. "To have greater employment, we must still further lower prices by the use of labor-saving machines." He conceded that such trends might create a "temporary displacement of men, but the eventual replacement will more than compensate."[14]

General Motors economist Stephen DuBrul took that argument further, saying that current labor requirements remained as high as those of ten years before, since auto manufacturers had added so many refinements to their models during that period. The work involved in adding fancy new accessories more than offset any displacement resulting from the mechanization of welding, painting, and other basic processes, he said. GM had hired many extra machinists, electricians, repairmen, and tool and die makers, while only a "comparatively few" manual laborers had been hurt. True, one set of production changes had made four departments of a single factory redundant, but 391 out of 490 employees had been shifted to other work in the company (DuBrul did not specify their exact positions and wages). Another thirty-nine workers had opted to leave, meaning that the company had laid off only sixty men (12% of the total affected), DuBrul reported. In short, he maintained, "Technological displacement is not a serious problem in the auto industry."[15]

The case of Milwaukee's A. O. Smith factory particularly intrigued Americans interested in issues of industrial change. Back in 1903, A. O. Smith had become the first United States firm to make pressed-steel automobile frames, constructing ten per day. Impressed by how other businesses had adopted mechanization to produce small items, managers at A. O. Smith resolved in 1916 to redesign their operation around technologies for the elimination of labor. "We set out to build automobile frames without men," president L. R. Smith later explained. In 1922, the company opened a new factory, one centered around their "quest for the 100% mechanization of frame manufacture." In that facility, conveyors and other systems carried rough steel plates through a series of specialized machines. At the start, inspection machines ran magnets along plates at a rate of nine hundred per hour, automatically pushing aside any that deviated from size or quality tolerances. Overhead electric cranes then took acceptable plates through washing and oiling machines. The plant needed two men to feed the clean plates onto conveyor belts, an "operation for which no satisfactory automatic apparatus has yet been devised." Once on the conveyor, plates ran through punching, pressing, assembling, and nailing machines, as well as a machine that could insert sixty rivets simultaneously into each frame, at the rate of 450 frames an hour. A set of finishing machines then rinsed, dried, spray-painted, baked, and cooled the frames before workers took them for final inspection and packing. The company boasted that aside from the few men needed to move frames between conveyor belts, its floor routine required almost no hand labor. With a staff of six hundred engineers and two hundred mechanics, plus supervisors, it could turn out as many as ten thousand standardized automobile frames daily, a feat that formerly would have required two thousand men.[16]

Observers agreed that A. O. Smith represented one of America's most advanced single-factory applications of mechanization, but, as in so many cases, different parties interpreted that concept in contrasting ways. Promoters of new production machines described the A. O. Smith setup as the bright hope for American progress, while those worried about technological unemployment regarded it as the symbol of disaster for labor. After visiting Milwaukee, Stuart Chase wrote that the company's "supreme" use of technology had created "a terrible specter" of job loss. True, conveyor belts and welding machines relieved humans of heavy or monotonous work, but when one man could perform the functions of ten, "technological unemployment looms." To Chase, the A. O. Smith factory had become "an iron bouncer." Boosters disputed that charge, maintaining that when A. O. Smith had first set up its mechanized operation in 1920, it transferred all the affected personnel to other operations. With the high demand for automobile frames, there remained plenty of work to go around. The firm had, moreover, hired dozens of supervisors, mechanics, and engineers to handle the new equipment and devise further improvements. Its research laboratory invented an electric arc-welding process which allowed the firm to start manufacturing oil pipes, a venture that promised to open entirely new avenues of employment.[17]

To A. O. Smith accountants, the workplace revolution generated pure economic gain. The firm had spent ten million dollars initially to purchase all the conveyor systems and sets of machines, but managers claimed that the investment paid off by eliminating high-skill, high-wage union workers. Executives had proven that with the savings in labor cost, removing men from the factory floor became "not only possible but also very profitable." Supervisors and technicians did not exactly come cheap, of course, but A. O. Smith's decision-makers approved of investing in research staff and a new laboratory building. To the company president, his plant represented a testimonial to the wonders of industrial engineering, the ultimate achievement in modernization. L. R. Smith either did not understand or did not care that his drive for complete mechanization, at a time when millions lacked jobs, might disturb some people. In a 1933 interview with *Good Housekeeping*, he apologized for not having fulfilled his goal of utterly abolishing manufacturing work. "We had figured that not a human hand would touch any part of these frames, but before we'd finished," Smith told readers, "we got interested in other things and left a few of the simplest operations to human labor." However, he continued, if the company really wanted, its engineers still "could easily fix it so that not a hand would touch metal" in the course of production.[18]

Smith did not explain the consequences of his goal: if his company could run with only fifty men on the floor to watch the machines operate, and if other manufacturers followed his lead, what would happen to all their former employees? Was it safe to assume they could migrate to alternative work? Where would those other jobs come from, and would the country have enough for everyone? For those who trusted that the natural course of economic growth would correct any problems of joblessness, A. O. Smith's quest for the manless factory heralded the future of a technological civilization. Julius Klein, Hoover's Assistant Secretary of Commerce, considered the company a great source of pride, proving that the United States led all other countries in industrial efficiency. He contrasted A. O. Smith's daily output of ten thousand frames with that of a Central European operation, in which two hundred hand workers reportedly manufactured just 31 frames each day. Superiority in mechanization, Klein suggested, would ensure that future Americans could continue to enjoy the world's highest living standards.[19]

Throughout the 1930s, promechanization forces echoed Klein's words of praise for A. O. Smith. The success of that company's experiment suggested that the nation must pursue labor-saving technology, without being deterred by any talk of unemployment. Just as mechanization had driven twentieth-century economic growth by making it possible to mass-produce automobiles and other new consumer goods, so future discoveries would give rise to entire new industries and provide plenty of opportunity for work. Adding up employment rates from the production of automobiles, radio, electrical machinery, airplanes, and sound movies, the NAM reported that "one out of every four persons employed in America today [1932] holds jobs depending on fourteen industries unknown in 1870." Inventive genius made it certain that displacement could never amount to more than a minor, temporary inconvenience, men such as former president Herbert Hoover maintained. Industry would soon perfect assembly-line methods for building homes, they predicted, and, in one swoop, solve the nation's housing problem and create construction jobs for thousands of men. Inventors and entrepreneurs had enough ideas just waiting for commercial application, Hoover said in 1939, "to put every one of the 11,000,000 unemployed to work in a few months."[20]

The Public Relations Battle for Faith in Mechanization

To Hoover, Kettering, and other supporters of corporate enterprise, common sense reasoning proved that advances in production technology must be good for civilization. Unfortunately, many complained, such a simple point tended to escape those ignorant of economic matters. Just as ancient civilizations had found it easier to create a myth about horses pulling Apollo's chariot through the sky than to master the complicated workings of the solar system, F. A. Merrick wrote, so modern Americans persisted in believing the myth that mechanization destroyed jobs. Throughout history, irrational fears had often led workers to raise a "clamour against improvements in manufacturing," said steelmaker Charles Hook. Though always "shown by subsequent events to be unfounded," he continued, "the old exploded theory, like Banquo's ghost, has come to life again wearing the label of 'technological unemployment.'"[21]

Talk about displacement seemed dangerous to many businessmen, who accused troublemakers of instigating a deliberate campaign to whip up resentment against new technologies and against employers who introduced them. One article on the subject by John Van Deventer bore the title "The Machine Has Been Libeled!" Herman Lind, manager of the National Machine Tool Builders Association, dismissed technological unemployment as a "bogyman [sic] paraded" by "pseudo-economists" and radical socialists. "For the last 75 years professional agitators have been crying out against the machine," he objected. The ranks of "agitators" included such men as Congressman Claude Pepper, who had reportedly commented, "Ever since machinery has come to be the real producing agency of the world, we have had economic maladjustment." Van Deventer immediately warned *Iron Age* readers that a "wave of anti-machine propaganda is being spread from high places." The editor waxed especially irate over proposals to institute a technotax, which he described as a direct strike against business, "a doctrine of the horse and buggy days if ever there was one."[22]

If demagogues had begun spreading unease about mechanization among gullible citizens, some members of the business community felt the time had come to fight back. The magazine *Iron Age* led the way in urging readers to defend themselves against the charge of having caused technological unemployment. Modern businessmen had been "suddenly set upon by literary phrase-mongers, cloistered professors, dilettante economists, grandiloquent politicians and 'buck-passing' financiers," Van Deventer complained. Overcoming such an attack would call for concerted effort, for example, in public relations. All too often, companies had gone about things the wrong way; one continuous-sheet steel mill had offered journalists special tours, only to have reporters ask, "How many men is this mill going to throw out of work?" Because spokesmen had not prepared to defuse that question, Van Deventer lamented, "hundreds of people went away with the firm conviction that technical progress means fewer jobs and less consuming power."[23]

An effort to convince people of the value of new workplace technologies would require plenty of resources, and organizations such as the NAM proved willing to commit the necessary support. The first measure on the agenda involved exposing the charlatans who linked job loss to the introduction of production equipment. The Technocrats drew fire, of course, and business leaders did not hesitate to attack politicians, even as high as President Roosevelt, who spoke of technological unemployment as a genuine issue. In regard to the "literary phrase-mongers," NAM officers helped lead an assault on Harold Rugg, the educator whose widely accepted textbooks encouraged students to discuss the consequences of mechanization. Critics painted Rugg as a Marxist sympathizer, citing his assertion that any gains from technological advances tended to go to the upper class rather than to ordinary workers. The Advertising Federation of America joined the attack, infuriated by Rugg's accusation that advertisers pushed consumers into buying things they did not really want or need. The American Legion published articles denouncing Rugg for daring to suggest that the American system might be flawed. Provoked by the scare campaign warning that Rugg's unpatriotic sentiments might corrupt youth, one Ohio community held a public burning of his books. Amidst the controversy, a number of school districts dropped Rugg's texts, and, by the forties, the steep decline in the textbooks' popularity led Rugg's publisher to halt further sales.[24]

For the NAM, *Iron Age*, and other business interests, the program to discredit talk of technological unemployment involved far more. The nation's most prominent executives wanted to emphasize their vision of a future centered around ongoing technical innovation. They promised a time when the power of machine production would allow enterprising capitalists to supply ordinary Americans with an ever-wider array of material goods, creating a potentially infinite path of prosperity which could provide as many jobs as necessary. National progress demanded a continuous expansion of consumerism, they stressed, and economic failure stemmed directly from the breakdown of purchasing activity. The common concept that "employment comes from employers" was ridiculous, Edward Filene scoffed. "It is on a par with the notion that milk comes from milkmen, or that water comes from faucets and money comes from banks. These notions are all true, but inadequate." Filene recast the issue of jobs in a way that shifted the focus away from how technological changes affected labor needs. In his analysis, employment depended primarily on the decisions that ordinary Americans made in their role as consumers rather than on any decision that businessmen might make to adopt machinery.[25]

Not wishing to simply wait for citizens to feel the impulse to buy new possessions, Depression-era industrialists turned to the advertising profession to convince Americans that they should spend their way into recovery. Advertisers joined corporate spokesmen in declaring that the wonders of a Machine Age economy entitled people to enjoy a world of consumer delights. If people just "stopped recounting our sorrows," said J. M. Mathes, they would observe an amazing "automatic world" at hand. The brilliant scientists and engineers at work in corporate research laboratories would soon create six-lane highways, aerial ferries, horizontal elevators, and other innovations, the demand for which must eliminate any question of unemployment. Supporting the effort at promechanization public relations, the Advertising Federation of America instructed its members in 1937 to act as "super-salesmen," disseminating positive impressions about the connection between new industrial machinery and higher living standards.[26]

Advertisements loudly proclaimed that innovations in technology always brought economic and social advantages, generating both employment and wonderful new consumer goods. In a 1937 ad, General Electric declared that its investment in science and invention had put "bread, butter, and jam" on Americans' tables. Thirteen million new jobs had come from "making or selling automobiles, radios, electric refrigerators or movie films . . . rayon or aluminum," the copy indicated. Demand for raw materials and transportation in those new industries supported millions of more positions, proving beyond doubt that the modern era had not and would not eliminate labor. Once people understood such simple facts, the ad suggested, they would start to appreciate how well off the Machine Age had left them.[27]

With a series of monthly advertisements running in popular magazines such as *Survey Graphic*, General Electric emphasized the second part of the equation, how the extension of mechanization yielded real gains for consumers. One ad, showing a picture of a woman proudly displaying two dresses, announced that thanks to recent improvements in clothing manufacture, today's consumer could buy two new garments for less money than her mother had paid for one. The copy stressed that researchers had developed new sorts of looms with special instruments to test and match colors, technologies of quality control which allowed the industry to offer consumers a better choice of clothing styles, colors, and fabric. While those Americans still out of work might respond skeptically to the idea that mechanization had brought ordinary people a superior standard of living, GE prepared to fight off such negative thinking. Another advertisement in the company's series carried a banner headline proclaiming that in the present-day United States, "millions of people are wealthy." With an illustration showing a couple enjoying a drive in a sporty roadster, the ad declared that for well under one thousand dollars, modern Americans could buy a car "far better than anyone owned even a decade ago." In fact, the text continued, "for what a leading car cost in 1907, John can now have, besides a better car, other things—automatic house heating, a radio, golf clubs. Mrs. Brown can have an electric refrigerator, a fur coat, and lots of new dresses." Glowing copy described this economic miracle of easy access to luxury as a gift wrought by the businessmen, engineers, and scientists who worked to improve production machinery. Such ads represented a design to legitimize mechanization, the strategy of countering popular concern about worker displacement by identifying technological change with consumer abundance.[28]

In addition to advertising, executives of major corporations had an entire range of public relations strategies at their disposal to proclaim their love for mechanization. In May 1934, General Motors president Alfred Sloan organized a special dinner meeting in the GM Hall of Progress at Chicago's Century of Progress exposition. Sloan invited three hundred leaders of science and industry to express their "tremendous confidence in ability of industry and science to evolve new enterprises calculated to create more jobs for more people," as well as "progressively higher standards of living." To reinforce the message, GM compiled a booklet of quotations from those famous men emphasizing that innovation had created jobs in the past and would assure even more employment in future. The emergence of electrical communication alone had offered "remunerative employment to hundreds of thousands," said Bell Labs president Frank Jewett, who felt "no scintilla of doubt" that ongoing research and development efforts would prove equally valuable. After all, as RCA president

David Sarnoff noted, significant "human wants remain unsatisfied"; inventors and industry would work together to devise useful new products and develop new industries which could absorb any workers displaced from older lines of employment. Jewett spoke of the need for improving telephones and telegraphs, and L. W. Chubb, Westinghouse lab director, promised advances in air conditioning. W. R. Whitney, GE vice president, offered the most exciting dreams of all, including television, news teletypes for the home, and trains that could run faster than two hundred miles per hour.[29]

The GM dinner gala, coming in connection with the World's Fair celebration of national identity, allowed business leaders a chance to establish the concept that mechanization represented a defining and inevitable characteristic of American life. Many participants welcomed the celebration as an opportunity to take a stand against talk of technological unemployment. Gano Dunn, president of the J. G. White Engineering Company, praised Sloan for the flag he was flying and added: "I lose patience with those who would check the application of science." Northwestern president Walter Dill Scott complained that despite the world's "constant acceleration" of progress, society had persecuted inventors more often than it exalted them. Workplace change helped advance civilization, GM's guests insisted, and any doubts about the need to introduce new equipment threatened to reduce the United States to a less forward-looking country. Mechanization of agriculture must proceed, said Cyrus McCormick, chairman of International Harvester. Whitney foresaw a future in which "all routine industrial and clerical jobs [had been] made wholly automatic." Through such comments at GM's Century of Progress event, business leaders were able to to display their faith in mechanization as the American way.[30]

General Motors gained a neat publicity splash by connecting its celebration with the Chicago fair, but in other cases business interests constructed their own events to spread the promechanization line. In 1940, the NAM put together a national trade show with the message "Fashions Out of Test Tubes." Publicity photographs showed a glamorous woman holding up a sample of garments made from marvelous synthetic fabrics. The accompanying message explained that "not so much fashions but the future of America and the jobs of tomorrow are being made by the research and invention of today." The NAM event carried a clear moral: by investing in research and development laboratories, far-sighted chemical corporations had laid the groundwork for devising novel products, consumption of which would ensure future employment. Critics charged that any increase in demand for synthetic material would simply subtract from sales of cotton and wool clothing, threatening jobs in those industries. Such arguments did not deter NAM publicists, whose campaign conveyed an air of excitement with the appealing image of new jobs flying out of a test tube.

Special events and advertising represented only the opening salvo in the business community's move to defuse fears of machine-based displacement. Business publications devoted editorials, articles, and columns to the argument that technological change did not really eliminate jobs. Such an aggressive campaign sparked some dissent. In 1938, the advertising magazine *Printers' Ink* ran a piece that cited employment, production, and price data from GM to remind readers that mechanization brought gains for labor and consumers alike. One subscriber, Thomas Baggs, protested that automakers' statistics appeared misleading and that in more typical areas of industry, such as the manufacture of tobacco products, employees had clearly been forced out "by the super-efficient machine." In short, he wrote, "Technological unemployment is the greatest single fact emerging out of all our economic backing and filling of the last few decades." In a reaction that underscored the importance business publishers attached to attitudes about mechanization, *Printer's Ink* prepared a special rebuttal to Baggs's letter and devoted four pages to the matter. Editors sarcastically observed that "if Mr. Baggs is right," then "industrial jobs must virtually have disappeared by now." Instead, factory jobs had increased from two million in 1870 to 8.3 million in 1937, a growth of over 300 percent. Until someone provided conclusive proof "that the machine actually is destroying us all," *Printer's Ink* declared, Americans could remain "serene in the belief that technology smooths the path of human progress."[31]

Visual imagery accompanying promechanization articles tended to lack the creativity exhibited in the numerous cartoons and drawings expressing alarm about job loss. Illustrators for the business press generally ignored the humanoid robot as a symbol of modern technology, perhaps because the other side had succeeded in vesting so much emotional power in the negative image of a robot towering over workers. One exception to that rule appeared in a 1934 issue of *Iron Age*, which featured a cartoon showing a humanoid robot walking arm in arm with a man dressed in overalls. The iconography conveyed a sense of harmony and advancement, implying that labor and technology could and should stride into the future as partners.[32]

For the most part, business magazines ran staple photographs of machines, with captions explaining how much they contributed to production, employment, and consumption. In a piece entitled "Labor-Saving Machines Make Jobs," *Nation's Business* printed a photo of a repairman adjusting a tire, with the note, "There are more jobs for garage men than there were for stable hands." A second shot, showing a gigantic parking lot filled to capacity, carried the line, "In addition to the thousands of workers making automobiles, some 11,000,000 make a living in industries dependent upon motor cars." A simple photo of a machine signified nothing about how it affected employment; such images revealed meaning only through the captions an interpreter chose to attach. With a photo showing rows of steelmaking machinery, *Nation's Business* printed a caption, "Since introduction of the continuous mill for mass production of tin plate, employment in the steel industry increased 28 per cent." One of Harold Rugg's texts included an identical image, with a note encouraging readers to notice how few workers appeared in the picture. Depending on the intended perspective, the same icon of modern production, the continuous-strip steel mill, could stand for either job creation or job elimination.[33]

The Depression-era publicity of AT&T took a completely different turn. Although company executives often remarked on how much pride they took in their laboratory's innovations and on how much new technology had improved customer service, their advertisements did not picture banks of dial switches. All through the 1930s, the Bell System continued to represent phone service in the old-fashioned way. Its ads featured drawings of young, pretty female operators, the very women who feared that the spread of dial technology endangered their jobs. That warm image of the "voice with a smile" conveyed an emotional depth, the idea that the company made a personal connection with each caller. Telephone offices still retained human switchboard operators to handle long-distance calls and special services, but the casual reader looking at advertisements would never realize how much the industry had committed to introducing dial equipment.

National business organizations devoted extensive time and energy through the 1930s to fighting off alarm over technological displacement. The American division of the International Chamber of Commerce declared that innovation "has failed to cause any appreciable diminution of employment opportunities." In fact, the group maintained, use of new machinery in the printing, electrical, and automobile industries had brought down production costs and thereby stimulated business so much that the number of jobs "has constantly increased." The National Industrial Conference Board similarly stated in 1935 that current levels of displacement remained "negligible." Such associations attempted to mobilize their membership, to turn each supporter into a booster for mechanization. The U.S. Chamber of Commerce published a special booklet setting out eleven different arguments against the idea that displacement posed a grave threat. The chamber urged its members and friends to consult those outlines whenever they needed to prepare speeches for luncheons, dinners, trade conventions, schools, or other public gatherings. The long months of depression had left "many sincere persons" confused about economic fun-

damentals, the booklet's authors suggested with dismay. Too many Americans
had observed that "the highest mark in unemployment is coincident with the
highest degree of mechanization" and then had jumped to false conclusions
about cause and effect.[34]

Precisely because technological unemployment was such a "plausible fallacy," the chamber maintained, businessmen had a "responsibility to help"
ordinary citizens start to "think along the right line of truth." To give members ammunition for such efforts, the booklet supplied tidbits of information
from prominent scientists and corporate executives. It quoted Walter Chrysler
as saying that without modern production techniques, automobiles would be
so prohibitively expensive that Americans would be able to afford only about
two thousand a year instead of the current annual figure, four million. Other
quotations from Charles Kettering, George Merck, Lammot DuPont, the
National Resources Committee, and many more sources offered a wealth of
ready-made arguments in defense of mechanization. Speakers who wished to
get more specific could refer to the chamber's outlines for talks on "What the
Automobile Means to America" or on "The Miracle of Industrial Chemistry."
In general, the chamber advised members to reiterate the idea that whenever
industry invested in the latest equipment, the public ultimately reaped both
"new employment and new wealth."[35]

Joining established groups such as the Chamber of Commerce and the
NAM in the effort to assert the economic and social value of mechanization,
machine-tool builders created a new association in 1933. Headed by former
U.S. Chamber of Commerce president John O'Leary, the Machinery and Allied
Products Institute immediately ranked among the nation's five largest trade
organizations. MAPI's officers made it one of their top priorities to organize
publicity campaigns attacking the idea of technological unemployment. Executives in the machine-tool industry felt especially victimized by talk associ-
ating job losses with mechanization and, for that reason, adopted a particularly
defensive stance. Herman Lind insisted that though the initial installation of
machines frequently eliminated a small number of workers, history proved
that employment grew fastest in industries such as automobile manufacture,
which led the way in embracing the latest devices. Labor needed to realize, he
declared, that "under the American plan, machines employ men."[36]

Through the 1930s, MAPI devoted considerable attention to publishing and
distributing a series of promechanization booklets. One of its publications,
Machine-Made Jobs, bore the subtitle "'But's' and 'And's' That Must Be Considered in Connection with Common Statements, Which on the Surface Appear
to Prove That Machines Cause Unemployment." The institute addressed its
comments directly to ordinary Americans, giving them what it called facts "to

think about before you swallow what appear to be good arguments that machines are doing all our work." Running down a list of occupations from automaking, steelmaking, textiles, and ice-vending to telephone work, office work, and farming, MAPI asserted that labor had never really suffered. The section headed "Goodbye, 'Hello Girls'? No, More of Them Than Ever," informed readers that even as the Bell System had installed dials in almost one-third of its system, the number of operators had risen from 190,000 in 1920 to almost 249,000 ten years later. Most companies installed new equipment not with the aim of displacing labor but in order "to improve quality and . . . sell a product for less money or to make an entirely new product," thereby stimulating employment. MAPI poked fun at the whole idea of technological displacement. Americans who objected that a steam shovel displaced 100 men with hand shovels, the group retorted, might equally well complain that a steam shovel replaced 10,000 men working with teaspoons.[37]

As the second part of its campaign, MAPI attempted to convince Americans that mechanization had brought new richness to consumer life. The group's publications defined material possessions as the primary measure of national happiness. Its booklets *Machinery and the American Standard of Living* and *Technology and the American Consumer* set out to show readers that production technology had helped slash prices. A set of simple pictographs indicated that the average person in 1914 had to work fifty-five minutes to afford a man's hat, but only fifteen minutes in 1939. Likewise, the amount of time needed to earn enough money for a pair of men's shoes had dropped from forty-five to fifteen minutes in those years. MAPI celebrated such supposed gains in consumer power as tribute to the success of innovation and free enterprise. By committing industry to mechanization, the institute insisted, far-sighted businessmen had raised the United States out of a primitive handicraft existence to a triumphant position as the world's foremost society.[38]

In 1938, the business publication department of McGraw Hill added further ammunition to the publicity war by producing a special twenty-four page magazine section which set forth justifications for mechanization. James McGraw Jr. explained that the company provided the material as an editorial service, intended to help businessmen fend off "mistaken and unfair" criticism. The publisher encouraged supporters to adapt the general line of argument as a starting point for their own public relations efforts, tailoring the specifics to suit their particular circumstances. Editors of *American Machinist*, one periodical that featured the McGraw Hill material, explicitly pushed readers to initiate such publicity campaigns. The single "biggest job confronting industry today" involved correcting the "many falsehoods" told about modern economics, the magazine warned, since so many people "have convinced themselves that machines have reduced employment." Just by looking at the familiar example of automobiles, McGraw Hill suggested, people could see how invention had opened new opportunities in manufacturing and selling cars, producing gasoline and oil, repairing vehicles, and building and maintaining roads. Over six million Americans, one out of every seven working men and women, "owe their employment directly or indirectly to the automobile industry," the writers declared. Another seventeen major industries developed since 1880 (airplanes, aluminum, asbestos, automobile supplies, calculating machines, cottonseed oil, electric equipment, gasoline, icemaking, movies, phonographs, photographs, radio, rayon, refrigerators, rubber, and typewriters) had directly created another 1,123,314 jobs and indirectly supported 25 percent of all working Americans. Such claims were not novel (though alternate sources cited different numbers), but McGraw Hill incorporated them into a slick publicity format it hoped to spread nationwide.[39]

As MAPI had done, McGraw Hill encouraged business spokesmen to play up the concept that the Machine Age created both jobs and tangible consumer benefits. To give that idea "vivid and personal" appeal, McGraw Hill constructed its argument, phrased in language ordinary people could appreciate, around the theme "What Machines Mean to Bill Smith." By defining their prototypical American in certain ways, publicists neatly avoided the issue of displacement. The material introduced "Bill Smith" not as one of the nation's millions of unemployed or underemployed but as a 48-year-old who had worked for twenty-four years operating a grinding machine. Descriptions of his workplace never hinted at the uncertainty real-life Depression workers faced. Instead, "Bill" praised mechanization for relieving him of heavy lifting, giving him shorter hours, and raising his wages. "Everything considered," the publicists wrote, the average worker had "a pretty good time out of life," thanks to changes in manufacturing technology. Bill Smith, his wife, and their children enjoyed a living standard that was "the despair and envy of other countries," McGraw Hill averred. The moral was clear: Americans must forget ridiculous worries about technological unemployment and be taught to feel, like the fictional Bill Smith, properly "grateful to and proud of the forces that have made America the most industrialized of nations."[40]

Ideals of Mechanization, Consumerism, and American Superiority

It was no coincidence that business interests placed so much emphasis on consumerism as a counterstrike against talk of technological unemployment. Their publicity campaigns defined national progress as an increasing standard of living, measured by sheer number of possessions (and, to a lesser extent, quality). Such an approach allowed them to argue that despite the painful sight of breadlines and apple-sellers, modern Americans still enjoyed the best of all times. Their ancestors had never had an opportunity to ride in an airplane or listen to radio. Electricity and the telephone had transformed countless households, while motion pictures offered an entirely new form of democratic amusement. Development of mechanization and the assembly line had transformed the automobile from a toy of the rich into a necessity for the masses. When life was seen in such a light, record levels of unemployment could be dismissed as a short-term glitch, an insignificant hiccup in the twentieth century's overall prosperity. The power of science and invention promised to yield still greater wonders in the future—providing that critics of mechanization could be prevented from derailing progress.

In case those readers whose family and friends had been hit by economic strain should prove skeptical, the business community hastened to reinforce the idea that mechanization had made American consumers the luckiest people on earth. On average, production in American factories used roughly twice as much machine horsepower as did the industries of Britain, Germany, or Italy, according to Hartley Barclay, editor of *Mill and Factory*. That pattern, he concluded, gave American workers a purchasing power twice that of their English counterparts, triple that of Germans, and almost four times as much as Italians. By committing themselves to making technological improvements, the Chamber of Commerce boasted, employers had "built up for Americans a standard of well-being unsurpassed in the history of other nations."[41]

In his booklet *Labor's Stake in the American Way*, Barclay equated consumerism with national superiority. Although Americans made up only 7 percent of the global population, he bragged, they used 70 percent of the world's motor vehicles, 75 percent of the silk, and 66 percent of all petroleum, coffee, and rubber. Depression-decade pamphlets produced by the National Industrial Conference Board and MAPI similarly declared that American consumers ought to be happy, since, under the laws of free enterprise, the savings from new production technology must eventually trickle down to them. With slick graphics, MAPI showed readers that Americans owned almost three hundred million more shoes, fifteen million more phones, and twenty-one million more radios than the British. Although such simple statistics did not factor in compensating variables such as population size or geographic area, they served their purpose, allowing boosters to paint American life as the all-time zenith of civilization. Modern Americans were privileged to enjoy "an automobile standard of living," while Europeans, once so triumphant in their era of world domination, must resign themselves to a "bicycle standard of living."[42]

Such material carried an implicit threat that if socialists and other agitators succeeded in rousing paranoia about technological displacement, American industry might easily slip from its pinnacle of progress. If the United States did not continue to embrace the latest advances in mechanization, it might slide down to join inferior countries. In boosters' rhetoric, China and the Soviet Union exemplified the state of primitiveness. M. L. Brittain, president of the Georgia School of Technology, found it "unthinkable that we are to Russianize American life, merely content to divide up jobs." Similarly, Malcolm Bingay, editor of the *Detroit Free Press*, declared that citizens of the United States must never be forced to accept limits on their standard of living. The "most successful nation" in history simply "will not be Chinafied."[43]

Even while the country's rebound from economic disaster remained shaky, even as government officials warned that the Machine Age might never again provide enough jobs to go around, the organized business community continued to assert that mechanization had brought unprecedented gains to the average family. Such language carried an underlying message that anyone complaining about technological unemployment was not only ignorant but ungrateful and even unpatriotic. Virgil Jordan, president of the National Industrial Conference Board, reproached Americans in 1936 for their readiness to condemn the corporate system as evil. The present-day mentality made businessmen into "eternal scapegoats," innocent parties "offered up for sacrifice to appease the wrath" of frustrated workers. Men and women had grown complacent, he complained, taking modern miracles for granted. Coasting along in the knowledge that their lights would come on at the flick of a switch, people never bothered to develop "the remotest notion" of how technology worked. If anything, Jordan seemed to suggest, Depression-era Americans needed a little more insecurity, to teach them to appreciate mechanization.[44]

Properly thankful workers, business leaders declared, would acknowledge that mechanization had allowed employers to give them shorter hours, higher wages, and better working conditions. Without the new power of production, Justin Macklin maintained, the United States might still be relying on child labor. Such an attitude overlooked the way unions had fought to win benefits, and it ignored government mandates on working hours and child labor. All gains for workers, it implied, had evolved naturally as a direct result of mechanization. Referring to the way women and children of the nineteenth century had worked fourteen-hour days in factories and mines, John Van Deventer suggested that Americans in the twentieth century should count their blessings. If only "we could give those who advocate a return of the good old days a taste of one week," he commented, "they would be glad indeed to return to the better days of 1931, depression or no depression." Van Deventer argued that while workers ought to credit their employers' history of technical innovation with making it possible to shorten the workweek from an inhumane sixty hours to forty, labor must also realize that the process had limits. An attempt to force working hours even lower would wreak economic havoc. *Iron Age* condemned the various union proposals for a thirty-hour week, commenting, "We cannot materially shorten the working day and still provide the quantity of goods and services which the American people aspire to consume."[45]

In fact, the business community suggested, unions' push for shorter hours and higher wages risked turning technological unemployment from a myth into a real problem. Harvard political economist William Ripley, former Interstate Commerce Commission member, urged railroad workers in 1931 to preserve their jobs by accepting a voluntary 10 percent pay cut. "Every increase in the labor cost of industrial operation immediately puts a corresponding premium upon the introduction of mechanical devices to take the place of manpower," he warned. If mechanization did seem like a threat, workers had only their own leaders to fault. The "phenomenal development of mass production methods since 1920 has been substantially stimulated by the sustained high wage level." Ripley observed. If employees would only limit wage demands and continue on a forty-hour week, the course of progress inherent in corporate philosophies of free enterprise would ensure prosperity.[46]

Advocates of machine-building insisted that job-sharing measures were not necessary, since even amidst economic downturn, the American system continued to present plenty of openings. New workplace technology did mean change, but the social consequences were not necessarily bad. Chicago bank official Franklyn Hobbs told a group of machine tool distributors, "True it is that, during the last thirty years, time-saving methods and tools have driven two workers out of every three away from the bench." However, he insisted, those men had subsequently moved "from that bench to other benches at an increased wage in every case." Business leaders must bring such facts to public attention, Hobbs continued. Once ordinary Americans finally learned the truth about how new technology had helped them, "the labor demagog [sic] who rants about machinery causing unemployment" would be forced to "find a small hole [and] crawl in."[47]

Belief that technical advances invariably created as many and better jobs than they destroyed allowed business to shift any blame for unemployment onto workers themselves. Cyrus Ching, U.S. Rubber's industrial and public relations director, explained in a 1932 radio broadcast that although industry wanted to mechanize processes that required high-wage skilled labor, the men involved would not suffer. Employers would generally choose to retain most of their knowledgeable and experienced personnel. A few might be transferred to less-skilled work with lower pay, but in a depression, they ought to feel grateful for the "opportunity." It was only the "inefficient or undesirable" man, the worker "who doesn't measure up," who failed to find a new position after being displaced. Any "adaptable and flexible worker has little to fear," Ching assured listeners. Business actually did employees a favor by updating equipment, since the process kept men versatile. Labor suffered most when a backward company let techniques stagnate, then panicked and introduced radical change overnight. In short, Ching implied, decent employees would find no difficulty in adjusting to mechanization and would not resent being temporarily shifted to a lower-rank job. Men unable to cope with the pace of change were by definition unsuited to the demands of efficient industry and did not belong there in the first place. Employers bore no responsibility for such casualties, Ching said, since their plight reflected "a social, not an industrial problem." Dealing with the ranks of inadequate workers should be a task for social workers, educators, or even eugenicists.[48]

Other business observers agreed with Ching that far from hurting good workers, innovation actually offered an advantage to intelligent, able, and "characterful" employees. Henry Ford argued in 1930 that, with rising demand for machine designers and builders, the American laborer enjoyed "a thousand chances" for success "where there was one in my day." A man who failed to seize one of those thousand chances might be presumed to be lazy, incompetent, or simply unlucky. Policymakers should not be worrying about technological unemployment as a permanent problem, Ford continued, they should be bracing for a shortage of labor. After the abnormal conditions of the Depression passed, employers might experience difficulty in finding enough men to meet production demands, to handle increasingly complex equipment. George F. Trundle Jr., president of a Cleveland engineering firm, declared that he had begun to witness a deficit of skilled and even semiskilled labor already. If industry could not get suitable men to keep machines running, he cautioned, economic recovery might come to a screeching halt.[49]

In the mindset of the prominent businessmen who spoke out, Depression-era joblessness represented an inconsequential detour from the historical inevitability of American triumph. The country must ignore misguided alarm over displacement to concentrate on achieving long-term economic gain through its embrace of new production technology. In the ongoing quest for wealth, Benjamin Anderson declared, "on no account must we retard or interfere with the most rapid utilization of new inventions." Future generations would expect living standards to climb ever higher, and industry could only meet such consumer expectations by expanding mechanization. Workplace mechanization represented the inevitable, the only possible way to attain national success. Westinghouse's J. S. Tritle announced that "instead of technological unemployment forcing . . . business to restrict its use of machinery, we will need machines in greater numbers . . . as we progress."[50]

Reassurances that technical innovation brought new employment and economic advantage over the long term failed to satisfy observers such as WPA administrator Corrington Gill. In Gill's eyes, John Maynard Keynes had summed up the problems with that way of thinking in his aphorism, "in the long run, we are all dead." Nevertheless, corporate leaders had mounted an impressive chain of public relations efforts. The gospel they preached, maintaining faith in technology as the key to national progress, rested in large part on the business world's alliance with science and engineering. Publicity experts promised that the laboratories of AT&T, GE, and other major companies would soon devise new products and industries, creating both new jobs and higher standards of living for the future. Business leaders effectively formed a partnership of common interest with America's scientific and engineering community, giving those groups the power and the incentive to work together on defusing public unease about the relationship between mechanization and work.

6

"Innocence or Guilt of Science"

Scientists and Engineers Mobilize to Justify Mechanization

As talk about mechanization and unemployment continued to escalate, the controversy drew in many of the nation's leading scientists and engineers. The scientific community heard a growing number of politicians, social scientists, labor leaders, and workers say that the modern development of science and technology had accelerated beyond the pace that society could handle. Such an attitude, the professionals feared, unfairly set them up as scapegoats for economic disaster. Still worse, if people became convinced that technology had grown too fast, they might start searching for ways to slow it down. The scientific and engineering communities remembered all too well that back in the 1920s, alarm over the increasingly destructive capacity of military technology had led to suggestions that the world needed a "science holiday." They worried that a popular scare linking machines to Depression-era job loss might rekindle the call for limiting new research and invention.[1]

Accordingly, prominent scientists such as MIT president Karl Compton and Caltech president Robert Millikan, along with engineering figures such as Charles Kettering and Ralph Flanders, rushed to defend the job-creating value of their work. The idea of technological unemployment remained purely mythical, they declared. Economic history had proved that mechanization always enhanced employment opportunity, consumer pleasures, and the well-being of workers. To spread that optimistic message, scientists and engineers turned to public relations. The country's most prominent scientific and engineering organizations mobilized their resources to protect their interests and defend their reputation as providers of progress. The American Institute of Physics, the American Society of Mechanical Engineers, the American Association for the Advancement of Science, the American Engineering Council, and other professional groups all created special events to downplay public fear of displacement and to celebrate mechanization. To bolster their argument, scientists and engineers constructed pet theories to interpret the entire world's past, present, and future. In their assessment, humans' mastery of science and technology had been the genesis of all civilization, starting the West off on a trajectory of progress. That route of improvement had inevitably peaked in the United States, whose citizens enjoyed an incredible material abundance. The country would realize even greater wonders in years ahead through its continued investment in research—providing that pessimists stopped frightening people with absurd talk about permanent technological unemployment.[2]

The fear that popular discontent with economic conditions might foster a revolution against science and invention was not completely unfounded. The idea of imposing a "holiday" on investigators continued to pop up, very occasionally, as the decade passed. A 1934 issue of *Rotarian*, for example, ran two articles debating the subject under the catchy title, "Do We Need Birth Control for Ideas?" In truth, though, most Americans expressed no interest in placing even a temporary moratorium on innovation during the Depression. William Green and fellow union leaders went out of their way to emphasize that they did not consider science and invention to be enemies of workers. Indeed, labor spokesmen generally supported the idea of intensifying the nation's research—with the stipulation that ordinary Americans must receive a proper share of the benefits. Editorial writers for the *New York Times* summed up prevailing opinion with the flat declaration, "The State dare not curb science."[3]

Nevertheless, as long as talk of technological unemployment persisted, scientists and engineers continued to fear a public backlash. During the 1930s, that all-consuming apprehension came to define their professional communities. The debate over mechanization affected how researchers thought about the meaning of their work and how they related to the public. Scientists, especially physicists, became increasingly assertive about assigning themselves a central role in modern society. Engineers devoted new attention to discussing economic issues and mused about their professional responsibility for how the introduction of technologies affected everyday life. National scientific and technical associations set out an agenda for defending themselves, a task that commanded the energies of some of their most prominent members. With a driving sense of urgency, these communities moved to associate mechanization with a faith in national progress.

Science as Scapegoat?

Scientists and engineers perceived talk about displacement as placing a certain blame upon those who had helped develop new machines. A 1933 lecturer at the Ohio Academy of Science, Robert Budington, spoke for many when he declared that he had begun to "vigorously resent" all the commotion. It seemed to raise insidious questions about "the innocence and guilt of science." No other area of intellectual "endeavor has been subjected to such a deluge of earnest opprobrium and unqualified reproach" as had science in recent months, Budington complained. Chemist Alfred Stock raised the rhetoric to a higher pitch, warning that when ordinary citizens failed to comprehend the value of intellectual life, it opened the way for a barbarian attack on researchers. Science could only continue in its "triumphant march," he declared, as long as "Archimedes escapes death by the rough hand of the soldier."[4]

Without question, talk of displacement did entail some criticism of science and engineering. In a 1936 speech to the American Society of Mechanical Engineers, Yale president James Rowland Angell threw down the gauntlet by referring to technological unemployment as the most serious of some "ill-advised consequences" stemming from present-day science and engineering. "One of the most conspicuous facts about mechanical inventions is that they may occasion large-scale dislocations of labor," he said. "The time has long passed when we can look upon these developments as simply interesting eccentricities exercising purely local effects." The twentieth century could not truly be considered successful, he told the engineering audience, since the very process of modernization left so many people miserable. "When we are willing to accept the benefits which engineering progress brings to us in the form of cheaper and better food and raiment and such like blessings, we must be willing to see to it that our neighbors are not compelled to pay in poverty and suffering for the advantages which we enjoy." Other observers similarly called for reforming scientific and technical education, to teach graduates how to bring their fields into harmony with social interests. The practice of science and engineering must give due weight to issues such as employment, AFL head William Green insisted. Engineering projects ought to incorporate "constructive consideration and counsel" on how the installation of new machinery would affect workers. "Science has been used without taking into consideration the fact that wage earners have an equity in their jobs," he explained. "Merely speeding up the industrial machine is not an unmixed good unless engineers realize that these changes profoundly affect human lives."[5]

Green and Angell never even hinted at the idea that society could or should force research to halt, but as concern about joblessness remained high, many voices called scientists and engineers to account for the potential negative social consequences of their work. Even President Roosevelt insisted that America's professional communities must face the evidence suggesting that mechanization hurt labor. In a May 1936 message congratulating General Electric chairman Owen Young upon being honored by the Society of Arts and Sciences, Roosevelt wrote, "I suppose that all scientific progress is, in the long run, beneficial, yet the very speed and efficiency of scientific progress in industry has created present evils, chief among which is that of unemployment."[6] The President's qualified phrase, "I suppose," amounted to a less than ringing endorsement of science, while his reference to displacement brought a harsh note to an evening of celebration.

Perhaps the most startling aspect of President Roosevelt's comment came in the combative directness with which he linked the pursuit of efficiency to modern social and economic disarray. Through preceding decades, American leaders in engineering, business, and government had held out efficiency as the ultimate virtue. From the 1890s, Frederick Winslow Taylor had led the way in preaching scientific management, a set of principles and tools for analyzing and reorganizing the workplace to eliminate waste. If experts could bring production systems to maximum effectiveness, the crusader believed, the benefits would spill over to labor in higher wages and a better standard of living. Taylor's disciples expanded on his approach and, as consultants, applied their techniques to restructure factories, banks, construction, and railroads. Colleges began training students in the new discipline of industrial engineering, and a fascination with Taylorism spread into popular culture. Reform-minded city managers, government officials, conservationists, teachers, and even home economists took part in a Progressive-era "efficiency craze." In the 1930s, talk of technological unemployment led some observers to ask whether the fixation on efficiency had gone too far, whether the quest for the "one best way" came at too great a human cost.[7]

The idea that no less a figure than President Roosevelt would condemn efficiency threw advocates into culture shock, though some critics felt that such talk was long overdue. Social scientist William Ogburn suggested that, if anything, society had been overly reluctant to assign science and engineering their fair share of blame for job loss. History had instilled an unquestioned respect for science and technology as forces that had helped Americans assert their independence and conquer the Western frontier. Moreover, Ogburn noted, people tended to associate unemployment with "moral" causes, assuming that anyone without work must be either lazy or stupid. It was "only recently that one would admit that a man was unemployed because a machine had destroyed his job," he wrote. For too long, an "unwillingness to admit the great role which so material a thing as technology plays in causing problems" had precluded serious discussion about industrial mechanization.[8] Though Ogburn believed that Americans possessed an ingrained love of science and technology, the Depression cast a new light on things.

Public debate forced scientists and engineers to work as never before to defend the value of their enterprise. The professions produced a flood of books, articles, speeches, and special events, generally following a three-part line of argument: To start, they flatly denied that technological innovations in production presented any major problems for labor. As a corollary, boosters argued that, by definition, advances in knowledge and technical ability would always promote economic and social welfare. Finally, they painted anyone concerned about technological unemployment as either an ignorant pessimist or as a scoundrel out to undermine the advance of civilization.

The professionals' discussion highlighted, among other things, a fairly firm consensus about the relationship between science and technology. A few scientists tried to emphasize the distance between pure investigation and applied research, to divorce themselves from any association with industrial mechanization. Those researchers who pursued knowledge for its own sake, in "the spirit of Thales," carried no responsibility for any consequences of production engineering, Robert Budington told the Ohio Academy of Science, just as "Llewenhoek is not accountable for the inhuman use of bacteria in war." Budington's view proved the exception, as most scientists resisted any impulse to escape the controversy by detaching their work from the realm of technology. Men such as Karl Compton referred to science and engineering as inseparable partners in the pursuit of prosperity and progress. Throughout the crisis, the nation's scientific and technical communities stood as a united front, with a shared interest in denying that mechanization brought any special trouble.[9]

Over and over during the 1930s, prominent scientists expressed a firm conviction that improvements in production methods had left Americans better off than ever. One fierce defender, Columbia University physicist Michael Pupin, lauded the Machine Age for having "made the physical side of human life ever more glorious than the life of the Olympian gods." Scientists and engineers had tapped into nature's powers, "which like a host of ministering angels are toiling for the good of this terrestrial globe," he told a 1932 audience. Yet, the world had not attained perfection, Pupin confessed, since "millions of idle workers are starving." Such problems stemmed from a failure of man's spiritual energy, he continued, rather than from any flaws in the divinely granted power of science and engineering. Technology was a "gift from heaven presented to man as a reward for his diligent study." Development of steam engines, electrical equipment, and manufacturing devices had created a wealth and material luxury which represented "the miracles of our power age." Though other scientists and engineers used less flowery and ecstatic language, the spirit of Pupin's argument held true for them. Charles Kettering described the whole notion of technological displacement as inconceivable, "entirely foolish." He found it impossible to "believe that an increasing in human information" about how to multiply production "can have any bad effect at all."[10]

One of those who devoted special energy to defending science was Karl Compton, president of MIT. Compton acknowledged that the "first effect" of installing machines "may be to throw people out of work through producing a given amount of goods with less labor." However, he maintained, such job loss could never amount to more than "local and temporary maladjustment." Compton agreed with those economists who believed that improvements must reduce a product's cost and thereby expand its market, creating a growth in demand which could employ "far more labor at higher wages." Adoption of machines for making light bulbs had initially eliminated many glassblowers, but had ultimately "made these lamps so cheap that they have become universal," Compton said, "providing again large employment in manufacture and distribution." Even though displaced glassblowers might not find any new outlet for their skills, even if new machines did cause some individual distress, Compton did not believe that such events warranted dismay. The United States could handle such cases by small measures, encouraging employers to offer retraining or pensions to the workers affected. Meanwhile, inventors must be left to work in peace.[11]

"Science really creates wealth and opportunity where they did not exist before," Compton maintained. Addressing the American Philosophical Society on the "Social Implications of Scientific Discovery," he quoted Louis Pasteur's description of science as "soul of the prosperity of nations and the living source of all progress" (a phrase that also became a favorite of Robert Millikan and many other scientists during the Depression). Compton cited figures showing that the United States, with less than 7 percent of the world's population, controlled 40 percent of the world's wealth. That lopsided distribution proved that national leadership in science had paid off in "unprecedented prosperity," he contended, "shared as never before by the great masses."[12]

At a time when unemployment hovered stubbornly close to 20 percent, asserting that Americans ought to thank science for unparalleled well-being might sound like elitist hubris. Yet Compton went further, announcing that the invention of labor-saving devices promised to transform the United States into an "analogue of Plato's republic." Mechanization would let every job be "performed as easily and quickly as possible" and produce enough wealth to gratify everyone's desires. Material paradise would guarantee happiness, and a country that achieved maximum consumerism would attain success. Many thinkers before Compton, of course, had hoped that science and technology could inaugurate a utopian era, but the public debate over displacement gave Compton's rhetoric particular importance. Instead of worrying about technology as a cause of job loss, he said, Americans ought to call for more mechanization as a way to hasten the evolution of that economic ideal.[13]

In Compton's view, lack of resources represented the sole constraint on the pace of intellectual achievement; once granted proper backing, investigators would speed ahead to greater discoveries. Pure research would automatically translate into applied insight, which would give birth to a host of new industries and provide countless jobs. Americans needed to recognize that present-day unemployment resulted in good part from the nation's failure to underwrite research over preceding decades, he said. "When a government official asked ... why science did not have more to offer now, I replied, 'The groundwork for what you want should have begun ten years ago,'" Compton recalled. Even without adequate support, research had helped fend off economic disaster, he claimed. If not for the scientific and technical progress made in the twenties, "unemployment would have struck us many years earlier." However, it was high time for Americans to learn that providing decent public and private funding for research "might well mean the difference between prosperity and economic catastrophe at no very distant date." Compton promised the investment would soon be repaid through the creation of new jobs—not least for scientists themselves.[14]

Compton's campaign revolved in part around the frontier thesis of U.S. history. Since twentieth-century America no longer had room for westward migration, he proposed, future expansion must seek a new frontier. In science, the "fields of exploration ... are probably unlimited," and the "thrill of discovery ... is no less keen" than in geographic adventure. Americans should hail the triumph of mechanization as "a turning point in the history of the world," he advised. Ancient Egyptians, Greeks, and Romans had built their empires through pillaging and taxation, while the British had expanded their territory by battle and colonization; Americans would achieve even greater prosperity through the more peaceful route of intellectual domination.[15]

Karl Compton spoke not just as a leading physicist and president of MIT, but also as the one-time chairman of the United States Science Advisory Board. Though the group had been formed in 1933 to study and guide government's scientific efforts, Compton seized the the opportunity his chairmanship provided to build a case for putting federal dollars into research. In a manifesto entitled "Put Science to Work: A National Problem," he argued that the country must grow into economic recovery by relying on the "well-known" fact that "science has created vast employment." The unfortunate rise of "depression hysteria" meant that research "is not being called upon . . . now to create new employment when this is desperately needed!" Other Science Advisory Board members joined Compton in expressing a desire to assert themselves. At their first meeting, director Isaiah Bowman urged the group to "assume the high duty of replying" to the "criticism leveled at science as one of the alleged contributors to the present instability." The wider scientific community supported that aim; at its annual meeting in 1934, the American Association for the Advancement of Science adopted a resolution crediting science for stimulating "enormous employment." Invention had pulled the United States out of past hardship; railroad growth had ended the depression of 1870 and expansion of the electric industry had halted the 1896 downturn, while the automobile's popularity led to the 1907 recovery, the AAAS asserted. Historical evidence had proved the folly of curtailing research support just "when properly directed scientific work is more than ever needed" to create jobs through "development of new products."[16]

In concentrating on the politics behind the Science Advisory Board's struggle for existence, historians have tended to overlook the extent to which the group participated in the national debate over jobs. To justify his ambitious plan of creating a sixteen-million-dollar fund for research in natural science, Compton promised it would alleviate unemployment. The boldness of his claims disturbed some observers. Fellow board member Frank Jewett worried that even sympathetic parties might have difficulty accepting Compton's extravagant demands; one of Jewett's acquaintances commented acerbically that he was "sorry to see that the Science Advisory Board has gotten into the trough with the rest of the hogs." Jewett worried that Compton had inflated estimates of how many jobs research could generate. It seemed "erroneous to assume that substantial new industries can be created in the manner suggested" over any immediate period. Frederick Delano and J. C. Merriam, members of the National Resources Board, also accused Compton of exaggerating America's neglect of science and of ignoring the practical complications that would stem from a massive influx of research money.[17]

Even after dissolution of the Science Advisory Board in 1935, Compton continued to insist that employment depended on America's commitment to research. Caltech president and physicist Robert Millikan echoed Compton's rhetoric and call to action. "The mere fact that all European countries now support four times the population than they had" in 1800, Millikan maintained, "is proof enough that in the long run science creates many times more jobs than it destroys." Like Compton, Millikan exhorted Americans to realize that science "has produced wealth and leisure, even in the midst of this depression." Perceptions made all the difference; Millikan envisioned invention as the source of utopia. "Call unemployment leisure, and you can at once see the possibilities." By applying machines to production, society could create abundance without toil. People would be free to pursue intellectual and cultural refinement, creating a "heretofore undreamt-of civilization," which would be "even better than the Greeks." High Athenian culture had relied on slave power to give its elites wealth and leisure, after all; the Machine Age would raise even common men to unprecedented heights. Millikan could not believe that a blind fear of technological unemployment would lead Americans to throw away such prospects. The "man who shouts that science is responsible for all our woes is just as intelligent as the fabled individual who killed the goose to get the golden egg."[18]

Millikan, Compton, Pupin, and many lesser-known figures all provided strong words in defense of science, yet the scientific community's approach involved still more. Just as the Chamber of Commerce and other business interests mounted public relations campaigns to associate mechanization with economic advance, so the scientific profession organized special events to refute talk about technological unemployment. Of course, professional activities were nothing new in themselves; for decades, scientific communities had held meetings, banquets, and public celebrations to highlight their latest accomplishments and to promise still greater achievements. During America's Depression, such efforts took on an added purpose, providing scientists with a platform to deny any responsibility for having contributed to joblessness. The AAAS assumed a primary role in asserting the link between innovation and prosperity. Its annual meeting in 1932 featured a special session devoted to defending the economic value of science and its applications. Recent discoveries had brought modern industry to a "high order of mechanical efficiency," said Charles Kettering, a fact which he maintained did not bother him at all. In fact, Kettering judged temporary job loss to be a sign of success. "If we did not have unemployment today, it would indicate the hopelessness of management in engineering," he said, since for the last "fifty years we have done absolutely nothing but attempt by machine design to displace labor."[19]

The AAAS session blamed economic downturn largely on the misdeeds of financiers and speculators, but Dexter Kimball, dean of Cornell's College of Engineering, sounded a discordant note. The problem of "permanent technological unemployment already exists," he asserted, and Americans would need to be even more "on our guard as industry becomes increasingly scientific." The likelihood of still more mechanization would force people to ask "how far we should permit the good of the majority to be advanced at the cost of suffering" for a minority. Yet beneath such foreboding, Kimball's remarks reflected a basic faith in the economic system's ability to balance out over the long term. As long as manufacturers who introduced new machines applied the gains to bring down consumer prices, rising demand must stimulate new jobs. Given such economic law, Kimball assured the AAAS, "we need not be troubled . . . as to the *final* results" of mechanization in modern life.[20]

"Science Makes More Jobs": Physicists Mobilize to Deny Displacement

In 1933, the American Institute of Physics undertook an especially ambitious effort to reinforce the link between scientific gains and social progress. Given popular misapprehensions, AIP director Henry Barton declared, physicists ought to mount "a directed campaign of educational publicity as quickly as it can be organized." Science had been thrown into a political battle, Barton told AIP chairman Karl Compton. Researchers who hoped to win funding from Washington must "enlist the really large memberships of our scientific and technical societies to bring what pressure they can upon the Congressmen." With that aim, Compton and Barton began planning to hold a special meeting under the title "Science Makes More Jobs," or "Science's Answer to Those Pressing for a 'Research Holiday.'" With the full weight of professional authority behind them, the two hoped to create "in the newspapers a counter-campaign against the 'moratorium for science' propaganda which now appears from time to time."[21]

From the start, the physicists geared their Symposium on the Fallacy of Scientific Progress as the Major Cause of the Present Unemployment Situation to achieve maximum exposure. The AIP arranged to hold the meeting jointly with the New York Electrical Society, a group that "always gets tremendous publicity"; Barton anticipated great benefits from being able to "hitch up the Institute and a good cause to their machinery." The director wanted to schedule the meeting either for February 11, "the birthday of both Edison and Darwin," or February 12, the day honoring "Lincoln, who founded the National Academy of Science." By choosing an auspicious date, Barton said, the AIP could "hook this meeting to claim the value of science onto the names of these great men." He settled for February 22, Washington's birthday, "an excellent publicity date because the morning papers after a holiday are sparse in ordinary news." As another advantage, Barton hoped to draw a large and prestigious audience from the scientists who would arrive in New York at that time to attend the American Physical Society's meeting.[22]

To gain attention, the AIP invited a list of special guests, including Orville Wright, Charles Lindbergh, Amelia Earhart, Henry Ford, and other celebrities associated with technological triumph. In a letter to Albert Einstein, Barton asked the world-famous scientist to come help oppose the "widely held" and "alarming" view that "scientific progress caused [the] troubled state of [the] world and should be stopped." In his response, Einstein praised the goal as an important one but refused to attend, pleading susceptibility to illness. The AIP sent another letter to Franklin Roosevelt, inviting the president to open the symposium via radio. The meeting would discuss "the part which pure and applied science should play in the new social order which lies ahead," encouraging professionals to "develop an increasing sense of the social values and responsibilities" behind their work. Though the White House declined to send a radio greeting, the president did put his name to a statement, the phrasing of which had been suggested by Burton. The "idea that science is responsible for the economic ills . . . recently experienced can be questioned," the message ran. It would be "more accurate to say that the fruits of current scientific. . . . development, properly directed, can help revive industry."[23]

Over the weeks of preparation, Barton and Compton grew increasingly excited about the potential impact of their event. To define their intent, the men issued a press release explaining that the "value to civilization of scientific . . . research has recently been questioned in uninformed but not uninfluential circles." Despite the "ridiculous" nature of such assertions, the statement continued, fear of technological unemployment had influenced politicians, universities, and business officials. To prove such concern baseless, the AIP promised to present evidence that scientific and technical progress had provided hundreds of thousands of jobs. In less formal terms, as Compton declared, the physicists intended to set up "a backfire against those . . . preaching such doctrines as 'science has had its day and made a mess of things.'"[24]

By the time the symposium finally took place, the AIP had worked up a major program. In the afternoon before the big talks, the New York Museum of Science and Industry hosted a gala preview of a new exhibit explaining "How Fundamental Inventions Contributed to Employment." The show centered around the theme that the modern applications of steam power, internal combustion, and other technology had created job growth "so extensive that it cannot be estimated." The museum distributed an information packet filled with charts and graphs to demonstrate that Thomas Edison's inventions alone, from the phonograph and movie equipment to the multiplex telegraph and lighting systems, had provided millions of jobs. True, the electric industry's introduction of machines had reduced the number of employees needed in the manufacture of light bulbs from about 26,000 in 1920 to 15,000 in 1930. The rise in efficiency brought down the average price of a light bulb from thirty-four to fourteen cents, however, and, over those same years, companies added another 10,300 workers to manufacture vacuum tubes. The course of invention would always compensate for labor displacement by expanding demand, the museum exhibit concluded. Scientists and engineers could rest secure in the knowledge that their research had led to "many millions of man-hours" of work.[25]

The evening session, held at the New York Engineering Societies Building, featured a series of speeches broadcast over NBC and CBS national radio networks. "The idea that science takes away jobs," Compton told listeners, "is contrary to fact, is based on ignorance . . . vicious in its possible social consequences, and yet has taken an insidious hold on the minds of many." While frequently innovations in production did "throw large numbers of men and women out of work" at first, their ultimate role in generating a net increase in employment must be considered "immensely more significant." The United States could not afford to reject change, Compton argued; if people at the turn of the century had blocked development of automobiles in order to protect horsemen and carriage-makers, they would have sacrificed an industry that came to employ ten million. The spread of "insidious and dangerous propaganda" threatened to cut off any chance of getting government to support research, a failure that would lead to "national calamity."[26]

Following Compton, Robert Millikan declared that on balance, mechanization had vastly improved life for ordinary Americans. The application of science and invention to industry had eliminated the "heavy, grinding, routine, deadening" tasks, while opening up "more interesting" forms of work. New jobs required greater intelligence and offered larger rewards than the ones that had been lost, he continued. Even if the process did not absorb every single man or woman who had been displaced, most would gain increased economical well-being and extra free time. As people began enjoying their newfound leisure, Millikan promised, employment opportunities in education and recreation would rise. In that analysis of long-term trends, he declared point-blank, "there is no such thing as technological unemployment."[27]

Looking back over his thirty years in industrial research, Bell Labs president Frank Jewett told the AIP, "I cannot find a single instance where a scientific achievement has resulted in a reduction in employment." Just the opposite, he said; the "benefits of scientific research flow to all classes," bringing "even the least competent" of Americans "a step further from the starvation line." W. D. Coolidge, director of General Electric's research lab, echoed Jewett. Over the next fifty years, physics and engineering would yield "new products, increased efficiency, shorter working hours, more pleasures." The "inevitable swift transitions" might cause unemployment problems, if statesmen and economists were not "wise enough to modify" America's economic system as necessary. However, Coolidge maintained, any such "disaster can no more be blamed on the scientist ... than the chemist can be blamed if his discoveries are diverted, by the crimes of political leaders, from their [peaceful] potential" to war. Coolidge painted the processes of discovery and application as entirely divorced from each other; researchers only created knowledge, with no influence on how it might be used. Such a picture allowed scientists to claim the best of both worlds, taking full credit for prosperity while disavowing responsibility for any uglier consequences.[28]

In sum, the process of innovation "not only was not the devil which caused depression," Owen Young concluded, "it is the most promising angel to lead us out of it." By enabling business to supply consumers with new and improved products at lower cost, mechanization became "the mother of obsolescence" and hence the father of wealth. Such declarations of faith thrilled Barton, who reveled in the resulting publicity. The AIP distributed copies of Compton's and Millikan's speeches, which many national scientific, technical, and business publications either reprinted or quoted extensively. One of a relatively few cynical notes came from the *New York Times*, which observed, "Neither the statistics nor the argument are new. Nor did any of the protagonists of the laboratory explain why there is poverty amid plenty." Editors concluded, "As yet, no one has devised the means of absorbing new technical developments with the least possible amount of distress."[29]

The "Science Makes Jobs" forum represented an impressive display of esprit de corps among leading scientists during what they perceived as a crisis. For all its publicity splash, the AIP's symposium failed to quell public concern about mechanization. Barton took it as a personal affront when Americans, from President Roosevelt on down, continued referring to technological unemployment as a real problem. To fight that "decidedly unhealthy" trend, Barton began considering new ways to force people to appreciate science. The very success of research had contributed to the difficulty, he mused. The "spectacular fundamental developments in physics the last forty years has unduly diverted attention from the applications." To drive home the lesson that research created jobs, Barton believed that physicists ought to increase emphasis on their applied achievements. "Chemistry is everywhere known," Barton complained, but "laymen do not know that radio, refrigeration, sound motion pictures, etc. are physics." William Buffum, head of the Chemical Foundation, echoed that notion, chiding the AIP, "You gentlemen do not advertise physics."[30]

To start self-advertising, the AIP set up a new Advisory Council on Applied Physics to undertake "missionary work," as Barton called it. "Far from being ... abstract and impractical," an official AIP statement declared, "physics is the basis of all ... technology," from electronics and x-rays to aviation and air-conditioning. To spread that message, the council proposed producing a popular book on applied physics, which would "emphasize how greatly ... civilization is influenced by ... science." As a tentative title, the group suggested *Physics in Overalls* or, with less subtle reference to the issue of jobs, *Putting Physics to Work*. Published in 1939 as *Atoms in Action*, the AIP-sponsored text cataloged both existing and anticipated results of research. Each chapter detailed a particular application, such as electric illumination (in the section entitled "Light for a Living World"), radio (in "Sound Borrows Wings"), or the spectroscope and x-rays (in "Eyes That See Through Atoms"). In every case, George Harrison wrote, investigators had uncovered products and techniques to improve everyday life and establish new industries. "Putting electrons to work has put men to work," one section informed readers. "That four great new industries—the telephone, the radio, the phonograph and the motion picture industries—rest directly on the vacuum tube ... is usually ignored when new devices are blamed for technological unemployment."[31]

Harrison celebrated some cases of innovation as a substitution of precise technology for fallible workers, without expressing any qualms about the fate of displaced labor. His chapter on glass included a paean to the marvelous machines that transformed a molten stream into hundreds of perfectly shaped, inexpensive light bulbs. "Development of automatic glass-blowing machinery shows how scientific control can achieve a result previously thought possible only by means of hand skill," Harrison wrote. "It would be incorrect to say that a machine which can do this is almost human. So far as glass blowing is concerned it is more than human—its abilities are those of 2,000 men." Similarly, he indicated, the telephone company naturally wanted to replace its female operators, however attractive and polite, with dial equipment "so reliable that fewer calls than one in a hundred go astray because of faulty apparatus." Engineers had conquered a tremendous technical challenge, constructing a switchboard with more than two million parts, connecting ten thousand wires and handling over two hundred operations a second. For Harrison, such a story represented pure triumph, making labor questions irrelevant.[32]

In other cases, Harrison disputed the whole notion that mechanization had caused trouble. While musicians protested that radio and talking pictures endangered their livelihood, he assured readers that mechanically reproduced music had actually fueled a burst of musical appreciation, which would create new opportunities for professionals to teach and perform. Harrison blithely concluded that "once the initial dislocation had been adjusted," radio and movies actually helped workers, "as happens in so many other cases of apparent technological unemployment." As researchers worked with businessmen to make air conditioning, television, and even atomic power common, future generations would find as many jobs as they needed. Such an upbeat conclusion pleased at least one scientific reviewer, who touted *Atoms in Action* as "prescribed reading for those pessimists who deplore the economic consequences of science."[33]

Atoms in Action interpreted history, economics, and sociology as the physicists wished to see them. Throughout the past, man had acquired knowledge and applied his power over nature to build greater and wealthier civilizations. The present, despite depression, testified to the fact that mechanization created advantages for workers and consumers alike. The future would demonstrate beyond doubt that research brought man employment, abundance, and happiness. In promoting that comprehensive vision of progress, the science boosters left no room for negative considerations.

Engineers Discuss Economics: Civilization as Mechanization

When physicists mounted their publicity campaigns to promote positive impressions of modern science and technology, they found powerful allies in the engineering community. Kettering, Jackson, Kimball, and many lesser-known figures joined scientists and business leaders in denying that mechanization could lead to any grave harm. Electrical engineer William McClellan dismissed job loss as an unfortunate but ultimately insignificant side effect of technological improvement. In language reminiscent of the previous century's Social Darwinism, McClellan argued that the historic course of economic evolution required an "inevitable" sacrifice of some individuals. "In the onrush of the crowd some are bound to be trampled on—some crushed to death," he wrote. "Sad as it may be, that is the price of progress."[34]

Those workers who got "trampled" might feel that mechanization was not worth such a price, but during the Depression, engineers, like scientists, were quick to resent any implied criticism. America's "current depression is really the first for which scientists and engineers have been generally blamed," AT&T chief engineer Bancroft Gherardi declared. "Not only have we had to stand our share of the grief of the depression, but, adding insult to injury, we . . . are blamed." In a 1933 speech to the Society of Automotive Engineers, Gherardi set out to relieve his fellows of guilt. Far from causing social and economic pain, he announced, engineers had eliminated the hardest forms of drudgery while giving everyone higher living standards and longer leisure hours. Engineers could honestly congratulate themselves on their service to mankind and observe growing breadlines with a clear conscience. Yale chemical engineer Clifford Furnas expressed even less patience with anyone who fretted about the pace of change in modern industry. "They look at our ten million unemployed . . . and say: There's your progress! and forget that they are not viewing the ultimate accomplishment but only a temporary case of science and engineering out of phase with sociology and economics," he wrote. "Since the world is not perfected in a few years they think its legs are tottering." Experts could repair such imbalance soon enough by reducing working hours and expanding the range of consumer products available. Meanwhile, Furnas instructed Americans to refrain from criticizing. "One could hardly be so naive as to suppose that all our social and economic troubles have been caused by technical changes," he wrote. "The Bible records 4,000 years of Hebrew trouble and there was not a worthwhile technical advance in the whole period." In fact, he continued, "all ancient history seems to reek with economic distress and there were no machines."[35]

Logically, pointing out that human society had faced problems in an age before mechanization could not prove that modern changes in production had not contributed to economic difficulty, but Furnas's fervor spoke for itself. In speeches, reports, books, and articles, leading engineers repeatedly stressed the "self-evident" advantages of mechanization. The United States contained one automobile for every five people, whereas the world's next four leading countries possessed only one car for every fifty. Dexter Kimball complained that the agonies of the Depression had made Americans forget how much they owed to technical ingenuity, but other observers resorted to conspiracy theories to explain how anyone could disagree with the seemingly obvious concept that mechanization multiplied well-being. Edward J. Mehren, vice president of the McGraw-Hill Publishing Company, accused radical subversives of having tricked workers into blaming engineers for job loss. "Communists, the Socialists [and] the professors of sociology in many colleges" had stirred up fear of displacement, he told the American Society of Mechanical Engineers. Mehren called on all intelligent, patriotic citizens to resist such misinformation, holding fast to a "faith in the fundamental soundness of American economic development."[36]

Like Mehren, a number of engineers felt that only a troublemaker or an ignoramus could doubt that technological change promoted happiness. McClellan complained that paranoia about mechanization had led to a flood of "maudlin discussion" about the perils of modern life, a discussion "bordering on asininity." However ridiculous that discussion might be, he worried that the Depression had made ordinary Americans uncertain about economic fundamentals, and so talk about permanent displacement nonetheless seemed to strike a chord. According to a 1938 issue of *Mechanical Engineering*, the subject of "publicity for engineering" had attracted significant attention at recent professional meetings. The editor praised such a trend, noting that business and trade associations had long promoted their interests and touted their achievements. Engineers similarly ought to "place [their] case before the public," in the hope that most people had enough "good sense and intelligence" to accept a well-reasoned argument. The "glib traducers of engineering and industry should be exposed," or else their "hostility to technology, a vital motivating force in modern society, may retard development." *Mechanical Engineering* cautioned readers to keep self-advertising within "realistic" bounds, since the "delicate" art of publicity would backfire if engineers tried "to lead the public to expect the impossible." Any attempts at "pulling rabbits out of hats should be left for magicians."[37]

To ensure that members did not give out the wrong impression, the engineering community policed itself on what was said about the topic of displacement. In 1934, McClellan was soundly rebuked by his associates for making a casual remark that "engineers are magnificent creators of unemployment." Lest he provide any ammunition to those who considered mechanization a source of economic distress, McClellan altered his comment to read, "Engineers are magnificent creators of leisure." That single-word change avoided the scary connotations of *unemployment*, stressing instead the pleasurable associations of *leisure*. Engineers commonly sounded a two-part theme of victory and victimization, characterizing themselves as tragically misunderstood heroes. Twentieth-century experts had made "dramatic progress" toward achieving ever-greater efficiency of production, Yale industrial engineering professor Elliott Dunlap Smith announced. Surely "no other profession has gone so far in the attainment of its goal, *and none has ever been so roundly abused for its success in doing this.*" Misleading terminology encouraged Americans to hold engineers responsible, Smith lamented. It seemed unfair that the "unemployment which has occurred because of the failure of the economist and the business man to devise means for effectively distributing the abundance which the engineer's technological skill has made possible, is called neither *distributive* nor *economic* unemployment but *technological* unemployment." Ordinary people proved simply ungrateful, Smith mourned. "Truly, the way of the progressor is hard."[38]

Just as Karl Compton played a particularly visible role in spurring the scientific world into action, certain individuals emerged to lead the engineering community's defense. In numerous speeches and articles, Ralph Flanders, the head of Vermont's Jones and Lamson Machine Company, condemned any talk of labor displacement. Speaking in 1935 as president of the American Society of Mechanical Engineers, Flanders urged his fellow members to fight such "shallow and inadequate thinking." True, the process of mechanization had forced many men and women into a "rapidity of adaptation" that proved "to some exhilarating, to others painful." But the nature of innovation was always "healthy and constructive," Flanders continued. Recent economic difficulty stemmed from the narrow vision of greedy "industrialists, workers, farmers, and financiers," not to mention "the imbecilities of our politics." Moreover, the end of westward expansion had weakened America economically, socially, and psychologically, Flanders said. To duplicate the old sense of frontier optimism, twentieth-century Americans must reach for a new challenge, raising living standards to new heights through the power of mechanization.[39]

With even more enthusiasm than the scientists, engineers wrote an entire construction of the past, present, and future in the process of defending their profession against blame for joblessness. Engineers explicitly promoted a specific type of historical analysis, based on the premise that centuries of technical advance had carried Western society to an ever-higher plane. MIT electrical engineer Dugald Jackson provided the most elaborate version of this celebratory history of technology in 1938, with his six-part lecture series, "Engineering's Part in the Development of Civilization." That assessment portrayed engineering as the essential component in all history, the *source* of social existence. Civilization first arose when "the sluggish primitive mind" turned toward "creative intellectual . . . invention of machines," Jackson indicated. As those "individuals of most active mind" developed stone tools and fire-making instruments, they allowed man to transcend a "beast-like, precarious" existence and establish community life.[40]

Later generations of scientists and engineers had led the world along an upward path of progress, from the building of the Egyptian pyramids to the construction of New England's classic white-steepled churches, which symbolized the gracious "community habits made possible by engineering." Most important of all for Jackson, Western society had begun directing the "dignity and power of the human mind" to discover ways of applying more and better mechanical power to the necessity of work. He skipped over details such as child labor, pollution, and health risks in order to paint the Industrial Revolution as an undiluted blessing, a time when steam engines and other machines assured ordinary people of "unremitting improvement" in life. By contrast, Eastern nations had made relatively little use of machines and so remained in a "miserable" state of affairs; for Jackson, the poor "coolie rickshaw pullers" embodied the fate of a world not enlightened by technological progress.[41]

His grandparents had somehow managed to remain happy in a less technological world, Jackson admitted, but earlier generations never knew what wonders they were missing. Mechanization had made the twentieth-century United States the greatest place on earth, one that gave its citizens greater leisure, more material possessions, more ethical communities, higher "mental alert- ness," extra stability of work, and hence, better "control of life," than anywhere else. Though unemployed Americans of 1938 might not feel especially lucky or secure, Jackson insisted that the country was rich overall. Two-thirds of the population remained well-fed, well-clothed, and well-housed, enjoying unprece- dented educational and recreational pleasures. The recent "defects" in eco- nomic performance arose from unsolved complications in public life rather than from any "faults in the fundamental ideal of . . . engineering." Citing Ogburn's cultural lag theory, Jackson blamed unemployment on society's fail- ure to adjust to power production. Once Americans brought their institutions up to speed with technical progress, he advised, continued mechanization would move them toward an ideal of leisure and wealth.[42]

Such an engineering-based account of civilization held a definite appeal for the profession during its decade of siege; *Mechanical Engineering* paid trib- ute to Jackson's work by reprinting his lecture series in full. More than that, the journal urged colleges to teach Jackson's type of history as a way of rem- edying the deplorable "lack of appreciation among engineer and laymen alike" for the cultural value of technology. Engineering departments should make Jackson's lectures required reading, to instill the next generation with a proper awareness of their profession's "dignity and significance," knowledge that would let them stand up to criticism. Nonengineering students should also learn technology-centered history, the editors continued, as a counterweight to the popular misconception of machines as job destroyers. Classes ought to incorporate the moral couched in the question, "If we have progressed so far from the times [of] . . . superiority of tooth and claw, shall we not proceed further through dependence on the intelligent use of technology?"[43]

Like Jackson, other Depression-era engineers reacted to the issue of labor displacement by depicting technology as the driving force behind historical progress. C. F. Hirshfeld, chief of research for Detroit Edison, told the Birm- ingham, Alabama, Engineer's Club that Western society had not only survived but also prospered under mechanization. Regardless of protests by Britain's Luddites and other groups of workers, "each time has lived to pass on to bet- ter living conditions." Between 1890 and 1929, American business had intro- duced more and more equipment, and still the total number of people employed had risen. Such "facts do not spell technological unemployment to me," Hirshfeld concluded.[44] National engineering and technical societies joined in manufacturing a preferred vision of the past in order to fight talk of job losses in the present. Organizations used the history of engineering as an intellectual weapon to serve two purposes at once: lecturing the public on a positive view of technology while reinforcing the self-confidence and spirit of their members. Engineering groups turned to meetings and events as a prominent platform for broadcasting their interpretation of history amidst intensifying public discussion of unemployment.

Mechanical engineers made the American Society of Mechanical Engineers' fiftieth anniversary celebration, in December 1930, into an excuse to revel in claims for the social and economic wonders of technology. In a pageant entitled "Control," the ASME literally dramatized the story of human existence as a record of engineering development, of progressive evolution from the age of cavemen to the development of modern automobiles. The script portrayed production technology as inherently superior to human workers, and, following that principle, the organizers chose to use recorded music rather than live accompaniment. "Inasmuch as the pageant celebrates the increasing triumphs of the mechanical engineer," a representative explained, "it has seemed wiser to do without a band or orchestra and to substitute electrical reproduction." In his speech to guests at the ASME dinner, Edward J. Mehren assured listeners that ever since the Victorian era, mechanization had made it possible to give people higher wages and reduced hours. Engineers had "sent the worker out into the sunshine with money in his pocket to enjoy his leisure." Inventors had released men from exhausting and dangerous jobs, and the next generation's engineers would solve any lingering problems of labor displacement in turn. Although Mehren drastically oversimplified the story of changing factory conditions, his account served its purpose for the ASME, giving members the sense of occupying a central place in history.[45]

Building on the success of their anniversary gala, mechanical engineers found other opportunities to make the argument for historical progress. In 1936, at the ninetieth anniversary of George Westinghouse's birthday, the ASME praised its former president not only for his inventions themselves but also for the way that his innovation had given birth to rich industries. Anyone concerned about recent economic distress, the society moralized, should learn that engineering advances ensured employment growth. That same year also witnessed the centennial of the American patent system, and, as at the 1930 ASME gala, advocates at this celebration used drama to convey a promechanization message. The Science Service presented a show entitled "Research Parade: Demonstrations of Scientific Achievements That May Become the Industries of Tomorrow." It conveyed the message that inventors, scientists, and engineers helped workers of the future by laying the groundwork today for developing television, air conditioning, and improved production machinery. The pageant's slant bore special meaning for the Depression; rather than emphasizing the patent system's legal or technical meaning, the writers chose to claim the patent's payback in jobs.[46]

The engineering community's consideration of the past connected directly to its judgment of the present. Depression-era engineers did not ignore the issue of how technological change related to current social conditions. Questions about the economic and social implications of mechanization absorbed many. Engineers did not regard discussions of technological change as an abstract philosophical matter but as a subject with immediate consequence for how they, as professionals, should think and behave.

In official pronouncements, major engineering organizations denied any connection between Machine Age engineering and Machine Age depression. In January 1931, the American Engineering Council organized a special committee, headed by Ralph Flanders, to analyze recent economic events. After two years of investigation, the group issued a statement, "Balancing of the Forces of Consumption, Production and Distribution," announcing that it had found no evidence of serious technological unemployment. The Depression had resulted from an ordinary recession combined with postwar deflation, farm trouble, and low investment opportunity; mechanization had played no role whatsoever, the AEC committee declared. Experts like Paul Douglas offered convincing arguments that the business system should absorb any workers displaced by machines, since labor-saving efficiencies in production would lower consumer costs and thus stimulate demand. Business owners might not always pass savings on to consumers, the AEC report acknowledged, and men and women might experience trouble shifting to new employment when mechanization terminated their old line of work. "At times the effect of progress does fall with crushing force upon individuals, business firms, entire industries and whole communities." However, the drive for mechanization created a significant number of positions in machine manufacturing, sales, and repair, the engineers concluded, and over the long run everyone should benefit.[47]

Many engineers welcomed the AEC's statement of faith, and some declared that it should be required reading for all political science students. Relatively few voices spoke up in dissent. One engineer criticized the committee for downplaying studies that proved how difficult displaced workers found it to change jobs. Such evidence, he suggested, raised a real possibility of long-term technological unemployment, something that might force a "sad goodbye to that dream of affluence for all." Such disagreements fueled further discussion. Mechanical engineers seemed especially intrigued by the subject, since, after all, their discipline had a special stake in the public's impression of machines. At some points in the Depression decade, the profession's interest in economics even appeared to overshadow its pursuit of technical knowledge. The January 1933 issue of *Mechanical Engineering* devoted six out of ten main articles to discussing how technical innovation affected employment and outlining an engineer's social responsibility. The ASME encouraged members to pay attention to such topics; the 1930 annual meeting featured a major series of addresses on "Engineering, Economics, and the Problem of Social Well-Being." One observer at the society's 1932 conference commented, "Economics and technology predominate." Still more prominently, when the nation's four leading engineering societies met in Chicago in June 1933, they convened two special joint sessions with the Econometric Society. The talks on "econometrics and engineering" were well-attended despite the summer heat, as attendees argued about exactly when and how technical men should get involved with public issues. Meanwhile, some engineers had reportedly also formed local groups to discuss economic matters.[48]

Many ASME members started from a premise that the country's present difficulty stemmed not from any problem with scientific and technical innovation per se but from the failure of community institutions. "We should not worry about the advances in natural science," W. D. Coolidge declared. "Our anxiety should be for the social sciences which are lagging far behind." The cultural lag thesis had an inherent appeal for many engineers, shifting the focus of debate away from mechanization onto social and political life. Rather than worrying about whether technical change had come too fast, they argued, the United States should concentrate on bringing economic science up to speed. Engineers did not take the cultural lag thesis as license to sit back and relax. Many spoke about the economic system as just another device, one that desperately needed repair. One issue of *Mechanical Engineering* ran a photograph of electrical equipment with the caption: "For the control of machinery, engineers have developed sensitive and automatic devices. Similar controls are sorely needed to regulate the fluctuations of economic machinery." In a letter to the editor, one ASME member compared the nation to an engine impaired by a defective flywheel, a faulty governor, and extreme swings in load. "The technological machine 'hunts,'" he wrote. "Adjustments do not take place until they are catastrophic." Who knew better than an engineer how to cope with balky machinery? "We must put on the overalls and service technology right out in the social field," the reader declared.[49]

By drawing an analogy between machine systems and economic systems, engineers made it sound natural and appropriate for them to become involved in social science. "Arising out of contemporary events has come a lively and inquiring interest on the part of engineers in economic problems," *Mechanical Engineering* reported in 1932. The editors commended that trend as a healthy one, and readers wrote to express agreement. In fact, some concluded, the very source of difficulty lay in the fact that economics had been left to economists! ASME members scolded mainstream economists for clinging to an outdated disciplinary framework; how could principles set up centuries ago, by thinkers who had never seen an automobile, possibly apply to the modern age? Engineering rested on a solid consensus of facts, whereas "economics is far from being a true science." Two economists might give completely different answers to the most straightforward question, Ralph Flanders complained, and the whole field was "swamped with vested interests and emotional reactions."[50]

Many engineers believed that they could certainly arrange matters better. Engineers possessed an "intimate knowledge of industry possessed by no other group," Dexter Kimball explained, and a man who seized the opportunity to work on economic problems could become an "important figure in public affairs." Technical classes gave engineers a superior education, ASME members asserted, an all-purpose background in rational analysis. "Engineers have a considerable advantage over other professions in trying to get at the root of things," one letter to *Mechanical Engineering* declared, because they had been "trained to reason carefully." Since even professional economists had proved to be mere "amateurs in the science of economics," a second correspondent suggested, the nation could no longer trust them with the task of restoring prosperity. Americans should look to engineers, men "unfettered by precedent or befogged by close personal interest," men who, "once awakened, are preeminently qualified to apply the acid test of logic to economic problems."[51]

It was "fortunate for society," a third ASME member commented, "that engineers as a class have suffered even more than other professional groups" in the Depression, since that would galvanize them to start considering national issues. "Engineers are, by nature and training, seekers after truth," he declared. "When engineers think, action is sure to follow." Such exuberance provoked one reader of *Mechanical Engineering* to complain about the "childish pretense that we are a superior class." Only false pride, he wrote, could lead engineers to believe they were "better qualified than anyone else to run the world's affairs." If anything, he said, college education left engineers much less able than other men to handle public matters, since technical courses dealt only with inanimate forces rather than with human behavior and institutions. "We can take a proper pride in our contribution to civilization without making the silly claim that we created the whole social structure," like the ridiculous "fly which sat on the wheel hub during the chariot race and exclaimed, 'Great Pluto! what a dust I am raising,'" the reader sarcastically remarked.[52]

Such "rank heresy," as the author himself described his comments, did not fluster those engineers eager to advance their claims of expertise. True, the challenge of analyzing economic activity was more "complex and dependent upon the vagaries of human nature" than the process of solving equations. Any "engineer who aspires to solve modern economic problems must expect to do an unusual amount of studying" to master the necessary groundwork in history, psychology, and sociology, Dexter Kimball cautioned. But, he quickly added, the "bewildering array of theories" only underlined how desperately economists needed engineers' analytic skill. By cooperating with the poor economists, engineers could transform vague speculation into a neat collection of data. Editors of *Mechanical Engineering* praised their readers for having "invaded the field" of economics with "characteristic vigor and confidence."[53]

ASME members hoped that invading economists' territory would help them defend their own interests. Even if incompetent social scientists and policymakers had mainly been at fault in failing to avert depression, the public seemed to hold engineers also responsible. In fact, some engineers judged that the Depression made it a duty for them to monitor the human consequences of mechanization. William Wickenden, president of the Case School of Applied Science, liked the idea of broadening the definition of engineering to include such a responsibility. In the future, "when an engineer or an inventor lays out a machine that will do away with men, he will have to show how he is going to take care of those men—by new jobs, or unemployed insurance, or pensions, or some other human method," Wickenden predicted. "Our engineers will have to have social inventiveness as well as machine inventiveness." Roy Wright concurred in the belief that public criticism, however undeserved, had been helpful in "forcing" engineers to expand their horizons. Present-day social and economic life had become more complicated than ever, he told the ASME in his 1932 presidential address. Technical experts must cultivate an "intelligent interest in helping to control all of the forces" involved, Wright declared. The world awaited a new generation of what he called "militant engineers," men who would "vigorously take the offensive" in leading public affairs.[54]

When phrased in terms of professional responsibility, the idea that engineers should devote more attention to the effects of technical development struck a chord. Popular criticism of mechanization had "surprised and disappointed" engineers, one letter in *Mechanical Engineering* said, but, the writer continued, such resentment might actually be somewhat justified. The public should hold engineers at least in part "morally responsible" when they devised new machines without providing for displaced labor, "as we would be were we to build boilers without safety valves, or steam turbines without governors, or speedy automobiles without brakes." Another reader adopted a different analogy. Just as observers would condemn someone who handed a boy a jackknife without supervising him or teaching him how to whittle without getting cut, Americans naturally blamed engineers for having ignored the issue of how their creations were to be used. Engineers had spent too much time on "sharpening the tools of industry and too little effort in giving instruction in the economic use of the processes of production," the ASME member wrote. "We have given so little thought to ultimate social utility that most of us do not even know how the clever methods and machines we have devised ought to be handled. Yet when they go wrong, engineers are blamed."[55]

C. F. Hirshfeld encapsulated such questions of responsibility in his much-noted lecture at the 1931 AAAS annual meeting, asking, "Whose Fault?" Society had failed to adjust to modernization, but the men behind technological change had been "guilty of a most unpardonable mistake," Hirshfeld said. "Science and appliers of science have brought about a very pretty mess because they were content to do the thing that was comparatively easy." Caught up in the rush to devise newer and more labor-saving machines, researchers had "almost criminally refused to give serious thought to the collateral results," Hirshfeld told the AAAS. "To my mind it is our fault that a civilization capable of producing a surplus" of leisure and material luxury had ended up in economic chaos. Hearing such a full-scale indictment felt uncomfortable, editors of *Mechanical Engineering* wrote, but engineers should listen precisely because it came from one of their own. Hirshfeld was no Communist or woolly-minded academic; he wanted discussion of how technical experts should handle popular perceptions, and such a debate immediately ensued. One member of the ASME protested that engineers bore no fault for any problems stemming from new technology, since the American system "denied [them] the right to control the economic effects of what they have created." When business owners ordered them to install machines, engineers were "forced to obey," just as "Lord Cardigan obeyed at Balaclava." If given free rein, he indicated, the "humanitarian" and "idealist" technical man would insist on reducing work hours and raising wages, to share the advantages of mechanization with labor. An engineer "may be the *deus ex machina*, but he is rarely the devil in it."[56]

Through the 1930s, *Mechanical Engineering* periodically printed articles and letters supporting ideas (mostly vague) for shortening the workweek and otherwise protecting labor. Some ASME members explicitly criticized profit-hungry businessmen for monopolizing an undue share of benefits from mechanization rather than applying them to raise wages and lower consumer prices, two developments that would support reemployment. But while engineers toyed with the idea of responsibility and sometimes advocated labor-friendly measures, they could not accept that unemployment problems might lead anyone to doubt the ultimate value of mechanization—past, present, or future. Just the opposite: once engineers had advised social scientists and politicians on how to operate the nation's economic machinery correctly, Americans would have no excuse to question further innovations in production. In this way, engineers felt they would provide the perfect defense for their profession.

Engineers walked a fine line, admitting that workplace changes posed real challenges to society but denying that such a fact contributed to any serious problems. To maintain that distinction, the professional community employed public relations tactics and rhetoric. By using metaphors of progress, engineers placed themselves at the center of civilization. The modern engineer was "as necessary to prosperity," McClellan declared, as "the medical man who keeps our bodies and minds capable of living." Such comparisons to healers placed engineering in a favorable light. Twentieth-century patients would not dream of abandoning hospitals and turning to magical treatments instead of medical research, engineers argued, and complaints about the latest industrial equipment must be equally preposterous. When a young man asked Charles Kettering about the idea of pacing technical innovation to match America's capacity for absorbing displaced labor, Kettering inquired whether the fellow felt a similar desire to curb doctors' ability to remove an appendix. Americans could not pick and choose, Kettering implied; voicing any reservations about mechanization equated to criticizing all modern knowledge.[57]

In another interesting turn of metaphor, Depression-era engineers (as well as businessmen) often employed the image of machines as substitutes for human slaves. The *Magazine of Wall Street* illustrated an article defending mechanization with a diagram of a nineteenth-century slave ship, explaining that a single steam shovel could perform as much excavation work as an entire cargo of slaves. Though such an analogy might sound crude and offensive to some ears, it served a clear purpose in the technological unemployment debate. By contrasting twentieth-century freedom to earlier barbarities, advocates aimed to reinforce faith in progress. Who could object to mechanization once they realized that it was engineers' commitment to innovation that had reduced society's labor needs and thus allowed the West to abolish slavery? Joseph Roe, a professor of industrial engineering at New York University, stated in a 1930 radio broadcast that modern machinery had supplied the United States with power equivalent to that of twelve billion slaves, one hundred for every person. Instead of fearing production technology, Roe instructed, Americans ought to welcome it as a source of liberty, happiness, and wealth. "To build the Great Pyramids 100,000 slaves worked thirty years, dying like flies. But the Panama Canal, as great a work, was built in a third of the time, by free men, well paid . . . using power and machinery, under conditions as healthy as those of New York City."[58]

Just as executives at General Motors took advantage of Chicago's Century of Progress exposition to pay tribute to corporate technical progress, so the nation's four main engineering societies organized their 1933 meetings around the same fair. By connecting their conferences to the celebration, leaders hoped to send a public message about the value of mechanization. They emphasized that all the wonderful products on exhibit in Chicago proved that the pursuit of new production technology had left Americans better-off, not poorer. Like the physicists, engineers hosted special public relations events. In May 1938, the American Engineering Council organized a day-long symposium in Philadelphia titled "Employment and the Engineer's Relation to It." Frederick Allner, chairman of the AEC's Public Affairs Committee and vice president of the Pennsylvania Water and Power Company, explained that the meeting was designed "to enlighten public opinion" by generating favorable images of technology. By virtue of their mathematical training and detached judgement, he said, engineers could provide a voice of sanity amidst popular confusion. He believed that the profession could still command respect, since "no engineer, either in literature, on the stage, or on the movie screen, has ever been cast "as a villain." American culture, he declared, tended to portray engineers as "poor but honest."[59]

The AEC conference emphasized that unemployment ran highest in industrial sectors that had failed to take maximum advantage of scientific and technical innovation; those businesses that had adopted the latest equipment kept people at work. The facts seemed plain enough to engineers; the problem became one of convincing ordinary Americans. Leonard Fletcher, a member of the AEC Executive Committee, complained that virtually every newspaper he picked up featured articles linking mechanization to job loss, perpetuating "economic and sociological untruths." He called on fellow engineers to fight back by describing "the constructive side of technological matters" in language "understandable to . . . the majority voter." The AEC should stress that in making production more efficient, mechanization opened the way for reducing work hours, Fletcher recommended. Anyone who valued leisure time should appreciate that the true definition of technological unemployment was "the unemployment of a Saturday afternoon." The nature of modern engineering handicapped attempts to highlight the advantages of mechanization, some suggested. The extreme specialization of engineers and their use of technical jargon discouraged communication with nonprofessionals. To overcome such obstacles, the head of the American Society of Agricultural Engineers suggested that the AEC "use its influence" to get big advertisers to provide sympathetic publicity. "If the matter were properly presented" to them, automobile companies might be willing to run advertisements saying "This car . . . has 25 percent more labor put in it than the one we sold 25 years ago." Such messages could reach millions of Americans, he advised, adding cynically, "We all know that constant repetition is accepted by the average person as truth."[60]

In addition to trying to swamp people with favorable images of mechanization, engineers set out to denounce any individual or group, such as the Technocrats, who expressed dismay over joblessness. Mainstream engineers and Technocracy agreed on some points, such as the inadequacy of established economics. Howard Scott predicted, however, that recent trends reducing the manpower per unit of production would lead to unprecedented technological unemployment. Although he emphasized that the United States could not solve the problem by halting or even slowing down the pursuit of innovation, Scott's words sounded dangerous to many engineers. Mobilizing against Technocracy in January 1933, the AEC passed resolutions to condemn such "exaggerated, intolerant, and extravagant claims" by troublemakers who had "capitalized [on] the fears, miseries, and uncertainties due to the depression." Though Scott brandished charts and statistics to support his argument, the AEC insisted that there was "nothing inherent in technical improvement which entails economic and social maladjustment. Indeed, technology offers the only possible basis for continuing material progress."[61]

After declaring that technological advances had created all of past civilization and present-day American wealth, engineers proceeded to promise that mechanization would likewise guarantee future progress. Just as physicists feared talk about a science holiday, so engineers worried that too much discussion of displacement might create a backlash against development of technology. In truth, virtually no mainstream politician or labor leader endorsed any idea of completely banning innovation, and public opinion had not turned against engineering as an institution. Nevertheless, the prospect horrified engineers. Like Karl Compton, leading engineers argued that future employment depended on a present-day investment in research. Just as machinery-makers urged businessmen to take advantage of lower prices to snap up the latest equipment, engineers suggested that the Depression represented a perfect time for companies to "acquire brains at bargain rates." The nation had a surplus of technical experts available "at ridiculously low figures," and it could not afford to let them sit idle. Companies must hire researchers to expand product lines or even develop entire new industries, "supplying work for thousands" and "increasing the wealth . . . and well-being of civilization."[62]

Charles Kettering, among others, contended that today's commitment to industrial engineering would assure work for tomorrow. Rewards could come from unexpected quarters, he observed; the most optimistic men of the twentieth century's first decade had not foreseen how the spread of automobile ownership would stimulate an enormous business in accessories and travel services. Similarly, experts of the 1930s could not determine exactly which avenues of development would yield the greatest return; like farmers, engineers could only "plant the seed and wait." Unlike Karl Compton, who promised that government funding would soon allow science to create jobs, Kettering warned that the payoff might be some time in arriving. Americans should not grow impatient, he continued. "If we tried to raise human children on the basis of the highly organized bookkeeping system which we apply to industrial children, a baby nine months old would have to earn its own living." Over the long haul, if the United States invested in a sufficient number of projects, "enough good will come out of it to make it worthwhile." If government could keep five or six hundred research laboratories running full time, he promised, "We would have 'help wanted' on every door of every factory in America."[63]

During the Depression, engineers strove to position themselves as agents of future well-being, creators of employment and wealth. The golden jubilee medal coined for the ASME's fiftieth anniversary celebration bore an idealized picture of an engineer, who the society described as "a thoughtful man of powerful physique, gazing intently . . . into the future," the blueprint for which lay "already within his grasp." By portraying the technical expert as a paragon of progress, the profession made labor's protests sound immaterial. How dare anyone distract engineers with complaints about technological unemployment, when they were busy "visualizing a world raised to higher levels," all "in the interests of humanity"? The extension of mechanization would render future generations of Americans increasingly dependent on engineering and move more engineers into influential public roles. True, recent "technological progress has created great problems," Elliott Dunlap Smith bluntly told the ASME in 1930. The spread of automatic looms had decimated textile employment, turning plants into a "sparsely inhabited wilderness of self-controlled machinery." Such cases, he declared, underlined the lesson of consequences; in bringing ideas from the drawing board to the factory floor, an "engineer not merely contrives a machine, but uproots the work and lives of men." Smith nevertheless situated engineers as the potential heroes rather than the villains of his tale, moralizing that the misery of textile workers only proved how desperately the country needed engineers' help. Those men, seemingly out to destroy hope, were really those best able to restore it. Smith promised that economic well-being would return once engineers took the reins of government and corporate management, applying their problem-solving aptitude "to the wholesome absorption as well as to the efficient construction of the machine."[64]

Roosevelt versus the Engineers: Controversy over Machine Age Education

Engineers' comments suggested that, in addition to engineering mechanization itself, technical experts could and should engineer society to fit a future of rapid changes in the workplace. Critics doubted, however, whether current forms of engineering training prepared students to address the human dimensions of innovation. Henry Wallace complained that too many colleges crammed technical courses into the curriculum at the expense of a liberal education, locking engineering majors into a state of "complete isolation from the economic and social world about them." In the interest of helping young professionals realize how their work affected everyday life, Wallace told the AAAS, "no great harm would be done if a certain amount of technical efficiency in engineering were traded for a somewhat broader base in general culture." Science writer Waldemar Kaempffert similarly urged engineering schools to teach the tough but momentous issue of how mechanization affected employment. The pursuit of innovation must become "correlated in the student mind with the social outcome of technical discoveries," he told a Carnegie Institute of Technology audience.[65]

As Kaempffert, Wallace, and others called for educators to steer would-be engineers into a new consciousness of public responsibility, two Ohio State University economists had published a text designed to accomplish exactly that. In *Economics for Engineers*, Edison Bowers and R. Henry Rowntree stressed that students pursuing technical careers must cultivate a working knowledge of social economics, with "special interest" in employment trends. In describing the way that machinery troubled musicians, coal miners, farm hands, and office workers, Bowers and Rowntree instructed readers to take talk of displacement seriously. If nothing else, they wrote, labor's sense of uncertainty could make an engineer's task more difficult. Employees who feared mechanization and distrusted a company's motives for introducing it might prove "reluctant to cooperate with management engineers in increasing efficiency."[66]

Technology-related job loss raised ethical problems, Bowers and Rowntree suggested. They presented the hypothetical case of a farm implement company that had increased its profits by two hundred thousand dollars per year by installing new production machinery, in the process displacing five hundred out of its one thousand workers. *Economics for Engineers* encouraged readers to consider whether the firm ought to be held responsible for the welfare of those thrown out of work. "Are engineers partly responsible for technological unemployment?" the text asked. "Should we not have a ten-year moratorium of invention and machine development to allow employment to catch up with science?" After all, even temporary displacement threatened to undermine the nation's economic soundness, wasting valuable human resources and contributing to a lopsided distribution of wealth. In the end, Bowers and Rowntree wrote, since the engineer was primarily responsible for developing new machines, he was "therefore responsible indirectly at least for the gains and losses accruing."[67]

Instructors adopted Bowers and Rowntree's book for use in numerous university classes during the early 1930s, but the arguments over whether and what engineers should learn about economics continued to rage. In 1934, MIT instituted a new five-year Bachelor and Master of Science degree program, which Compton described as a combination of technical and liberal arts courses. The "new adventure in technological education" would help students master both the technical and the social sides of engineering, Compton promised. Its graduates could show the world how to cope with mechanization, inaugurating a new "cooperative approach to economic problems by the engineer and the economist, similar to the fruitful cooperation of the engineer with the physicist."[68]

To skeptics, educators still had not moved quickly enough to teach students about technological unemployment, and in the fall of 1936, President Roosevelt thrust the question of engineering education into newspaper headlines. In a letter sent to more than one hundred college presidents, accompanying a pamphlet on soil and water conservation, Roosevelt indicated that recent environmental and economic problems, including the persistently high rate of joblessness, suggested that engineers must learn to "cooperate in designing accommodating mechanisms to absorb the shocks of the impact of science." The president called on engineers and university leaders to discuss ideas for reforming the technical curriculum to give students a more evenly "balanced" background. Young engineers must acquire "vision and flexible technical capacity," Roosevelt wrote, so that they might "meet the full range of engineering responsibility" in an age of mechanization.[69]

Roosevelt's advocacy apparently stemmed, at least in part, from the initiative of Rural Electrification Administration official Morris L. Cooke, who had long campaigned for reforming professional education. The widely distributed

"Little Waters" letter attracted significant attention, especially after some college presidents and leading engineers responded in a decidedly cool fashion. Nonetheless, all endorsed the general principle that engineers ought to pay more attention to modern economic and social problems. Robert Doherty, president of the Carnegie Institute of Technology, called it indefensible for engineers to "disregard, as they have tended to do in the past, the social consequences of their work, leaving these for others to worry about." Representing the AEC, Purdue's engineering dean Andrey Potter reassured Roosevelt that professors felt "fully appreciative of the responsibility of the engineer in bringing about a better balance between technological progress and social control."[70]

Though educators might agree that social issues had a place in engineering life, many perceived comments from outsiders, even the president, as unwarranted interference. In criticizing engineering curricula, Compton protested, Roosevelt failed to acknowledge how much the discipline had already expanded to incorporate "a notable increase in attention to the study of economics and social science." Lehigh University president C. C. Williams concurred, declaring that over the past decade, professional engineering societies had been hard at work on course reform. H. H. Rogers of the Polytechnic Institute of Brooklyn worried that schools could not make much more space for nontechnical courses without subtracting from the time devoted to fundamental requirements. In any case, he added, nobody knew precisely what to tell students about economic matters anyway, given the confused state of the social sciences.[71]

The heat generated by Roosevelt's letter did not stem just from the question of educational reform per se but from the president's continued references to technological unemployment as a serious problem. Rushing to defend their profession from the perceived criticism, engineering professionals repeated the argument that even if changes in production methods caused occasional temporary displacement, new technologies added jobs and improved living standards over the long run. Karl Compton of MIT and Potter Adams, president of Norwich University, both publicly chided Roosevelt for feeding Americans' baseless fears about mechanization. Adams scolded the president for displaying a distressing "lack of faith" in the safety of Machine Age progress.[72]

Depression-era alarm over labor displacement created a dilemma for the Taylor Society, the organization devoted to promoting and elaborating on the theory and techniques laid down by Frederick Winslow Taylor. Some unions extended blame for job losses to the philosophy of scientific management, accusing its proponents of promoting an "unbalanced efficiency." In the single-minded quest to multiply per capita output, the AFL declared, industrial engineers had ignored the economic importance of maintaining stability in both employment and consumption. "Ability to improve methods of production without ability to stabilize it," one labor publication editorialized, "is like building a powerful machine without regulator, brakes or indicator gauages [sic] and turning it loose on the highway." The device might have a tremendous power to shoot ahead, but would "certainly wreck itself and do appalling injury" as a result.[73]

The accusation that scientific management had helped drive up unemployment instead of creating an economic utopia caused consternation within the Taylor Society. Members began asking whether, in espousing the drive for efficiency, they had unwittingly embraced a socially and economically short-sighted ideal. At the group's annual meeting in December 1929, Leo Wolman warned, "We unfortunately know too little about the natural history of invention and of technological changes." The prospect of achieving further efficiency gains and cost savings in automobile manufacture seemed slim, a fact that could limit the industry's future growth and its capacity to absorb workers displaced from other avenues of employment. Although optimists liked to believe that the ongoing course of economic development would provide an automatic safety cushion, rapid mechanization might at some point exert a price too great to be overlooked. Humans simply did not possess enough intelligence "to control this enormous flood of power that we have loosed upon the world," Stuart Chase told the group. "Machines seem to be displacing men faster than jobs can be found." Scientific management must start to "pay particular attention to the question of technological unemployment," factoring labor issues into every analysis of efficiency and productivity.[74]

At the Taylor Society's 1930 conference, discussion of a possible link between scientific management and technological unemployment grew intense. Paul Douglas maintained that the aim of raising productive efficiency remained a sound means to ensure a long-term expansion of jobs. As an illustration, he proposed the hypothetical case of a publisher who used scientific management to double efficiency and then cut journal prices in half. If the 50 percent savings induced customers to purchase twice as many copies, the company could retain its entire labor force; if demand were not perfectly elastic, some workers might lose jobs. In such an event, buyers could apply the money they saved on the journal to other purchases, theoretically stimulating employment elsewhere. In practice, such ideal economic relationships might have broken down over the past decade, Douglas admitted. Per capita productivity had grown 45 percent over the 1920s, but, since managers had not passed on the savings to consumers, demand and employment both stagnated. As Americans lost confidence in the job market, Douglas warned, fear of displacement could pose a major obstacle to further efforts at rationalization or even provoke a back-lash against industrial engineering.[75]

In a pessimistic follow-up, Harvard economist Sumner Slichter stated that displacement had been rising for the last eight or nine years and that the labor market could not compensate. Once a company had replaced experienced workers with machines, the job loss could never be recovered; even if managers slashed prices and buyers responded, a mechanized firm would never rehire the old hand labor. To make matters worse, economic benefits of technological change inevitably lagged behind the cost. Firms that introduced machinery tended to eliminate redundant workers immediately, but any bounce in consumer demand and resulting reemployment would take longer to show up. Such delays imposed substantial hardship on American families, Slichter concluded, and scientific management experts must refocus their attention on ways of relieving such pain. Industrial engineers had refined methods of moving material through a factory, but the country had "just as much need for careful planning to adjust . . . labor to the ever-changing needs of industry." Slichter proposed organizing Taylorites, businessmen, union leaders, and government representatives into a Federal Labor Board, which would investigate how workplace innovations might affect jobs and recommend steps to minimize harm.[76]

During the ensuing discussion, Henry Kendall, Taylor Society president, agreed that industrial mechanization had accelerated displacement. In referring to technological unemployment as a "tremendously serious" problem, Kendall effectively endorsed the concept that the society bore an obligation to address such matters. Prosperity had made it easy to portray efficiency as an inherent good, to hold up the dream of ever-higher per capita productivity. The Depression threatened to change Taylorism from virtue to vice, but Slichter offered an escape clause. Industrial engineers did not have to retire in disgrace; the country needed them more than ever, to conquer the unemployment problem they had accidentally helped cause. Harlow Person, the society's managing director, declared that the crisis posed a "new challenge to scientific management." Practitioners had formerly studied individual companies and subroutines of work, but they should now concentrate on the "organization and control of industrial society conceived as an organic whole."[77]

Person's talk explicitly put forth the notion that "operations of industrial society are not yielding the greatest possible good to the greatest number of industrial citizens." By considering the evidence that decades of efficiency gains had not added up to an economic and social miracle, the Taylor Society hinted at flaws in its founders' principles. During the Depression, arguments that would have amounted to heresy a decade before suddenly became thinkable.

At the Society's 1931 meeting, Commerce Department statistician Robert McFall criticized the rising use of labor-saving devices and efficiency schemes, which might seem "essential to the well-being of individual industries," but "may be subversive of the general economic well-being." An ill-considered fad for mechanization had aggravated unemployment, he declared flatly, and scientific management experts should be doing their "utmost to discourage" inappropriate extensions of technology during an economic downturn. "Too much of our thought has been spent on the mere saving of labor," McFall maintained; genuine wisdom came with the skill to identify "where labor saving is not a true saving."[78]

McFall effectively redefined the goal of scientific management to mean achieving a healthy "balance between labor saving and labor utilization." Such statements show how much the popular concern about displacement influenced the Taylor Society, which nonetheless did not relinquish its claims to expertise. To the contrary, concern about technological unemployment led leaders to advocate an even more central role for scientific management in promoting a socially enlightened view of efficiency. Addressing a radio audience in the early 1930s, Person conceded that old-style scientific management had been an incomplete and hence a cruel vision, ignoring the importance of employment. Industrial engineers had encouraged businessmen to abuse their power, "applying mechanization in such a manner as to deny a share in the results to an ever increasing part of the population." Taylorites stood ready to confess their sins and reform their ways, Person announced, to help society establish greater job security and a broad-based economic welfare.[79]

Person and others in the scientific management field took a radical step by acknowledging problems of technological unemployment, yet at the same time they asserted their claims to professional importance more strongly than ever. Especially among industrial and mechanical engineers, popular alarm over labor displacement spurred episodes of soul-searching. Even as members of those professional communities debated amongst themselves about their responsibility for the social and economic implications of mechanization, their organizations began circling the wagons, preparing to defend their interests against assault from outsiders. The ASME, AEC, AIP, AAAS, and the Taylor Society insisted that present-day turmoil only proved how urgently the country needed to rely on technical expertise. Even as workers revolted against mechanization, those men who had developed and promoted new production methods coaxed Americans to trust them. In a desire to paint themselves as national saviors, scientists and engineers paid special attention to public relations. When pure research seemed too esoteric to be relevant, physicists redefined their image to emphasize the valuable results of applied research. Engineers built technology into the central feature of civilization, constructing a usable past from which they might draw reassuring analogies to the present. Popular science and technology magazines bolstered such a view, echoing the line that talk of displacement had been misrepresented. One article in *Popular Science* used the eye-catching title, "Will You Lose Your Job Because of a Machine?" and answered, "no!" Ongoing research would perpetuate an "endless chain" of new jobs and consumer goods, the journal assured readers, so that instead of worrying, Americans ought to "pat the machine on its shining back."[80]

As part of the mission to shore up the equation between mechanization and national well-being, technical and scientific groups effectively formed a partnership with the corporate community. While a few scientists and engineers might occasionally blame managers for failing to use technology well, their societies generally cooperated with big business to advance a common argument. Prominent executives attended the scientists' and engineers' celebrations of technical progress, and vice versa, projecting a mutually sympathetic vision of national destiny. Together, those groups interpreted American history as a march toward progress, promising that the force of innovation, fostered in company research facilities, would open new horizons of material abundance and employment opportunity. Corporate advertising incorporated images of men in white lab coats as a visual shorthand for that promise, while Karl Compton campaigned for bringing industry into education.[81]

Such an alliance had not been inevitable. Earlier in the twentieth century, engineers had engaged in internal debate over where their allegiance should lie. Their desire for professional independence competed with the lure of corporate money. By the 1920s, the progressive reform impulse had started to die out, shifting the engineering community toward a conservative orientation.[82] The Depression offered a chance to reopen that discussion, posing questions about whose interests science should serve and about the proper direction of engineering. In the end, group dynamics pushed scientific and engineering organizations to reassert the faith that workplace innovation did not pose any real danger. Scientists and engineers drew rhetorical and symbolic links between technological change and the notion of American progress in the past, present, and future. Talk of technological unemployment turned what had started as a partnership of convenience between science, engineering, and the corporate world into an alliance of mutual defense and a powerful joint force of public relations.

"What Will the Smug Machine Age Do?"

Envisioning Past, Present, and Future as America Moves from Depression to War

WITHIN A DECADE of uncertainty, the idea of mechanization became a lens through which people observed and interpreted the past, present, and future. For all the twentieth century's vaunted advances in knowledge and wealth, too many citizens faced "enforced idleness and cold and hunger," Silas Bent wrote. "What will the smug Machine Age do about it?"[1]

Such pointed questions compelled attention, especially since forecasters predicted that inventors would continue extending the force of mechanization into new areas of industry, agriculture, and the services. The spread of photo-electric eyes, business equipment, and other devices promised to create an entire chain of consequences, reshaping work and redefining its meaning in modern life. Commentators frequently expressed a conviction that as production technology allowed employers to keep turning out more goods with less labor, the United States must ultimately slash working hours to avoid mass unemployment. But opinions were divided over the impact of such an unprecedented leisure world; optimists anticipated a utopian era which freed men and women to pursue self-enlightenment and community improvement, while pessimists worried that the death of an old-fashioned work ethic would only foster laziness and delinquency. The sheer acceleration of technical innovation carried breathtaking implications, leaving plenty of room for philosophical speculation about whether and how human beings could cope with such revolutionary potential.

In envisioning the effects of workplace mechanization, Americans similarly had to think about how expanded production power might transform the market and consumerism. Through the 1930s, corporations, scientists, and engineers countered talk of displacement by launching public relations campaigns that celebrated mechanization as a guarantee of wealth. In an alluring promise of abundance, manufacturers put the latest automobile models, washing machines, and even experimental television sets on display at World's Fairs for a Depression-weary audience. Emphasizing that ordinary men and women could share the wonders of scientific genius by using exciting new dial telephones, AT&T effectively diverted attention from the fate of switchboard operators. Exhibits defined consumerism as the primary standard of American well-being, measuring personal happiness and national civilization by material possessions. Business interests constructed a clear message that those who fretted over job loss simply failed to comprehend the fabulous potential of corporate-sponsored technical ingenuity.

The symbolic power of the great New York World's Fair proved critical in promoting the equation of mechanization and progress. By 1939 and 1940, the debate had been raging for roughly a decade. Even as hundreds of thousands of visitors thronged to fair exhibits, other Americans traveled to Washington, D.C., to participate in the Temporary National Economic Committee (TNEC) hearings on technological unemployment. Statisticians with the WPA and other government agencies testified that their research showed the persistence of disturbing economic trends; a parade of labor representatives told Congress how their occupations had been hit by workplace change. Countering such evidence, executives from the nation's largest corporations offered their own graphs and accounts to demonstrate that critics had greatly exaggerated any such incidents. Business leaders and the other promechanization forces eventually lost the argument on Capitol Hill. The TNEC investigation concluded that Americans had good reason to feel concerned about the social ramifications of change, warning that future increases in mechanization seemed likely to intensify problems of displacement. But the real battle did not take place in Washington; the business, scientific, and engineering communities had deployed their best weapons at the World's Fair in New York. Moreover, by 1940, international events had begun to alter Americans' perceptions and priorities. The tide of world conflict would soon rise to sweep aside the subject of technological unemployment.

Historians' Interpretations of Mechanization as an American Trait

Throughout the 1930s, scientists, engineers, and businessmen propounded versions of history that defined gains in scientific awareness and technical capacity as the central component of Western progress. During those same years, mainstream historians also grappled with various ways of interpreting the record of mechanization. Historians' analysis of the past was informed by the present-day debate over labor displacement, and the public fear of job loss appeared as a subtext through their writings. One of the assessments most sympathetic to mechanization came from Charles Beard, who regarded "progress" as a faith in power, especially in the power of technology to improve man's place on earth. He considered the benefits of science and engineering self-evident, arguing that the spread of knowledge had fostered democratic principles, raised the standards of health and longevity, and kept people supplied with both the necessities and luxuries of life. Beard defined civilization by its embrace of technology and science, asserting that it was the trust "that mankind is advancing" which set the Western world off from the "fatalism of the Orient and the other-worldliness of the Christian Middle Ages." In Beard's mind, such belief in scientific and inventive triumph was more characteristic of the United States than Europe, assuring its ultimate economic, intellectual, and cultural preeminence. His model made America the natural home of progress; unnecessary paranoia about mechanization risked destroying that achievement and sending modern culture backward to subsistence agriculture and primitive medicine.[2]

Roger Burlingame seconded the view that the modern world's "incessant movement" toward further mechanization represented "a natural, not a perverse" trend. His 1940 work *Engines of Democracy* linked all economic and social growth in the United States to technological triumph, reading the country's entire history as "fundamentally a history of invention." After outlining the way nineteenth-century Americans had derived great benefits from the spread of steam engines and railroads, Burlingame turned his attention to more recent developments. He praised the dial telephone as a technical masterpiece and accepted AT&T arguments that switchboard operators had not really suffered. New phone equipment contributed far more economic good than harm, since over the long run it "multiplied employment in a hundred spheres by increasing speed of production and the tempo of business." Labor itself bore the blame for any job loss, Burlingame wrote, since unions' push for higher wages and shorter hours only gave business owners incentive to reduce their need for workers. Mechanization did "nothing but what man intended," helping "to take severe burdens off human shoulders . . . and give the spirit a chance."[3]

Following William Ogburn, Burlingame declared that the source of trouble lay in humans' failure to adapt to mechanization. The United States must "reduce that lag of society behind technology which is responsible for most of our disease today. That men released by mechanisms should find themselves without occupation for their newly freed spirits shows a profound failure in social invention." After Americans had learned to redirect their surplus time and energy into education, art, and city planning, they would see it was "well that machines have been made to perform these lesser functions so as to release men's minds for higher ones." He recommended that rather than waste time bemoaning the loss of jobs to cotton-picking machines, African Americans in the South should seize the opportunity to form choirs, touring the country to perform concerts of spirituals. Bethlehem, Pennsylvania, residents displaced by continuous-strip steel production could similarly earn a living through the musical skill exhibited in their famous Bach Festival. If only steel "puddling could be done by machines and the singing made a full-time job," the nation would "have no less steel and far more delight to our senses." Burlingame recast technological displacement as a cultural asset. For him, agricultural and steel laborers were dispensable; the only irreplaceable workers in modern life were engineers. Once Americans could come to "accept the belief that we need more rather than less machines," he forecast, "we shall need more engineers" than ever.[4]

Harold Rugg provided a drastically different diagnosis of the past, present, and future relationship between technology and society. In his judgment, the original Industrial Revolution represented a period of unusual expansion, when workers could move into new avenues of opportunity as mechanization closed off old ones. But starting around 1900, Rugg declared, the world had entered a second Industrial Revolution, a time of economic limitations. Once the arrival of larger and more automatic machines had accelerated the "permanent ousting of labor from industry," Americans had good reason to fear for their well-being. Rugg predicted that, while people during the first Industrial Revolution had concentrated on perfecting new devices, the second Industrial Revolution would force them to think about better social management of change. Only by learning to "apply the scientific method to Man-Man relationships as well as to the Man-Thing relationships," he concluded, could humans achieve a more stable distribution of wealth, work, and leisure.[5]

Lewis Mumford similarly considered that the accumulating effects of change should and would lead people to regard technology in a new light. In the past, he wrote, the "gains that the machine has brought have rarely been balanced up against the losses." The Depression had finally driven Americans such as Stuart Chase to start calculating that balance, revealing just how "uncertain such an estimate must be, once one drops the comfortable Victorian notion that all change is progress and all progress is beneficial." True, modern machine power had raised civilization beyond dependence on human energy, defying Aristotle's prediction that society's need for labor would make slavery a permanent phenomenon. In practice, however, Mumford complained, the rise of mechanization had merely converted people from slaves of production into slaves to consumerism. The advertising industry pushed "wanton multiplication of fake wants" just to keep the market moving, while workplace changes left communities "burdened with chronic unemployment, a curse and not a benefit."[6]

In his 1934 work *Technics and Civilization*, Mumford argued that the Industrial Revolution, the "paleotechnic" stage of historic development, had devastated labor, the environment, and economic stability. Nevertheless, the very prevalence of technological change made people into passionate believers in progress; questioning the advantages of mechanization amounted to unthinkable heresy. "What paleotect dared ask himself whether labor-saving . . . thing-producing devices were in fact producing . . . enrichment of life?" Mechanization allowed industry to turn out more and more goods, but simultaneously eliminated the value of human beings as participants in production, fostering a culture that rewarded profit-making at the expense of workers. It appalled Mumford to see that even after 1929, when the displacement effects of dial telephones, linotypes, and other machines became seemingly obvious, some observers persisted in believing "that maladjustment to the machine can be solved by . . . introducing greater quantities of machinery." He ridiculed the Hoover administration's *Recent Economic Changes* report as an example of such "classic fatuousness." Mumford judged it the height of idiocy to suggest that Americans must learn to "conform our living and thinking to the antiquated ideological system" that had distorted priorities in the first place.[7]

Mumford anticipated that the course of history itself would correct such blind worship of mechanization. For him, the Depression represented the painful evolution of civilization from its destructive "paleotechnic" form into a more mature, balanced "neotechnic" state. As part of such a transformation, workers would no longer sit passively through radical changes imposed on their lives. Rather. they would seize the power to evaluate developments for themselves, rationally deciding whether to accept or reject various forms of technology. By taking control over the direction of mechanization, people would make the "social unemployment of machines" in a future neotechnic civilization "as marked as the present technological unemployment of men." Machines would lie unused if necessary, once society had come to value the laborer as "quite as important as the commodity he produces." Neotechnic countries would end the "dissociation between capitalism and technics" by reinventing the economy along more well-planned, cooperative principles; through more equitable distribution of the gains of mechanization, all citizens would enjoy improved living standards while working less than twenty hours a week. That efficient yet humane world would maximize both human and engineering potential, conquering the problem of technological unemployment for good.[8]

Writers of the time superimposed their ideas about present-day mechanization onto interpretations of the past, then extrapolated that analysis to prescribe proper terms for America's social and economic future. Rugg and Mumford agreed that the selfishness and short-sightedness behind the Industrial Revolution had devalued workers, creating a serious problem of labor displacement which society could only correct by reordering its priorities. To Burlingame, Beard, and Dugald Jackson, the Industrial Revolution represented a central step in the upward path of Western civilization, inevitably due to peak in the United States. A writer's assessment of contemporary conditions both dictated and depended on his broader philosophy about how civilization evolved. The Depression-era controversy over machines' impact on labor created a unique climate within which historians produced a special set of literature, one characteristically focused on technology in the past, present, and future.

In reviewing the history of the Machine Age, Americans continued to reevaluate the meaning of their immediate social and economic position. The sudden shock of collapse from a culture of prosperity to one of poverty suggested that the United States had arrived at a momentous turning point. Racing at full speed into the twentieth century, the nation had crashed head-on into urgent questions about labor displacement. Joblessness signaled a shift in eras, the start of an age when, more than ever before, mechanization represented the single most important factor defining the foundation of national life. Workplace technology had seemingly become the determining variable that would shape employment conditions, public debate, and financial relationships for generations to come. Though observers might agree on the historic significance of mechanization, they were often split over its human implications. Public discussion revealed both an uncertainty about what a future Machine Age would look like and a deeper ambiguity about whether change would come for better or worse. "We are living in one of the most interesting periods" in history, Ralph Borsodi commented in 1933. "Industrial civilization is either on the verge of collapse or of rebirth on a new social basis."[9]

Curse or blessing—regardless, many Americans feared that they had no power to influence events. Machinery had already been set in motion, moving the country into an era defined by ever-more-rapid technological change. The effects of workplace revolution seemed to pile up and create a momentum for further innovation. Engineers and businessmen promoted this sense of a predetermined fate, arguing that no matter how much people worried about displacement, the United States could not turn away from embracing new technologies. "The Machine, and the Age which it has moulded, is as inescap-

able as it is unpreventable," Ralph Flanders declared. "Like other manifesta-
tions of natural law [it] can be neither ignored nor combatted." Such statements fit well with sociologists' cultural lag theory; given the assumption that
technological change must be permitted to develop along a natural path,
humans must simply figure out how to adjust. Flanders continued, "Our only
safety lies in adaptation. The rapidity of adaptation required is to some exhilarating, to others painful."[10]

Such a vision left Americans no escape route from the technological revolution. Lucky individuals might adjust quickly and benefit from change, but
men and women in less advantageous situations could lose their jobs. This
analysis raised the question, how long would the most stressful economic conditions persist? Henry Ford insisted that the transition to "the machine age
is barely started now"; Americans had only begun to experience the new reality. However, he promised, as workplace innovation advanced further, it would
clear up the present "noise and confusion," setting the stage for a glorious
future. Dexter Kimball, on the other hand, reassured Americans that the traumatic pace of upheaval would not continue indefinitely. Mechanization would
proceed far more slowly over the next thirty years than in the previous three
decades, he predicted, thanks to the law of diminishing returns. Employers
would discover inherent limits to the payoff from mechanization, and some
might even opt to discard equipment in the hope of gaining flexibility. Such
trade-offs would halt the race to mechanize, Kimball said, restoring a natural
balance in the labor economy and keeping the problem of displacement down
to manageable levels.[11]

The principle of diminishing returns had already started to take effect,
noted Sumner Ely, an engineer from the Carnegie Institute of Technology.
Business managers had implemented the simplest technological changes first,
and further innovation would arrive more gradually. Workers had therefore
just about survived the worst phase of disruption, Ely concluded, although
mechanization would continue shrinking manpower needs down to "an irre-
ducible minimum." Industrial designers would design more and more factories to run with virtually no men on the production floor, but the laboring
class "will adjust itself as time goes on." People would settle into new employment opportunities, since even the most mechanized processes would still
demand human engineering and supervision. "When we have once become
thoroughly mechanized," Ely rationalized, workers at least could "no longer be
affected by the machine."[12]

Though such words might not prove reassuring, there seemed little anyone
should or could do to defy the trend toward mechanization. Most unions did
not desire to wage such an enormous battle and, even in the midst of fear about

displacement, did not usually destroy equipment or rail against science and engineering. Despite the nightmares of Karl Compton, Robert Millikan, and their fellows, few advocated placing a moratorium on research and invention. The idea that the United States would follow an inevitable path of technological development, to which human behavior and cultural institutions must adapt, proved a powerful assumption. The 1933 Chicago Century of Progress exposition captured that sense of mechanization as a predetermined force in its famous slogan, "Science Finds—Industry Applies—Man Conforms."

Mechanization and Values: The Leisure Age versus Work and Consumerism

If men and women had no choice but to conform to a future determined by increasing technological change, what would be the shape of that emerging world? Through the 1930s, educators and sociologists debated how Americans seeking employment should approach a rapidly evolving workplace. C. A. Prosser, a Minneapolis industrial educator, warned that because of mechanization, every person in any line of work would encounter at least temporary unemployment more often. In modern times, he wrote, "the job which the typical worker holds is likely to be either completely abolished by technological change or to be so greatly modified . . . as to become virtually a new job." Rather than teaching students specific skills targeted to particular occupations, Prosser suggested, vocational counselors ought to give youngsters the practical advice and psychological reassurance they would need to face a constantly changing job market. Business could help prepare workers both physically and psychologically for the possibility of displacement by rotating them between a number of different tasks, management expert John Garvey suggested. Employees accustomed to switching work would gain self-confidence and begin "welcoming changes as happy experiences rather than dangers to avoid." Americans should learn to appreciate the healthiness of continual change, Garvey concluded, comparing the shift between jobs to the "movement of your blood through your own body. . . . When your blood becomes sluggish, your actions become slow," he cautioned, and no one in the rapid Machine Age could afford to be slow.[13]

As mechanization proceeded along its inevitable course, educators agreed, the sheer pace of change placed the burden on workers themselves to keep up. Flexibility became the key to survival in the face of mechanization; anyone who could not adapt would invariably be left behind. Young people must avoid putting all their eggs in a single basket, advisors indicated, and boys certainly should not count on following their fathers in the family's traditional work. A person could no longer even be sure that a particular line of employment would last throughout their lifetime, R. B. Cunliffe of Rutgers pointed out, since "with the coming of the machine and power age, an occupation may be born, reach maturity and die within a decade or two." Education could no longer stop in the teenage years; as the expansion of mechanization continued to alter work processes, adults would need ongoing training and retraining to brace for the perpetual threat of job loss. Such intensive programs could prove expensive, raising the issue of whether the federal government, states, localities, unions, business, or workers themselves would assume primary responsibility for financing and facilitating classes. And yet, educators argued, the country had no choice but to ensure that its human resources could keep step with its technical power. Spencer Miller, secretary of the Workers' Education Bureau, summed up the problem as one of "re-education to the machine age."[14]

If adults would devote more time to continuing education in the future, many Americans rationalized, they might also come to spend less time on the job itself. The invention of increasingly powerful equipment would change the labor-intensive nature of production, they reasoned, thus altering the nature of work itself. Ultimately, workplace technology would redefine the social, economic, and psychological importance of employment. The most dramatic impact could come in the amount of time spent working; looking at the power of mechanization, some observers predicted that the workweek would soon shrink to twenty or thirty hours. Others anticipated a workweek under ten hours, while still others believed technological change would eliminate virtually any need for labor at all. Some commentators welcomed such a prospect, hoping that the end of employment would finally grant ordinary Americans the ability to realize their full human potential. Setting out an ideal of well-invested free time, those optimists envisioned a world of traumatic joblessness transformed into a utopia of virtuous leisure. Others felt less sanguine about the prospect that the modern economy would create a superabundance of idleness. Employment played an important role in people's lives; how would individuals and society react if that disappeared?

Labor leaders argued that it only made sense to cut the workweek to compensate for mechanization, which appeared to put a strict upper bound on the amount of labor required for production. Under the maladjusted present-day system, such limits meant the economy simply did not provide enough work to go around; a record number of people stood in breadlines, while others toiled long hours. By shrinking the standard workday, the nation could redistribute available work among everyone who needed it. Such a move represented the natural course of history, union leaders philosophized. Factory owners in earlier centuries might have felt compelled to keep employees going from before dawn till after dark, but the modern age, with its unprecedented access to machine power, had no excuse for such inhumane expectations. William Green especially hailed the prospect of a shorter workweek as a necessary and desirable step to counter present and future technological unemployment. Such a move would benefit employers as well as labor, he promised. Research surveying what happened when businesses cut hours showed that when workers felt less stress and fatigue, companies enjoyed higher morale, greater productivity, and lower absenteeism. More than that, he contended, every American deserved the right to a healthy balance of labor and leisure. A two-day weekend would give people the chance to pursue personal fulfillment and meet community obligations, thereby enhancing democratic politics and raising civilization to new heights. "Leisure makes possible broader views of life," Green proclaimed.[15]

Chemical engineer Clifford Furnas agreed that a reduction of work hours must become inevitable as geniuses of the not-too-distant future invented devices that "will make even Rube Goldberg run for cover." With machines substituting for men, he predicted, the economy could provide everyone a rising standard of living while imposing an average workday of just two hours. Furnas condemned the "kill-joy" pioneer spirit which made moralists assume that "hard, grinding labor is the means and object of all life." Americans would only enjoy true liberty, he wrote, when they turned over all routine work to brainless robots and began to define their identity through hobbies and personal interests rather than by occupation. Furnas envisioned a revival of the lost art of letter-writing and a flowering of amateur science. Such options might not appeal to all, he admitted. "The blasé and the bored individuals will probably always be numerous, but if they are harmless they can be allowed to roam at large." Sports activities might absorb the energy of such people and prevent them from causing trouble, leaving the rest of the population free to indulge in a new renaissance.[16]

Although they approached the subject from different perspectives, Furnas and Green both forecast a shorter workweek paradise. Such a hedonistic view disturbed other observers, who challenged the assumption that modern progress would entitle everyone to leisure opportunity. Engineering contractor Thomas Desmond warned that the common laborer was neither intellectually equipped nor mentally inclined to make productive use of spare time. Those workplace technologies already installed had not fostered any notable cultural or spiritual improvement of the masses, he scoffed. Ordinary people spent the weekend in mindless pursuits such as speeding through the countryside in automobiles. "Do they stop in the day time to go out into the fields to study geology or botany? Do they stop at night to gaze at the galaxies?" Reducing the workweek could only lead to catastrophe, Desmond maintained. Freed from the discipline of long hours, men and women would waste time or use it for troublemaking. One *New York Times* reader predicted that as mechanization shortened hours, the country would witness an "ever-rising tide of crime." During a 1933 sermon in St. Patrick's Cathedral, the Reverend Thomas Graham insisted, "If men have more leisure, sin will flourish." After quoting the old adage, "The devil finds work for idle hands," Graham suggested that perhaps advocates of reduced hours "expect no difficulty from that source. Maybe they have made a deal with Satan."[17]

Speculations about labor forces being in league with the devil certainly opened up new avenues of debate. The vision of a future based on mass leisure represented a distinct break from the traditional principle of civilization built around the Protestant work ethic. Although plenty of Americans continued to argue that industriousness strengthened character and virtue, theological opinions during the Machine Age varied. *New York Times* editorial writers noted, "One of the Ten Commandments enjoins a day of rest from labor. It cannot be irreverent to suggest that its scope should be extended to include the blessing of longer periods of leisure" made possible through modern technology. Sociologist L. P. Jacks worried about the sheer rapidity with which the prospect of mass leisure had come upon modern society. Americans were "not prepared either biologically or by education" for an abrupt gain in free time. Twentieth-century civilization was "threatened by a surplus of leisure," he cautioned, "of which the present unemployment is a foretaste." The very survival of the human species might be at stake; without "a life of skillful activity" and rewarding outlets for energy, people would "begin to degenerate biologically." In short, Jacks warned, the "evils of enforced leisure" could prove "almost as bad as the evils of enforced labor."[18]

If past experience had not prepared men and women to handle the Machine Age's gift of freedom, social scientists stood ready to intervene. Self-proclaimed "leisure experts" could guide Americans in the proper management of spare time, to ensure that boredom and frustration did not generate foolishness or vice. Under professional oversight, workplace mechanization would open up harmonious social order rather than chaos and misery. In light of such thinking, Depression-era concern about technological unemployment stimulated a burst of interest in the science of leisure. Although counselors did not provide any magic recipe to alleviate the current uncertainty, they envisioned a remarkable transformation of life. Future generations would know no fear of displacement, only the luxury of recreation. *A Guide to Civilized Loafing* touted the idea that workplace mechanization could create a new age "in which we can at last begin to call our souls our own." In that book, H. A. Overstreet defined "civilized loafing" as the use of free time for mental improvement, physical exercise, or community involvement. He approved specific forms of social activity, such as singing or acting in local amateur troupes; canoeing, walking, and ice skating helped with "building the skilful body," while pottery, woodworking, and similar handicrafts allowed "the fun of handling materials." In endorsing such pursuits, Overstreet indicated that modern civilization imposed a duty on everyone to apply free time in the most rewarding directions. "The man who plants his garden, or plays his violin" will help "transform life into the delightful and adventurous experience it ought to be."[19]

Merely recommending acceptable outlets for energy did not guarantee that people would choose wisely, and such a crucial matter could not be left to fate. A 1934 handbook, *The Challenge of Leisure,* produced for the National Recreation and Park Association, took it for granted that technical changes, which could not and should not be stopped or slowed, had started to remove the value of work. Mechanization multiplied efficiency, Arthur Newton Pack wrote, but threatened to deaden the senses. When industry placed machines above human labor, men and women could only restore meaning to life through a creative use of leisure. "What the American people do in their spare time henceforth will largely determine the character of our civilization." Without supervision, citizens could slide into a rut, watching movies and otherwise killing time. The top minds must go to work on shaping the approaching leisure society, steering people toward more self-improving habits. Social critic Paul Frankl concurred that once the country had conquered present problems of technological unemployment, constructive use of the new machine-made leisure would become the burning social question: "Our claim toward civilization will have to be proved in the use we make of our idle time rather than by the size or capacity of our factories." The United States ought to begin coaching children on how to respond to the inevitable course of change, teaching them to pursue the most responsible ways of filling free time. Just as adults superintended toddlers' playtime to prevent accidents, experts must guide ordinary Americans' use of leisure to avert chaos.[20]

Authorities in the 1930s thus portrayed adaptation to the Machine Age as a matter of paternalistic leisure education. As part of the "Social Living Series" of junior high school textbooks, teachers Mabel Hermans and Margaret Hannon produced a volume called *Using Leisure Time.* "People now living remember when almost everyone had to work at least twelve hours a day," the 1938 book advised, but modern life had become much easier. Hermans and Hannon hinted at the links between workplace technology and lifestyle, asking students to consider the likelihood that "in the future there may be still more leisure.

Can you tell why?" To illustrate the worth of leisure, the authors quoted William Henry Davies' poem "Leisure," which opens, "What is this life if, full of care, We have no time to stand and stare." But ironically, standing and staring would not count as proper use of free time; the younger generation must pursue more "active," "stimulating," and "wholesome" forms of recreation like softball games. The twentieth century made leisure a community responsibility, Hermans and Hannon declared. Towns should invest tax money in organized public recreation, while schools could help spread the gospel of worthwhile leisure by preparing displays and booklets describing local pools and playgrounds, libraries and museums, parks and camping facilities.[21]

At the federal level, New Deal projects provided money and initiative to support that goal of wholesome leisure. The Civilian Conservation Corps, the Works Progress Administration, and other Roosevelt-era agencies not only provided jobs but also represented a social investment in community recreation. According to a 1941 report of the American Youth Commission, the federal government spent $1.5 billion between 1932 and 1937 on enhancing recreation services and social infrastructure. WPA teams built ten thousand swimming pools, tennis courts, and other public facilities, not to mention repairing and improving nine thousand existing centers. Such efforts reinforced the message that by preparing a new generation to make the right choices for a world in which technology reduced working hours and drastically reshaped life, the United States could minimize if not avoid a future crisis. Such an approach would in effect reinvent the old-style work ethic for an age in which mechanization would have made jobs less important. Instead of cultivating virtue through paid labor, youngsters would have an obligation to refine their character through beneficial recreation.[22]

In creating a model for this novel leisure culture, government officials and social scientists ironically reverted to a distinctly nostalgic ideal. The experts explicitly rejected twentieth-century pursuits such as moviegoing and automobile riding in favor of old-fashioned community singing, sandlot baseball, and other nontechnological pursuits. Future economic and labor conditions might be defined by a headlong race toward mechanization in the workplace, but private life, in this vision, could and should remain stubbornly human-centered. The technological imperative might force Americans to adjust their employment expectations, but in their expanded spare time, people might preserve a slower way of life.

Leisure utopians waxed poetic over the prospect that technical advances might counter the twentieth-century trend of women seeking employment outside the home, freeing them for a paradise of relaxation and nurturing. L. P. Alford, editor of several technical journals, compared modern engineers to medieval knights, honoring women through their work at a drafting board and a conference table rather than in chivalry and tournament. By giving up long hours in factories and offices, women might return to the hearth and enjoy a deserved round of bridge games, shopping, and socializing. Yet in real life, evidence suggested, reducing women's working hours did not automatically bring them into a life of ease. A Women's Bureau study of one factory that switched from three eight-hour shifts to four six-hour shifts showed that most of the affected female employees approved of the shorter workday. Out of 224 women surveyed, however, only forty-nine reported using the extra time for recreation and just sixteen for self-improvement. The majority devoted their new "leisure" wholly or in part to housework. Marital status appeared to be the determining factor. Eighty percent of the single women used their free time for sports, automobile outings, and even, in one case, flying lessons, but 55 percent of married women applied the time to laundry, gardening, and child care. The twentieth century had brought new washing machines, electric appliances, and other "labor-saving" equipment into the home, relieving the backbreaking strain of chores like laundry and ironing. Technical innovation could not, however, reduce the labor-intensive nature of child-rearing and family obligations, a factor that would limit the promise of leisure for women.[23]

In light of immediate unemployment and poverty, some Americans voiced skepticism about all the talk of workplace technological change as a force for leisured well-being. C. E. Kenneth Mees, research director at Eastman Kodak, created a stir when he criticized mechanization during a 1931 symposium (sponsored by the major engineering professional societies) on the topic of "Engineering Progress." In front of that audience of technological enthusiasts, Mees uttered the heretical sentiment that he would prefer to have lived in ancient Egypt or the Golden Age of Athens rather than in the present-day United States. Engineers and businessmen had promised that improvements in production would yield abundance and security, but so far Americans had seen just the opposite. Life had been happier four thousand years ago, he asserted, offering "more leisure, less pressure, more opportunity for the exchange of ideas, less emphasis on material things." Mees's comments elicited scorn and ridicule; the *New York Times* devoted an editorial to reminding him that ancient civilizations had supported only a small leisured elite, their privileges maintained through reliance on slave labor. The modern American "citizen has free time for everything that the Periclean period or the eighteenth Egyptian dynasty offered—and a marvelous world besides, of which neither had ever dreamed."[24]

Such talk about the joys of mechanization still left room for doubt, as some observers questioned whether a society built around leisure could keep the economy moving. Virgil Jordan, president of the National Industrial Conference Board, complained in 1936 that already millions of people had grown "physically or psychologically sick, lazy, untrained, uneducated, incompetent." Mechanization threatened to reinforce such unhealthy tendencies, to the point where generations raised in a "push-button . . . paradise" would take abundance for granted. Invention and engineering had made the United States the wealthiest country on earth, but Americans must also realize they remained just "a few steps from the jungle of primitive savagery." Only "incessant labor and intelligence" could prevent society "from slipping over the edge," Jordan declared. A world based on leisure would not survive; Americans must keep on with "more work, harder work" to maintain any level of prosperity.[25]

Jordan's remarks reveal a fundamental truth about the business community's attitude toward the prospect of mass leisure. In order to dispose of the goods being turned out so efficiently by mechanized workplaces, the economy would need to maintain consumer activity. William Green promised that the very act of reducing the workweek would stimulate consumption by giving people the opportunity to pursue hobbies, home repair, and other activities that required special supplies. Many business leaders worried, however, that past a certain point, an increase in leisure might send purchasing behavior into a tailspin. A modern Athens in which everyone spent the day reading or singing did not represent their ideal. Their market-oriented vision required a steady cycle of consumerism, the desire for new possessions. Accordingly, rather than stress the vision of mass leisure, the business community of the 1930s chose to present a dream of mass luxury.

Some Depression-era commentators criticized the economy of consumerism for promoting the concept that happiness came primarily from material things. Stuart Chase blamed advertisers for cultivating an obsession with novelty, a pointless fad that "drained away the savings conceivable in a machine economy whose sole objectives were the abolition of poverty and the increase of leisure." Advertising multiplied wants, transforming the bath from a simple place to wash into an opportunity for marketers to sell back scrubbers, sprays, and fancy soaps "nicely adjusted to every part of the anatomy." Machines could indeed speed up production, but piling up goods did not automatically qualify as progress, Chase indicated. He feared that businessmen were convincing Americans to trade leisure for consumption, throwing off the healthy balance of life. Families might "be happier with a five-hour day and rather less in the way of cosmetics, Hollywood films, overstuffed davenports and electric refrigerators," however much such a principle appalled advertisers.[26]

Humorist Robert Littell poked fun at the basic assumption that modern products necessarily made life better. Littell, the Andy Rooney of his time, complained about ice cube trays that froze solid and about fancy telephones that allowed salesmen to call at the most inconvenient moments. The consumer-products industry had at least created employment for repairmen, he remarked. Whenever a refrigerator broke down, the mechanic called to the scene "finds the problem is too much for him, and calls in the second assistant engineer specializing in number 31-b magneto brushes, all resulting in a bill for forty or fifty dollars." In a more serious vein, Ralph Borsodi too questioned whether a rush to consumerism really represented social progress. The more capitalism kept people dependent on daily wages, the more they remained vulnerable. "Insecurity is the price we pay for our dependence upon industrialism for the essentials of life." Displacement posed an ever-present threat in the Machine Age, Borsodi declared, and Americans could only avoid any danger of falling victim by breaking free from the industrial system. His own family had used a few hundred dollars to set up a self-sufficient homestead. Opting out of the commercial system did not mean reverting to a primitive existence, Borsodi stressed. Far from rejecting all technology, his family used a tractor, sewing machine, and other modern equipment to produce their own food and clothing, a strategy that promised a "permanently better way of living for every man, woman and child now struggling for happiness in our industrial civilization."[27]

Needless to say, the prospect of many families separating themselves from established production-consumption networks would not have thrilled business executives and advertisers. William Green and other labor leaders reassured people that fear of job loss need not drive them to abandon the mainstream economy. If managed carefully, with an eye to workers' welfare, the introduction of new production technology offered the prospect of genuine advance, Green maintained. Corporate management hastened to reemphasize the most positive spin on mechanization, the promise that improving production could guarantee a wealth of consumption.

World's Fairs as Weapons in the Technological Unemployment Debate

Charges of technological unemployment had thrown the business, scientific, and engineering communities on the defensive, but those interests continued fighting to link mechanization with a faith in progress. Ever since London's great Crystal Palace Exposition of 1851, national and international public fairs had provided opportunities for manufacturers to impress observers by promoting their latest products. During the 1930s, public expositions took on an added purpose. In an atmosphere of economic crisis and amid fears of technological displacement, companies turned their displays into a glittering vision of how the mechanization of production could provide unparalleled consumer wealth.

Of all the decade's expositions, the 1939 New York World's Fair best demonstrated the importance decision-makers attached to reinforcing trust in mechanization. From the beginning, the civic leaders, business executives, architects, artists, and planners responsible for articulating the fair's direction pursued the theme "Building the World of Tomorrow." In their eyes, pure knowledge could not stand as sufficient cause for celebration. "Mere mechanical progress is no longer an adequate or practical theme for a world's fair," explained Michael Hare, member of the fair's Board of Design. Advocates constructed the fair to convey specific assumptions about the economic and social role production technology played in the country's past, present, and future. Regardless of recent employment difficulties, their imagery suggested, Americans must embrace workplace mechanization as the necessary course of history, fulfillment of a new manifest destiny for the twentieth century.[28]

The New York World's Fair represented cultural theater, an assurance that for all the pain of yesterday's economic reality, adaptation to tomorrow could prove easier. The film *The City*, commissioned for the fair by the American Institute of Planners, dramatized the promise of a culture that smoothly assimilated the implications of change. Opening scenes show women sewing quilts and farmers using small tools, hand labor symbolizing a preindustrial stability and harmony of being. Industrialization and urbanization had upset that idyllic world, the movie indicated, as people began inventing more and more "machines to make machines." To suggest the way technology kept accelerating the pace of life, directors Ralph Steiner and Willard Van Dyke filled the screen with images of traffic jams, auto accidents, and crowds gulping down meals at lunch counters. The nation could yet escape such horrors, the planners continued, by creating a new, better-designed city that put technology in its proper place. "Here science serves the worker, making machines more automatic . . . and those who [handle] them more human." To men and women worried about their jobs, making production still more automatic might not seem like the best solution, but *The City* implicitly preempted questions of technological unemployment by defining well-planned mechanization as an integral part of the perfect community. Along similar lines, the "Democracity" display created by industrial designer Henry Dreyfuss featured movies showing men and women employed in agriculture, manufacturing, and service, purposefully converging on the twenty-first-century city. Those future workers would have no difficulty finding their place in that perfect economy, an America in which extensive use of labor-saving machines would pose no obstacle to labor.[29]

Such exhibits did not detail exactly how society could go about adjusting to the impact of increasingly rapid workplace innovation. As the *Official Guide Book* informed readers, the fair merely presented "the materials, ideas and forces" which Americans might employ to shape the future. "You are the builders; we have done our best to persuade you that these tools will result in a better World of Tomorrow; yours is the choice."[30] *The City* similarly told audiences that they must decide on the living environment for future generations. Despite such rhetoric of empowerment, most exhibits did not encourage visitors to reflect too deeply about recent conditions. In reality, the New York Fair only presented one economic and social philosophy, that of Machine Age consumerism. Displays emphasized that men and women enjoyed a glorious freedom to select their style of car and kitchen appliances, but the choices really did not extend further. The fair vision left no room for people to reject or even affect the course of technological change. The exposition's interpretation of American history as a continual movement, from horses to trains to airplanes, from country to city to a modern planned community, conveyed a sense of inevitability that could not easily be gainsaid. "Despite every restriction that can be placed on it by so-called 'reformers,'" Henry Ford told potential World's Fair visitors, "the quest will continue—invention will go forward."[31]

Historians have long recognized the 1939 "World of Tomorrow" as a celebration of consumerism, but such an assessment tells only half the story. The fair did not enshrine consumption just for its own sake but also as a counterweight to criticism of mechanized industry. In defining the future as a period characterized by wonderful revolutions in production, exhibitors effectively excluded discussion of any accompanying cost to workers. In many ways, the New York fair was meant to serve as the ultimate argument against fear of technological unemployment, the culmination of almost a decade's worth of debate about the relationship between mechanization and progress.[32]

That perspective came through most explicitly in comments by John Van Deventer, editor of *Iron Age*, who encouraged business leaders to treat the World's Fair as a great opportunity for public relations. Companies would never enjoy a better chance "to disabuse the public of the impression that mechanization destroys jobs and to show Mr. and Mrs. Average Man how it multiplies them." Unless exhibitors consciously designed displays to fight the idea of technological unemployment, they might accidentally end up reinforcing it. Van Deventer worried that the standard type of corporate promotion, showing off the latest advances in production, might backfire. Exhibits that highlighted "the ability of modern mechanized industry to turn out an almost unlimited quantity of products" without intensive human effort could leave Americans more nervous than ever. After learning how far industrial engineers had taken mechanization of steelmaking, oil refining, and bottle-making, people could hardly be blamed for saying that technology had become so wonderful that they worried about the future. Visitors might tell friends back home, "I saw the most amazing automatic machines turning out great quantities of products without . . . human hands" at the New York Fair. "Even cows milked by machinery. No wonder we have ten million unemployed."[33]

The executive community could not afford to neglect public impressions, Van Deventer wrote, since heightened resentment might give rise to legislation restricting business practices. "Let even a comparatively small percentage" of visitors "go away from the Fair with the belief that the machine is destroying jobs," he warned, "and American industry will soon find productivity wearing a muzzle." Van Deventer believed that critics had succeeded in creating an atmosphere of paranoia, with talk of technological unemployment being "spread from high places, preached from platforms and pulpits as well as from soap boxes." World's Fair exhibitors should avoid anything that might "feed fuel to the fire of the powerful anti-machine propaganda already under way in this country." Writing in February 1939, Van Deventer regretted that the time had passed when business interests could have built a "Hall of Machine-Made Jobs" for the fair. Exhibits could have taught visitors that continuous-strip steel mills and other new production technologies ultimately created more work than they destroyed. It was too late to create any central focus for that important message, he wrote, but individual companies still had time to prepare displays explaining how their pursuit of mechanization had enhanced employment opportunity. Advocates could even prepare special pamphlets or films with titles such as "How Science and Invention Multiply Jobs and Wages."[34]

Picking up on Van Deventer's message that they dare not ignore ordinary Americans' misgivings, the fair's image-makers set out to deny that mechanization raised any serious social or economic difficulty. An article published under Henry Ford's name in a special "World's Fair Section" of the *New York Times* called it "astounding" to "realize that there are still in this world men who actually believe that machinery is a menace." Though new production technologies had occasionally caused "temporary dislocations of employment for numbers of men caught unawares by the new time," the piece asserted, most soon found new work. Victims themselves bore some blame for job loss; a shortsighted laborer who insisted on clinging to "comfortable unprogressive habits" would necessarily encounter trouble in a revolutionary environment. Americans must accept that their country had "scarcely entered upon the machine age as yet," that soon many more operations would become fully mechanized, "reducing employment on many jobs still further." Continued economic growth would prevent disaster, ensuring that "there will be more jobs in industry than ... men to do them." People who anticipated developments would enjoy plenty of opportunity, and, thanks to such mental preparation, technological unemployment "is now a disappearing term. A generation such as the present one which is technologically alert will always be employed." In any case, Ford maintained that talk of displacement missed the main point, that mechanization made business efficient enough to supply consumers with more, better, and cheaper possessions. Skeptics who visited the fair could "learn at first hand how every scientific advance in the production of goods and services contributes" to both individual happiness and "the advancement of civilization."[35]

Many prominent exhibits at the New York World's Fair sent that precise message, that mechanization justified itself by creating an abundance of stunning new products. Beyond just tantalizing visitors' dreams of wealth, the parade of consumerism could effectively negate any worries about labor displacement. Planning for the exposition, after all, took place in the very months when David Weintraub and other federal officials were producing new economic data revealing disturbing trends in employment. Companies could use the World's Fair to counter such troublesome statements. Displays designed to entertain visitors with the ingenuity of modern production, to overwhelm them with a tempting vision of Machine Age mass consumption, might manage to bypass the entire question of jobs. Many businesses actually featured the latest forms of production technology in their exhibits, giving ordinary Americans a chance to see the machines at the heart of debate. The American Tobacco Company, for example, displayed new cigarette-making machines with the dictum that such equipment helped keep Lucky Strikes affordable. Crowds could marvel at the sight of machines turning out row after row of perfectly rolled cigarettes. While admiring the industry's technical ingenuity, visitors would absorb the moral that they all, as consumers, reaped the benefits. Such an angle neatly evaded any question of what had happened to the hundreds of skilled workers who had once rolled cigarettes and cigars by hand.

The Borden Company similarly avoided the issue of labor displacement by playing up the "gee-whiz" aspect of its equipment. Its "Dairy World of Tomorrow" featured the "Rotolactor," a device for simultaneously milking five cows on a rotating platform. Borden's model dairy, which offered guests free milk samples, touted the Rotolactor as embodying the latest hygienic principles "that may be used universally in the future for the benefit of mankind." The Sheffield Farms display also portrayed dairy mechanization as an asset to consumers. Thanks to technology, the firm boasted, it could keep hundreds of thousands of families supplied with "protected milk from selected farms." Advertisements assured mothers that machines handled the entire task of bottling and sealing milk, bringing it into the home "all untouched by human hands." The Sheffield and Borden exhibits thus reframed impressions of production technology, making milking machines into a force that preserved family health, rather than one that imperiled agricultural labor.[36]

Considering how much controversy had arisen in recent years over the way the introduction of dial equipment affected switchboard operators, the telephone company had a particular interest in trying to redefine public perceptions. AT&T aimed to elicit appreciation for its technical sophistication, to get audiences thinking about the scientists and engineers of Bell Labs rather than about its "hello girls." Communications would break down without continued innovation, the firm emphasized, given the sheer "magnitude and complexity of the operations involved in interconnecting telephone subscribers." An exhibit entitled "What Happens When You Dial" used paths of flashing lights to show how dial machinery placed calls faster and more accurately than even the best human operators. Publicists proudly cited one audience member's comment, "The average individual should be able to appreciate the vast improvement that the dial system brings." Other viewers reportedly declared that while they had always preferred dealing with human operators, suspecting that the mysterious new equipment overcharged for local calls and busy signals, the display had convinced them to trust dial technology. AT&T attracted fairgoers by allowing some to place complimentary long-distance calls, reinforcing the message that technology contributed to customer satisfaction. Switchboard operators, who suspected that management really had introduced mechanization to break their union, had no such prominent venue to dispute the notion that machines inherently provided superior service.[37]

It was Westinghouse's showmanship which offered New York crowds the most diverting technical possibility of all, the humanoid robot. Throughout the Depression, artists and cartoonists had identified the image of the robot with labor displacement, making it a visual indicator of distress. Science fiction writers had constructed entire plots around the robot as the ultimate job stealer, ruin of the human race. Westinghouse provided an instant counterweight to such impressions with its exhibit starring "Elektro" the "moto-man." When Elektro stepped out on stage, he did not start assembling cars; his accomplishments were limited to moving forward on command, counting on his fingers, and other simple tasks. But Elektro made for terrific public relations; newspapers ran photographs and articles raving about his act. Westinghouse spokesmen touted Elektro's capacity to amuse people as proving the robot's nonthreatening nature. With every captivating performance, with every publicity photograph, Elektro deflated the issue of technological unemployment and supplied seeming evidence that modern science and engineering could work wonders.

To drive that point home for Americans unable to attend the festivities, Westinghouse made a special movie, *The Middletons Visit the New York World's Fair*. Opening scenes show a boy complaining about his generation's poor employment prospects. "Maybe it's difficult, but it's worse to be a quitter," Bud's father replies. "You've heard all the talkers," the father continues, "now I'm going to show you the doers." At the exposition, the Middletons receive a personal tour of the Westinghouse building from an old friend, Jim Treadway, who praises his employer for proving that "nothing is impossible." When the son expresses reservations, Treadway replies, "Open your eyes, Bud, the proof's all around you! Why, this whole Fair is a product of research." Corporate-guided science and engineering kept technological unemployment away, he promises, and in the future, "industry will make so many jobs, there won't be enough people to fill 'em." When Bud again evinces skepticism, Treadway tells him, "You're liable to hear anything these days [if] you're willing to let a lot of self-appointed leaders do your thinking for you." After Treadway has shown off Elektro and let Bud play with experimental television equipment, it appears Westinghouse has converted Bud into an unabashed enthusiast. Meanwhile, his mother and grandmother admire the company's demonstration kitchen and watch the "Battle of the Centuries," the slapstick dishwashing race in which "Mrs. Modern," whipping plates through the latest model dishwasher, leaves poor "Mrs. Drudge" awash in her own sinkful of suds.[38] The movie could not have carried a clearer moral: average Americans should dismiss pessimists' ridiculous talk about technological unemployment and focus on real heroes, the researchers and businessmen busy devising wonderful things like television.

A cartoon film made about the New York Fair similarly encapsulated the message of mechanization creating a material utopia in which the issue of labor displacement did not matter. *All's Fair at the Fair* tells a humorously affectionate story of a country couple who gaze with thrill at the exposition's industrial exhibits. Without a single worker in sight, without even a person around to press the start button, production machines whittle logs into clothespins and assemble perfect little houses. When the pair goes to buy refreshments, an automated watering can sprinkles a pot full of soil, which instantly sprouts an orange tree; automatic shears cut off the fruit, automatic juicers squeeze it, and automatic hands insert two straws. In the fair's barber shop, mechanical hands remove the man's jacket, push him into the chair for

Fig. 6. In order to promote its exhibit at the 1939–40 New York World's Fair, Westinghouse produced a special movie, *The Middletons Visit the New York World's Fair.* These panels, taken from the accompanying campaign of print advertisements, show an all-American family enjoying the company's spectacular displays and learning all about the technical wonders of mechanization. With the opportunity to play with steel-handling machines or see Elektro the robot, who would be ungrateful enough to complain about a temporary loss of jobs? *Country Gentleman* (July 1939): inside cover.

a shave, and present themselves palms-up for a tip afterward. At day's end, the couple purchases an instantly assembled roadster from a handy vending machine and drives off with their old horse sitting in the back seat.[39] The cartoon derives its comedy, of course, from its very exaggeration; though some restaurants had adopted pancake-making machines in the 1930s, human waiters and cooks still existed. Behind the fanciful humor lurked a serious message, one which conveyed the inevitability and the incredibility of mechanization. With a future in which machines served up food, houses, and new cars at unbelievable speed and at virtually no cost, who would worry if employment evaporated?

The proconsumerism, promechanization slant of New York's World's Fair appealed to the nation's beleaguered engineering community. *Mechanical Engineering* praised exhibitors for presenting material "that should strengthen the faith of men in their ability to create and satisfy new wants," thereby eliminating any problem of job shortages. The exposition offered a powerful set of "emotional, aesthetic and educational experiences that put the scoffer to shame," (à la Westinghouse's fictional Bud Middleton). When displays held out the thrilling possibilities of mass consumption, the journal editorialized, Americans would stop worrying about displacement and finally show the proper gratitude for improvements in production. Once annoying talk about labor problems had been dispelled, researchers and inventors could proceed without criticism to develop even more powerful machines. The American Society of Mechanical Engineers held its 1939 annual meeting in New York, enabling members to make a pilgrimage to the fair's cathedrals of corporate glory, where exhibits encouraged people to worship consumerism and reaffirm their faith in technological progress.[40]

The TNEC Hearings: Talk of Machines in a Nation Moving toward War

The decade of dispute over the relationship between mechanization and work had not come to a close by 1939. Unemployment had not yet fallen back to satisfactory levels, and even as hundreds of thousands gawked at Elektro and Detroit's latest automobiles, the federal government continued to investigate the human consequences of changing production technology. In 1939 and 1940, Congress undertook a major set of hearings on the concentration of economic power in the United States. As part of that broad mandate, the Temporary National Economic Committee (TNEC) held fifteen days of hearings in April 1940 focused squarely on the issue of machines and jobs. A parade of witnesses came to Washington; executives of the nation's largest corporations, union leaders, economists, sociologists, and statisticians each described in turn their perspective on the question of whether mechanization caused significant social and economic distress for labor.

TNEC chairman Joseph O'Mahoney opened by posing what he called the "all-important riddle of our time": in an age that possessed a greater domination over nature than any civilization before, why was it that "we still have not learned to apply the wonders of technology . . . to provide decent jobs for the millions of idle"? From the outset, O'Mahoney disavowed any antiprogress intent, declaring that he hoped "to do everything in the world to aid technology." Nevertheless, he emphasized, rising manufacturing output could not count as beneficial unless some employment security came with it.[41]

During the first days of testimony, TNEC's economic advisor Theodore Kreps drew on figures from the WPA's National Research Project to emphasize that though new production technologies had raised living standards, they had also imposed significant costs. Businesses could discharge their superfluous employees, but the country itself could not dispose of surplus citizens. Ideal economic laws suggested that rising efficiency of production ought to bring price cuts and market growth which would stimulate employment, but that model had broken down in the Depression. Over recent years, the economy had employed just 91 new workers for every 100 displaced, Kreps told Congress. Redundant laborers could not simply waltz into new positions; many needed retraining or other assistance from somewhere before they could rejoin the workforce. Nonetheless, it seemed Americans had no choice but to accept the pressures imposed by mechanization. Any steps to restrain research and invention would be "both unwise and impossible," Kreps continued. Therefore, the United States must conquer the cultural lag problem, balancing the unprecedented rate of progress in production with compensating social accommodations. "Change in one must be synchronized with changes in the other, just like the front and back wheels of an automobile." The years ahead might well bring still more remarkable developments in production technology, and for Kreps, the most promising strategy entailed promoting consumption. If the nation could reshape itself as "America Unlimited," problems of labor displacement might indeed vanish. Science writer Watson Davis and General Motors vice president Charles Kettering enthusiastically endorsed such a notion, assuring Congress that scientists and engineers stood ready to produce an America Unlimited.[42]

Subsequent testimony revealed that the practical details of adjusting to mechanization remained most contentious. The TNEC took up specific industries and economic sectors in sequence; business spokesmen usually testified first, followed by representatives from the related labor union. Such a format underlined just how far apart the two sides generally stood. During discussion of conditions in automaking, for example, the Ford Motor Company's official statement reported that managers had "no knowledge of any technological improvement that has resulted in the permanent displacement of workers." Improvements in manufacturing might have reduced the number of men needed in particular functions, but all "willing workmen" should be absorbed by job growth created through still more technological change—at least "under normal conditions." Ford engineer R. H. McCarroll maintained that by adopting the latest equipment, the firm had been able to lower prices and thereby maintain sales. That forward-looking attitude also led Ford to add new features and accessories to its cars, a step that raised man-hours and thereby helped absorb excess labor.[43]

Testifying later the same day, United Auto Workers president R. J. Thomas agreed that such a constructive feedback cycle had indeed existed in automaking between 1923 and 1929. Technical innovations in the industry had increased productivity by 134 percent during that period, he said, but surging consumer demand simultaneously had stimulated a 212 percent leap in production which sent employment up 30 percent. In the decade since 1929, Thomas continued, the relationship between productivity and jobs had turned sour. Auto companies continued to pursue labor-saving techniques; figures from the National Research Project showed that even during the Depression years of 1929 to 1938, technological change in the industry had resulted in a 17 percent gain in per capita productivity. Massive displacement would have resulted, Thomas contended, if the UAW had not succeeded in spreading out the work by reducing hours from an average of 46.8 to 35.8 a week. Contrary to industry claims, he added, mechanization did not always serve consumers' interest by improving the product. When a company eliminated 250 men by substituting automatic paint sprayers for hand finishing, cars came off the line with an imperfect "orange peel" surface rather than a smooth coat. The UAW supported the ideal of engineering advance, Thomas concluded, but mechanization would not bring economic health until employers offered labor a fairer share of the gains from higher productivity.[44]

A different argument arose in the hearings covering railroads, in which management agreed with labor that trends in recent years had caused sizeable job loss. J. H. Parmelee, from the Association of American Railroads, acknowledged that automatic train control and signaling systems, paint-spraying machines, and office mechanization had displaced people. Competitive pressures had forced railroads to adopt such equipment, he argued, since the government unjustly favored road and ship traffic by subsidizing highway and waterway construction. Parmalee also blamed unions for worsening technological unemployment. By demanding higher wages, he said, workers had priced themselves out of the market; as the cost of labor rose, management found it easier to recoup their investment in labor-saving machines. If railroads could only eliminate those twin problems of high labor costs and oppressive competition, J. J. Pelley promised, the resulting boom would lead them to hire many more men, even given the labor-saving effects of technology. In response, Byrl Whitney, speaking for the Brotherhood of Railroad Trainmen, denied that labor costs had driven management to adopt new equipment. Unions had agreed to a 10 percent wage cut in 1932, he reminded Congress, and over the years since, the Chesapeake & Ohio at least had paid more money in dividends than in wages. Road and water transport had grown, Brotherhood of Railway Clerks president George Harrison agreed, but had not pushed railroads into truly desperate straits. Executives had seized on competition as an excuse for mechanization, concealing their real desire to weaken union power.[45]

Later TNEC proceedings examined the relationship between technological change and employment in the telephone and telegraph business, office work, textile manufacture, coal mining, and agriculture. Those sessions followed a clear pattern: industry spokesmen stressed that machines not only improved service but also created jobs over the long run. When pressed, they admitted that temporary displacement might have occurred but denied that it had reached any serious level. In response, union spokesmen claimed that unemployment statistics should supply more than sufficient proof of trouble. Philip Murray complained that for over a decade, business leaders had counseled workers that if they waited patiently, technical innovation would bring new jobs and lower consumer prices. Labor had kept its part of the bargain by muting its objections, only to see the savings from mechanization pour into corporate profits. Millions remained trapped in "the craziest national economy ever recorded in the history of any civilized nation."[46]

TNEC hearings finished with testimony from WPA officials Corrington Gill and David Weintraub, along with Commissioner of Labor Statistics Isador Lubin. Those three had led the Roosevelt administration's efforts to collect solid evidence on how technological change affected employment. Gill repeated the WPA's key conclusion that over recent years, productivity simply had not grown fast enough to make up for technical advances, leading to declining job opportunity. The system allowed businessmen to write off the costs of installing machines, thereby giving them all the benefits while externalizing the costs. That approach amounted to "bad social bookkeeping," Lubin declared, since "we haven't recognized the fact that progress . . . involve[s] hazards for somebody."[47] In short, those speakers reinforced what many social scientists inside and outside government had come to believe: the problem of Machine Age displacement was too large to be ignored but too complex for easy solution.

Complementing the outpouring of statistics, charts, and arguments over months of hearings, the TNEC produced forty-three monographs covering antitrust law, consumer economics, and business activity. In number 22, *Technology in Our Economy*, TNEC staffer John Blair offered detailed analysis of the displacement question. Defying classical economic predictions, he wrote, the new efficiency of mechanization had not translated over recent years into proportionately rising wages, reduced prices, or shorter hours. Employment growth in previous decades had come from the ripple effect of first railroad and then automobile development, but Blair judged it unlikely that the immediate future would witness the arrival of another such megaproduct. Americans should not assume that the future must always generate enough jobs to absorb displaced workers; new industries would keep pursuing labor-saving mechanization, after all, and purchasing could not expand indefinitely. Many new products simply substituted for existing goods, siphoning off consumption and employment from older industry rather than creating a net gain. It seemed "technology will continue to increase labor productivity" and displace men, Blair concluded, so that "economic and social distress may be expected to accumulate."[48]

Such pessimistic sentiments did not please the National Association of Manufacturers, which condemned the TNEC monographs for revealing a clear "hostility to corporations and individuals of wealth." The committee's work represented a waste of public money at best, the NAM said, and at worst, "evidence of a deliberate design" to prepare the way for "government control of private activity along virtually Nazi and Fascist lines." In evaluating monograph 22, the NAM found it incredible that anyone could still believe in technological unemployment as a major problem. While it might be "easy and natural to feel pity for the few who have lost their jobs" to machines, all good workers should be able to adjust without undue difficulty. The NAM dismissed WPA research on displacement as mere "economic sophistry" and referred to the TNEC report as "an all-time low in economic reasoning." At the most fundamental level, NAM writers objected, Congress had missed clear evidence of progress in its eagerness to "paint a dark picture and dwell on the hardships" of mechanization. History showed that over the long run, "diffused benefits for the millions may outweigh the hardships of the few who are injured" by change. Just as growth of railroads and automobiles had created both employment and higher standards of living, so in future the creation of new industries would absorb any workers thrown out of older economic sectors.

Blair's failure to admit such a likelihood merely proved his lack of vision and faith, the NAM said. Those who ignored how much mechanization had improved American life were nothing less than "economic perverts."[49]

With such insults, the NAM set itself up as defending the notion of progress against those parties terrified by modernization and out to destroy free enterprise. The TNEC hearings had brought out reams of statistics and charts, but they had not exactly brought out consensus. In that way, the committee's experience encapsulated the whole decade-long argument over technological unemployment. Though WPA economists and other observers had collected substantial evidence on recent trends in mechanization and work, the numbers left more than enough room for disagreement. In the Depression era's emotionally charged atmosphere, arguments weighing the costs of technological change against the benefits grew heated and heavily symbolic.

As far as the ultimate turn of the technological unemployment debate, the TNEC hearings really did not matter. By 1940, rising concern about the international situation had started to affect the way Americans regarded mechanization. As an impending danger of war spread across the world, U.S. officials began emphasizing the need for readiness. With national security possibly at stake, production technologies suddenly acquired new meaning. To the extent that mechanized equipment could help manufacture weapons and military supplies while releasing manpower for other vital functions, it became invaluable. In the Depression, labor-saving machines seemed to present a threat; in a time of global crisis, they came to represent a virtue.

Psychologically, as the United States moved toward an emergency, people needed to take refuge in the notion that American technical brilliance would guarantee wealth and happiness. In a 1941 essay, "The American Century," Henry Luce defined Freedom (which he capitalized) as the use of science and technology to achieve "the more abundant life." During the 1930s, those who talked about the United States as the ultimate consumer utopia had hoped to distract troubled people from thoughts of labor displacement. For the 1940s, the idea of American culture as economically and socially superior took on new importance.[50]

Coming out of the 1930s, the American business community welcomed prospects for expansion. Defense-related growth could provide the final justification for adding new equipment and would prove once and for all that talk about permanent labor displacement had been wrong. "With its moving parts working at capacity, mechanized industry can do more to maintain a high standard of living and provide jobs than by any other method," *Nation's Business* observed in September 1940. According to census data, the writer continued, "more American people are now engaged in the production of machinery than were employed in all manufacturing eighty years ago." The same conveyor belts that moved shoes or boxes of cereal through a factory could carry gas masks, tanks, and other military items equally well.[51] For war-conscious readers, the moral seemed clear: mechanization enabled modern business to turn out everything from canned food to weapons more efficiently, all representing the necessary and desirable course of progress.

As it turned out, of course, World War II did solve many immediate economic problems. As the military absorbed able-bodied young men and as production expanded to meet war needs, official unemployment rates fell to just over 1 percent in 1944. In a desperate search to keep defense plants running around the clock, management scouted all possible sources of labor. During the 1930s, fear that mechanization cut the total number of jobs available had fostered resentment of working women. In the 1940s, "Rosie the Riveter" became a force for the survival of democracy. The entire context for talking about workplace technological change had shifted, an evolution shown even in America's film world. Moviemakers in the Depression had wanted to document the imbalance between men and machines, and by decade's end, Pare Lorentz had started shooting factory scenes for *Ecce Homo!* Lorentz left his grand work on technological unemployment unfinished, however; events in Europe distracted attention and diverted investors. Lorentz had shot some footage inside factories, intending for the scenes to show how drastically machines had eliminated labor. Ironically, some of that material ended up in Office of War Information productions, where it underlined a more positive message about how the power of manufacturing technology could win the war.[52]

Willard Van Dyke's film *Valleytown* similarly ran head-on into the changing climate. The movie had succeeded in capturing how deeply mechanization frightened some Americans, but by the time of its release, that concern no longer seemed relevant. *Valleytown* "didn't get any distribution because it dealt with unemployment and there was no unemployment during the war," Van Dyke remembered, "so it seemed irrelevant . . . out of step with its times." As a matter of fact, Van Dyke had done his best to keep *Valleytown* in touch with the times. His final scenes conveyed a sense of industry in wartime motion, showing workers busy learning how to run the latest equipment for manufacturing airplanes. An optimistic conclusion declared that the United States would remain safe and prosperous as long as it heeded the human element in work, retraining people to "keep their skills as modern as the new machines." After *Valleytown*, Van Dyke dropped his inclination to make "films of protest" and began directing government wartime movies that treated the subject of technology quite differently. While *Valleytown* had focused on displacement,

Van Dyke's 1943 film *Steeltown* glorified industrial machinery as a source of patriotic pride. The political climate and popular culture had made talk of technological unemployment outdated.[53]

Why was talk of displacement so easily and suddenly superseded? The reasons ran deeper than just the change in times. The topic of workplace mechanization had raised significant alarm among labor interests, social scientists, government officials, and the general public all through the 1930s. But concern about displacement, while substantial, had never escalated into a full-scale challenge to technology or corporate management. The nation's scientific and technical community feared a backlash, but in reality, few Americans expressed any desire to suppress technological change. Scientists and engineers worried about being made into scapegoats for the Depression, but in the end, those professions came into the 1940s with their power and prestige largely intact.

The major reason why talk of displacement did not build up to a thorough rebellion against the system may have been the difficulty of overturning the entire ideology of technological progress, past, present, and future. Americans liked to believe that their country had been built around invention and discovery. From the mythology of railroads conquering the nineteenth-century western frontier to the twentieth century's love affair with the automobile and airplane, people wanted to associate technology with excitement and virtue. During the 1930s, even as labor leaders and other critics questioned the present-day impact of mechanization, they never denied its historic importance.

The business, scientific, and engineering communities took advantage of those positive associations to reinforce Depression-era faith in technology. Their skill in public relations proved a valuable weapon. At best, most skeptics offered only vague hints about how they would reform and control the evolution of technology. In contrast, corporations and their allies in science presented a clear vision of progress focused on consumer goods, one in which material possessions defined freedom and personal happiness. Advertising and World's Fairs linked the pursuit of mechanization with wealth. Despite the stress that unemployment, bank collapse, agricultural failure, and other economic crises had placed on so many Americans—or indeed *because* of that stress—the dream of consumerism carried real power. People did not, after all, want to relinquish their automobiles, radios, or other products of Machine Age manufacture. Concern about job loss raised deep questions about the nation's direction, but Americans were ready to believe in the promise of abundance.

Scientists, engineers, and businessmen had presented a powerful defense of technological determinism, the notion that innovation could not and should not be restrained. In their view, the continued development of new production machinery was foreordained, a force best managed by technical and business experts alone. Labor leaders and other critics never succeeded in articulating an alternate vision that would give ordinary Americans some input into the adoption of technology, some choice over workplace conditions. No one was really prepared to fight the myth of technical inevitability. Americans had been guided to accept the prospect of accelerating mechanization and were told that they could do nothing else. At the outset of the Depression, sheer alarm over unemployment appeared to offer a chance for people to reevaluate the relationship between technology and society. Corporate, scientific, and engineering leaders rushed to head off such a challenge and ultimately succeeded in reinforcing the ideology of mechanization as progress. With World War II, the window of opportunity for rethinking such fundamental questions had closed. And yet, Americans had never settled the dispute over how mechanization affected labor and what should be done about it. With the return of peace, still more powerful and versatile forms of workplace technology would soon appear, leading the country to pick up the discussion where it had been interrupted.

8

"Automation Just Killed Us"

The Displacement Question in Postwar America

Dᴜʀɪɴɢ ᴡᴏʀʟᴅ ᴡᴀʀ ɪɪ, as full employment returned and shortages and rationing constrained purchasing for the duration, Americans looked forward to an expected burst of peacetime consumerism. Advertisers interpreted the war as a fight to defend not only political but also economic freedom, especially Americans' right to enjoy material plenty. Postwar prosperity did indeed accelerate personal consumption, especially after the baby boom generated new family expenses. During the 1950s, consumer spending jumped 38 percent in real terms. By 1958, John Kenneth Galbraith defined modern America as an affluent society, one in which material abundance and economic opportunity had greatly reduced poverty. The country had apparently made the transition from an old economics of scarcity to a new economics of wealth, though racial discrimination and financial disadvantage conspired to limit some people's access to the American dream. The postwar years brought more families close to realizing the futuristic ideals set forth at the 1939 New York World's Fair: neat suburban living in a modern house equipped with television and the latest kitchen appliances, larger and more powerful automobiles, multiple-lane highways, and relatively dependable employment.[1]

The period significantly changed the political and institutional context for scientific and technical research. Vannevar Bush's 1945 report, *Science, the Endless Frontier*, defined research as America's key to power in the new global arena. As Cold War tensions intensified, the Defense Department poured unprecedented funding into physics and engineering departments at MIT, Stanford, and other universities so that academics could help improve aircraft, rockets, and nuclear technology. The future of democracy in the atomic age seemed to rest on whether the United States would keep ahead of the Soviet Union in science education, research, and development. Richard Nixon announced during his 1959 "kitchen debate" with Nikita Khrushchev that ordinary American families should feel proud knowing that free enterprise and a free political system gave them a superior standard of living. The economic commitment and technical knowledge that produced missiles could also produce a better dishwasher, and during the Cold War consumer choice came to symbolize success of the American way of life.[2]

Beneath the surface of affluence, lingering economic and social concerns, including technological unemployment, continued to haunt many in the United States. While the onset of World War II had abruptly disengaged attention from machines and jobs, that did not mean the issue had been resolved. Quite the contrary; as postwar scientists, engineers, and businessmen developed new techniques of automation and computerization, the familiar discussion about the fate of labor reopened almost immediately.

The Fascination of Automation: Cold War Perspectives

With the military's urgent need for weapons and supplies, World War II itself had provided a new rationale for reinventing production equipment. Under pressure to expand defense business, some contractors experimented with ways to take mechanization further than ever. The W. F. & John Barnes Company began developing a high-explosive ordnance factory that would use four-station robots, handled by one female operator, to manufacture shells. Conveyors would move material between rows of automated furnaces and machines, loading and unloading without continuous human intervention. The plant did not reach fully operational status before the war ended, but automation advocates hailed the effort as indicating that future factories would break new technical bounds. In modern manufacture, "any feat that hands and body can perform can be duplicated automatically," said Barnes Company president William Barton.[3]

Just as engineering and management experts had hailed construction of Milwaukee's A. O. Smith plant for both its technical and its symbolic importance during the Depression, so the Barnes plans signified the future of labor-saving technology during the war. Once peacetime arrived, representatives of business, engineering, and science quickly built up automation into an obsession, a new gospel of postwar economics. During the 1930s, individual employers such as A. O. Smith had introduced new types of production equipment, but each instance of mechanization still had seemed unique. By the late 1940s, promoters extended the concept of automation into a universal ideal, promising it would revolutionize every area of industry.

That hope appeared most vividly right after the war in a November 1946 *Fortune* article, "Machines without Men." Physicists Eric W. Leaver and J. J. Brown articulated a theory of how to design a completely automatic factory for virtually any type of manufacture. Transport units and conveyor belts would start by unloading raw materials, then carry the emerging product through automatic lathes and assembly machines up to the final stage of automated packaging. Master controls would initiate operations and monitor each machine constantly. Photoelectric cells, thermocouples, and other testing devices could inspect parts, automatically comparing their quality to specifications encoded on perforated rolls of paper. Such a perfectly automated system was not a technical stretch, the authors informed skeptics; engineers had already developed most of the different machines and control units, so gathering them into one automatic factory became primarily a question of vision. Leaver and Brown had faith in that vision, and, for them, the beauty and power of automation lay in its versatility. Factory owners could apply the basic principles of automation to produce everything from cars to telephones, transforming the entire manufacturing sector into a virtually labor-free enterprise.[4]

To help readers appreciate such possibilities, *Fortune* set two illustrations side by side; one photograph of an old factory showed a shop floor full of workers, while a sketch for a prototype automated plant did not contain a single human. More than that, *Fortune* ran diagrams that cleverly captured how machines could duplicate every human capacity. The artists drew a human eye next to the photoelectric cell, a nose next to the gas detector, an ear next to the microphone, and a human brain next to the factory's master "electronic brain." In every case, Leaver and Brown emphasized, the mechanical sensor operated more reliably and efficiently than its "makeshift" biological counterpart. The two concluded blithely, "Nowhere is modern man more obsolete than on the factory production floor."[5]

Given that automation could perform better than human senses, *Fortune* told readers, dispensing with factory labor offered "obvious" advantages. Humans felt tired, distracted, or angry, but machines never demanded more pay and were "always satisfied with working conditions." Technology could operate around the clock, turning out an ever-greater volume of goods, and there, Leaver and Brown acknowledged, lay the potential snag. The automatic factory "would be worse than useless" if owners could not sell the enlarged output; therefore, in the interest of maintaining a mass market, the country would need some way to reemploy the workers made obsolete. Leaver and Brown did not worry long; though the "automatic factory may well loose waves of temporary unemployment," they wrote, "new machines will force the issue, force society to find a better use for men than to make them mechanical operators of machines." Automated factories would still need technicians to arrange and reset machinery, plus engineers to prepare control tapes and literally oversee operations from balconies suspended over the factory floor.

Leaver and Brown did not address the issue of retraining or whether the new automation economy could provide enough jobs to go around. For them, automatic manufacturing could be justified by the inherent superiority of machines over men and by the promise of virtually infinite production capability.[6]

CAUSE OF BREAKTHROUGH TOWARD LIFE OF PLENTY

IMPACT OF AUTOMATION begins when automated machine (upper left) produces goods quicker and cheaper, which are bought by more people and enrich the manufacturer. Manufacturer invests in still more automation, producing even more goods for material abundance and producing more leisure (upper right). But automation also displaces workers (left, center). Some of them get new jobs operating and servicing the new machine. Some switch to non-automated industries. Others enter service industries like the Laundromats. A few remain unemployed.

Fig. 7. This complex little cartoon captures a sense of idealized assumptions about how workplace mechanization could set off a virtuous cycle of positive economic relationships. Technological changes in production, represented in the late 1950s by the boxlike computer mainframe, would keep moving people, products, and money around. A couple of displaced workers, shoved to one side by the force of change, might end up at a dead end, but most would fare well. Some would head for nonautomated factories or the service sector, while others would pick up a wrench and begin servicing the new office machinery. Meanwhile, American consumers, busy enjoying their neat suburban houses and large automobiles (complete with the latest tail fins), could rejoice in the flood of goods. All's for the best in this best of all possible worlds, where technological unemployment appeared a minor detail amidst prosperity and progress. Illustration by Erdoes, *Life*, December 28, 1959, 36.

Advancement published a whole series of promotional pamphlets promising that the miracle of new engineering could generate jobs, eliminate drudgery, and churn out enough consumer goods to create a capitalist utopia.[11]

Cynics scorned the hype anointing automation as the latest wonder of the world; the *Baltimore Sun* at one point labeled automation "the Cliché of the Year." Nevertheless, advocates of "the automatic factory" had already extended their ambitions to include "the automatic office," and new technical developments seemed to bring that prospect home. In 1955, the Bank of America unveiled its "Electronic Recording Machine-Accounting" system (ERMA), which could track transactions for fifty thousand checking accounts, sort checks by magnetic code, and print monthly statements. With just nine operators, ERMA reportedly performed the functions of fifty bookkeepers. Management assured employees that rather than discharging superfluous clerks, the bank could retrain them for new positions as operations expanded. In fact, the president hoped that ERMA might have enough excess capacity to let the bank open a sideline handling payrolls, taxes, credit charges, and other accounting for factories, department stores, small businesses, and airlines.[12]

Thanks to machines such as ERMA, companies could foresee the first stages of a "coming victory over paper" (not to mention the ranks of clerical workers once needed to handle that paper). Automation had not yet become sophisticated enough to dispense with the human element entirely, John Diebold lamented; offices still needed clerks to feed information into machines. Engineers were well on the way to perfecting equipment that could scan documents and automatically turn them into code, he added, so that business would soon "eliminate the need for the typist." Such prospects alarmed Walter Reuther, who forecast that automation of insurance firms, inventory operations, and other white-collar work might bring layoffs for thousands of secretaries, clerks, and bookkeepers. The United States could not keep stretching the employment system's capacity to compensate for change indefinitely, he warned, and the incredible power of automation threatened to displace people too fast for even a growing economy to absorb.[13]

Just how many workers might automated factory and office equipment displace, and how many new jobs would the technology open up? As Depression-era researchers had ruefully observed, the multidimensional relationship between technological change, economic activity, and employment proved inherently complex. Postwar observers complained that they had virtually no statistics or other reliable evidence on which to form a basis for discussion. Once again, government agencies launched investigations to clarify the effect of machines on work. One 1955 study by the Labor Department's Division of Productivity and Technology investigated modernization at a large radio and television manufacturer. Where employees had formerly wired, soldered, and inserted electronic components by hand, the company introduced automatic machines "operating like large staplers." Assembly work proceeded "without human intervention," and by switching to mechanically printed or stamped circuit boards, the firm soon eliminated more hand labor. Though purchasing, installing, and running this new equipment cost millions, executives still planned to extend automation further, since the savings on labor allowed them to lower prices and thereby expand sales. The Labor Department congratulated the company for its efforts to retrain some displaced workers for higher-paid jobs operating machines and to reassign others to nonautomated tasks such as parts testing, tube installation and adjustment, final inspection, and packing. Meanwhile, the business had more than doubled its ranks of skilled machinists, jig-and-fixture workers, repairmen, and engineers. Researchers lauded managers' enlightened approach to employee relations, concluding that "good planning" (combined with a favorable environment for market expansion) could facilitate "an orderly transition to automatic technology."[14]

In an equally upbeat assessment of office automation, the Labor Department reported on how one large life insurance company had made "exemplary" efforts to prevent layoffs as a result of modernization. After the firm had installed a computer system to track policies and process changes, clerical needs in the main office dropped from 198 to 85 people. Management assigned nine displaced workers with experience and mathematical ability to operate the new computer and transferred eighty-seven others to different jobs (reportedly paying at least as much as old ones). Other workers either retired or resigned, but researchers noted that the gradual nature of the transition offered leeway for personnel adjustment. It took two years to arrange for and install the computer, during which time the firm experienced a natural rate of departure among lower-grade female clerks. While admitting that not all computerization would occur under such favorable conditions of market growth and high labor turnover, the study saw "no reason to believe that this new phase of technology will result in overwhelming problems of readjustment."[15]

Such reassuring evaluations by no means ended discussion. In a striking parallel to the 1939–40 TNEC investigation, the 1955 Congress held two weeks of hearings on the economic and social effects of automation, with special attention to "possible and probable displacement of personnel." Reportedly, Henry Ford II, Douglas MacArthur, and other prestigious figures declined invitations to testify, but, even so, the Joint Committee on the Economic Report heard plenty of opinions. Opening the "Automation and Technological Change" investigation, chairman Wright Patman declared that Americans' interest had compounded weekly "as the newspapers, Sunday supplements and magazines report ever new and startling developments in automation." In language directly echoing that of the Depression, Patman said it was time to ask whether automation represented a "'blessing' or 'curse.'" While the general thrust of debate covered old ground, experts in the postwar period phrased it in unfamiliar language. The committee asked speakers to clarify the strange word *automation* (not yet included in major dictionaries). In fact, experts had not yet even agreed on terminology, speaking variously of "automizing" or "automatizing" as well as "automating." Though engineers and businessmen offered technical explanations of feedback, continuous processing, and rationalization, Patman declared that he had "never attended a hearing where I knew less about the subject matter."[16]

Technical concepts such as feedback seemed straightforward compared to disagreement over the philosophical definition of automation. Advocates maintained that though the word sounded novel, it really only described a natural continuation of old-style engineering. Ford spokesman D. J. Davis told Congress, "The first use of automation I can remember was perhaps little David when he slew Goliath with the slingshot." Other speakers claimed that the flour-mill system Oliver Evans had developed back in the colonial era represented the root of modern automation. Nineteenth-century inventors had pursued increasingly bold ideas, promoters said, making the twentieth century's transition from mechanization to "supermechanization" a logical and inevitable extension. By contrast, labor representatives declared that postwar technical knowledge had jumped to an entirely different level. Machines had always supplanted human physical labor, but new engineering aimed at replacing humans' mental ability. In short, electrical union leader James Carey considered automation "new and revolutionary," while Ford representative D. J. Davis termed it "just another evolutionary phase" in mass production.[17]

The wrangling over calling automation "evolutionary" versus "revolutionary" represented more than a word game; it cut to the heart of whether the pace of change allowed individuals and institutions time to adjust. As their counterparts had done in the Depression, Davis and fellow business spokesmen complained that radicals had whipped up resentment of technology with an alarmist campaign containing "90 percent emotion and 10 percent fact." One poll conducted by a Detroit radio station supposedly showed that in that manufacturing town, listeners' level of anxiety about automation came second only to their concern over Communism. When public hysteria ran so high, William Barton said, authorities must step in to calm things down. He recommended a broad-based educational campaign teaching Americans to appreciate the need for "orderly growth of more and ever more automation."[18]

Economic science had proven that improvements in production brought prosperity and employment over the long term, making any fear of permanent displacement false, W. F. Barnes told Congress. Updating Say's Law with an atomic-age metaphor, General Electric president Ralph Cordiner insisted that mechanization encouraged a "chain reaction of economic growth: more productive machines reduce costs and prices; this increases volume of business, creating a need for more workers." Switching analogies, he suggested that recent trends revealed a "'bow wave' theory of technological *employment*." Just as a speedboat pushed water ahead of its path, automation propelled "a wave of new employment opportunities." As in the 1930s, absence of hard evidence did not weaken such faith. The fluctuating nature of the automobile market meant Ford "frankly" could not "trace in precise detail" how automation affected jobs, Davis admitted, but he felt sure that engineering meant "change always for the better."[19]

While promising that automation would raise efficiency to unbelievable levels, executives simultaneously reassured Congress that such innovation posed no danger to labor. Engineers could not work instant miracles, and automation would not be "applied overnight to every activity of man." At least in the near term, Diebold testified, automating did not make economic or technical sense for small-volume manufacturing, mining, construction, and agriculture. The fields currently "ripe for automation" (chemicals and textiles, petroleum refining, printing, and communications) employed just 8 percent of the nation's labor force. Little harm would result even if automation displaced half those workers, Diebold concluded, since postwar conversion had proven that the system could readjust up to 2.5 million people. Moreover, even in fields where automation appeared technically feasible, practical and economic constraints limited the rate at which companies adopted new equipment and so contained the human impact. The very novelty and complexity of the technology meant that once a firm made the decision to automate, planning and installation still took months, if not years. Companies could use that lead time to adjust employment, often through natural attrition. The country's workforce "continually reallocat[ed] itself voluntarily," NAM director Marshall Munce told Congress, with 8 million Americans entering, shifting, or quitting jobs each month. That fact made human resources "highly flexible," inherently able to absorb any impact of automation.[20]

Providing workers stayed flexible, they would find plenty of opportunity in an age of automatic production, advocates reassured listeners. Americans should not be intimidated by news reports that described chemical facilities run by just two operatives, Diebold emphasized. "Automatic factories will not be workerless factories. Many hundreds of maintenance men will be required," as well as supervisors, clerks, salesmen, distributors, and professionals. Automation itself would create entire new job categories. Cledo Brunetti, engineering research and development director at General Mills, predicted that the next decade would bring 15 million positions for machine operators, electronics technicians, key-punch operators, computer programmers, and systems engineers. Workers displaced from old assembly-line functions could easily take up automation-related jobs, Diebold insisted. Operating and maintaining complex equipment did not require advanced education, just a "desire to do good work, and ingenuity." Companies actually did labor a favor by introducing devices that cut off older lines of work, he told Congress. Men and women should welcome displacement as a blessing in disguise, since it opened the way for them to learn new skills and switch to more interesting jobs, gaining self-respect and higher wages.[21]

The power of automated production not only upgraded work in established occupations, promoters maintained, but also enhanced employment by creating entirely new industries. While Depression-era business leaders had repeatedly stressed that the automobile brought more jobs than horses ever had, postwar executives dwelt on the promise of television. Just as Henry Ford had developed an affordable car and offered workers the great "five-dollar day," General Electric and Sylvania claimed that their investment in automation would make television into a source of employment and a product for the masses. The industry had already realized tremendous savings by adopting automatic machinery for making tubes and screens, GE's Cordiner told Congress. Thanks to such innovation, consumers in 1955 could buy a 21-inch set for less than a 12-inch model had cost five years before. Engineers had started developing machines that could automatically apply color to tubes, and once they accomplished that, Sylvania chairman Don Mitchell promised, pent-up demand for color television would create "hundreds of thousands of jobs" in manufacturing, selling, and servicing sets.[22]

Future consumers could enjoy color television and other such terrific items, the business community maintained, if and only if manufacturers introduced more and more automation. In fact, they said, given estimates that the country's working-age population would grow by only 15 million over the next two decades, Americans should stop worrying about technological unemployment and start bracing for labor shortages. Lack of workers would cripple the economy faster than any problems of displacement could, industry leaders told Congress. Without the saving grace of automation, a labor crunch might send productivity into a nosedive and create "a situation in which the ordinary man cannot buy automobiles and other products we now regard as necessities." But with automation, ever-rising consumer expectations should foster unparalleled economic expansion, ensuring once and for all that technological unemployment would not "constitute a social or human problem of even a minor nature." In that light, simple demographics made labor-saving technology seem an imperative.[23]

Speaking on behalf of organized labor, Walter Reuther agreed on "the desirability, as well as the *inevitability* of technological progress." Though critics had spread a "great deal of propaganda" blasting unions as antiprogress, he told Congress, the CIO actually "welcomes" automation as key to both a shorter workweek and higher living standards. However, economic events usually proved more complicated than automation optimists pretended. He accused the National Association of Manufacturers of approaching the issue with "irresponsible complacency." Its pamphlet "Calling All Jobs" featured the line, "Guided by electronics, powered by atomic energy, geared to the smooth, effortless workings of automation, the magic carpet of our free economy heads for distant and undreamed of horizons." In real life, Reuther said, "economic expansion does not arise simply because we desire it." Consumers could not keep stretching their budgets, especially if companies refused to pass on the savings in manufacturing, and once market growth stalled, the output of automated production lines would pile up disastrously. The United States had experienced nasty downturns as recently as 1949 and 1953–54, showing how easily the modern economy could be "shoved off balance."[24]

Reuther charged that in their eagerness to push automation, industry spokesmen risked ignoring real hardship caused by even temporary job loss. He challenged assurances that automation would come gradually enough to prevent harm. Promoters themselves predicted "that automation probably will make almost twice as much progress in the next five years as . . . in the past ten." One Labor Department study showed that job growth simply did not keep pace with such rapid productivity gains; a 275 percent jump in electronics production between 1947 and 1952 had brought just a 40 percent rise in work. Once business started "automating the automation factories," Reuther cautioned, reemployment could never catch up with displacement. Ford spokesmen declared grandly that the "hand trucker of today replaced by a conveyor belt might become tomorrow's electronic engineer," but Reuther doubted whether retraining would come so naturally, especially for older workers. The NAM might talk about a magic carpet economy, but Americans could not trust in a "laissez-faire belief that 'these things will work themselves out.'" The country could only prevent automation unemployment through proactive steps; instituting a thirty-hour workweek, for instance, could serve as a much-needed "shock absorber."[25]

Following up on Reuther's arguments, other union leaders told Congress that automation had already caused problems for their members. W. P.

Kennedy declared that the Brotherhood of Railroad Trainmen had been hurt by the introduction of new automatic signals, tracklaying equipment, and automated yard systems which moved cars using automatic switching, speed controls, and tape storage of routing information. One such system in North Carolina allegedly had displaced 64 of about 150 yard conductors and switchmen, while a Pittsburgh version eliminated roughly 250 workers. In general, Kennedy claimed, "robot yards" tended to take over 10 to 90 percent of old yard jobs. Railroad employment had fallen by 485,700 between 1947 and 1954; while some of that decline had been because trains lost business to competing forms of transport, the brotherhood calculated that technological change had been the reason behind 195,875 of those lost positions.[26]

Where the TNEC hearings had investigated changes in telephone technology, 1955's discussion revealed that "switchboard girls" had already become objects of nostalgia. "*If one thinks at all about such things,*" chairman Patman mused, "one is forced to wonder about what happened to all of the friendly, efficient telephone operators in large and small towns who used to handle our calls." Clifton Phalen, president of Michigan Bell, reassured the chairman and told him not to upset himself over the fate of operators, who had just not been efficient enough to keep up with demand. AT&T had extended dial service to 85 percent of its operations by 1955, Phalen told Congress, and its payroll simultaneously had soared to an all-time peak. Since 1940, he explained, the number of phones in use had grown from about 17 million to 45 million, doubling jobs from 300,000 to over 600,000. Far from eliminating humans, Bell still relied on 237,000 operators to handle information requests, collect calls, and other special needs. Though AT&T anticipated extending dial equipment to over 95 percent of phones by 1960 and reducing operator assistance on long-distance calls, Phalen assured listeners that the company's aggressive sales tactics would multiply demand for service and so keep up employment.[27]

Speaking separately, Joseph A. Beirne, president of the Communications Workers of America, acknowledged that "despite intensive mechanization of local telephone calls there are over 150 percent more people employed in the telephone industry today" than twenty-five years before. He doubted, however, whether AT&T could sustain its astounding rate of customer growth. While acknowledging management's "genuine attempts" to help workers through the transition, Beirne mentioned that distressing incidents occurred nonetheless. During the 1949–54 conversion to dial, employment had fallen 80 percent at some Michigan exchanges, from 1,414 to 273 workers; over half the total, 761 people, had reportedly been laid off. Remaining workers lost any sense of security; Beirne described "women crying in restrooms, improperly prepared for new methods and fearful of losing their jobs or being pressured into unwanted, early retirement." Paeans to automation covered up such distress, "entire lives shaken up by so-called progress." Dial technology merely represented the tip of the iceberg; automatic testing equipment seemed likely to reduce the need for maintenance workers, while centralized message-accounting machines could eliminate the white-collar work of tracking calls and preparing bills. Two hundred thousand Bell System jobs might disappear by 1965, Beirne concluded, unless the company compensated for automation by reducing the workweek.[28]

Such arguments between business and labor would have sounded familiar to both sides fifteen years before, but the 1955 hearings also brought distinctly postwar twists to the discussion. In 1939, record job loss had been fresh in people's minds; the early 1950s brought the United States close to full employment (with joblessness fluctuating between about 2.5% and 4.5%). While unions worried about displacement in specific occupations, general prosperity and economic confidence removed any sense of emergency from the debate. Although the TNEC members had interpreted labor's prospects in the most pessimistic tones, their postwar counterparts ultimately concluded that, given current fiscal well-being, automation posed no real danger. To cover all bases, the committee warned that if the economy took a turn for the worse, serious technological unemployment might still arise. Prosperity had not banished the threat of mass displacement, but it had definitely made the possibility appear more remote.[29]

Cold War politics and tensions also shaped the 1955 perspective on automation. America's engineering prowess and its industrial capacity had become vital factors in the superpowers' contest for global position, which bolstered arguments for pressing full speed ahead with workplace technology. Even amid 25 percent joblessness, Depression-era critics had not generally endorsed limiting innovation; fifteen years later, the concept of a "research moratorium" felt even more unthinkable. John Snyder, president of automation company United States Industries, warned Congress, "The enemies of freedom can continuously improve their industrial processes" and so "might seriously menace our ascendancy." Soviet machine-tool makers had reportedly made "astonishing strides" lately. The United States seemed to have no choice but to push on, since automation might give its industry a decisive edge in manufacturing arms for some future "hot war." Fear of Communism created the sense of an automation race, relegating labor questions to secondary importance. Cold War political posturing also provided a second justification for automation: if American engineering could guarantee consumer abundance, it would provide visible evidence for the inherent superiority of the free enterprise system. As Reuther said, automated production could become even more valuable than the H-bomb as a way to "prove that the Communists are wrong."[30]

The Cold War atmosphere, combined with feelings of economic confidence, made challenging automation ever more difficult. 1955's discussion of machines and jobs proved less tense than the TNEC hearings. Reuther and Kennedy called for shorter hours and more support for displaced workers, but they stopped well short of demanding a ban on automation. The cultural lag theory remained influential, but union leaders expressed faith that ultimately "our economy can adjust to the challenge of automation." Americans still approached technological issues with a powerful sense of historical determinism, a feeling that the country had been set on an inevitable course. Both business and labor portrayed automation as an autonomous force racing ahead, while economic relationships, social institutions, and workers could only try to keep pace.

Beirne alone questioned the premise that anything that technically and economically could be done, must be done, with utmost speed. Democratic freedom meant giving people the choice to embrace only those options that represented true progress and to oppose any that imposed too great a cost, he dared suggest to Congress. Referring to spread of the dial system, Beirne asked, "What does it really matter if it takes 30 seconds instead of only 15 to complete most long-distance calls if we gain this speed at the price of unemployment and . . . misery? Why the headlong rush into mechanization if slower movement gives us time to contemplate what we are doing and where we are headed?" Such radical sentiments vanished into thin air, however, lost amidst talk about technological change as an unavoidable fact. Congress, business spokesmen, and most labor leaders agreed: the United States was entering an era of automation triumphant.[31]

Technological Unemployment as a Fear in Postwar Popular Culture

So was the question of technological unemployment finally closed? Not from the evidence of 1950s popular culture, which, as in the Depression, translated political and economic discussion of machines and jobs into public images and entertainment. On the very day Congress opened automation hearings, the comic strip "Blondie" showed Mr. Dithers inspecting a new device where "you put the figures in here, then merely press the button and you get the results here." Unfortunately, Dagwood had just decided, "This will be a good time to ask the boss for a raise." After Dithers told him, "You know you can be replaced by a machine, don't you?" a shaken Dagwood walked away muttering, "That *wasn't* such a good time to ask."[32]

Catchy headlines in the popular press underscored the possibility of displacement. Accompanying a report on automation in Ford engine plants, *U.S. News & World Report* announced, "Push-Button Plant: It's Here—Machines

Do the Work and a Man Looks On." The magazine's artists put statistics into visual form with a diagram comparing employment levels. The "before automation" side of the chart showed twenty-nine figures, representing the number of men formerly needed to drill holes in crankshafts, while the "after" side includes only nine silhouettes. In stark terms, writers labeled one photograph of conveyor belts, "NO MEN WANTED." Emphasizing labor's marginality in modern production, the caption on a second photograph notes that the lone worker shown was "there just to watch the colored lights on a control panel."[33]

The link between automation and workplace revolution stimulated imaginations in Hollywood, where producers of the film *Desk Set* transformed concern about office computerization into light comedy. Set in the headquarters of a national broadcasting company, the 1957 movie opened with management's decision to make operations more efficient by adopting the Electro-Magnetic Memory And Research Arithmetical Calculator (EMMARAC—a clear pun from the real-life machine ERMA). EMMARAC's inventor, "methods engineer" Richard Sumner (played by Spencer Tracy) knew that move might upset workers and grumbles, "Every time I mention what I do, people seem to go into a panic." Sure enough, Sumner's presence and rumors about bosses reevaluating personnel files rouse suspicions in the company reference library, where a four-person staff fields inquiries about geography, nature, literature, and baseball statistics.

> Peg Costello (Joan Blondell): "That Richard Sumner is . . . trying to replace us all with a mechanical brain! He's under special assignment . . . to see if EMMARAC can be adapted to this department. That means the end of us all!"
> Bunny Watson (Katherine Hepburn): "Peg, Peg, calm down! No machine can do our job!"
> Costello: "*That's* what they said in payroll . . . and as soon as it was installed, half the department disappeared!"
> Watson: "[That's] just a calculator. They can't build a machine to do our job, there are too many cross-references. . . . I'd match my memory against any machine's any day. . . . The worst thing that can happen is for us to get panicky, so let's keep this between you and me and not tell Ruthie and Sylvia."
> Costello: "Not tell them! They're down at union headquarters right now to see if there's a law against this!"

Talk of displacement hangs over the office Christmas celebration as well. Costello remarks, "If we do get canned, we won't be the only ones to lose our jobs because of a machine," to which another staffer adds, "I understand thousands of people are being replaced by these electronic brains." Resentment intensifies once EMMARAC appears; handing the operator a pile of punch cards, Watson remarks, "The complete history of the American buffalo—it too is becoming extinct." The image of computers' infallibility, however, quickly disintegrates in the film; mishearing a simple question on Corfu, the harried operator sets EMMARAC to printing pages of irrelevant data on curfews, including the eighty-stanza poem, "Curfew Must Not Ring Tonight!"

Outraged that the librarians had received notice, Sumner contacts the firm's president only to learn, "The whole darn building's been fired! That crazy fool machine of yours in payroll went berserk this morning and gave everybody a pink slip!" Once that error has been sorted out, Summer promises Watson that researchers' jobs will be safe. EMMARAC "was never intended to replace you! It's here merely to free your time for research, it's just here to help you." Given prospects of a network merger and a heavier workload, the company might well need to hire extra staff. That perfect solution resolves any question of technological unemployment; *Desk Set* presented the perfect case for automation as a source of jobs. But underneath the romance and upbeat conclusion, the film also conveyed a vivid impression of labor's anxiety. Hollywood could ensure a happy ending; in real life, Americans enjoyed no guarantee that change would turn out so smoothly.[34]

While *Desk Set* promised that automation would ensure more and better work in the future, Kurt Vonnegut was creating fiction in which automation made unemployment the American norm. *Player Piano* imagined the Faustian bargain in which a society desperate for manpower embraced technology, never to back away. "Production with almost no manpower" had proved the "miracle that won the war," but in peacetime, veterans soon realize that such a system offered little chance for reemployment. Authorities quickly suppress a wave of frustrated attacks on machinery, clearing the path for further automation by steering superfluous men into the Army or the Reconstruction and Reclamation Corps. Known even among themselves as "Reeks and Wrecks," those men kill time filling potholes and bemoaning the day when "machines took all the good jobs."[35]

The engineering described in *Player Piano* closely paralleled real postwar research at places like General Electric. Record-playback systems captured an expert worker's every movement on tape; when replayed, that technology allowed automatic lathes, presses, and drills to duplicate human skill. Such symmetry between fictional and actual automation was no coincidence; Vonnegut had worked in GE public relations during the late 1940s, experience that lent plausibility to his description of automatic factories. In his novel, vending machines dispense everything from nylons to legal documents and medical diagnoses, taking over the functions of professionals and service workers alike. A nervous barber tries to reassure himself by thinking about all the complex motions in hair cutting that couldn't be automated, only to visualize ways of doing exactly that; in the end, he himself creates the machine that makes his job obsolete. In an obsession with innovation, one manager throws himself and seventy-one other employees out of work by inventing devices to automate his whole petroleum terminal. Suddenly a full "job classification has been eliminated. Poof." In the engineering world's view, such a sacrifice served a larger purpose, realizing the long-awaited golden age of efficiency and wealth. Engineers express incredulity that corporations had ever entrusted work to humans, who would all too often make the "stupidest mistakes imaginable." Like their real-life counterparts, Vonnegut's automation proponents measured social well-being strictly in material terms. Consumerism justified automation, allowing even "Reek and Wreck" families to live in fancy houses. Some foolish people refused to admit their happiness; one housewife admits she liked having the ultrasonic washer broken, since hand-laundering gave her something to do besides watch television.[36]

Echoing the real-life way that the ASME and other scientific and engineering societies had fought talk of technological unemployment, Vonnegut's novel includes a scene in which advocates ceremonially reaffirm the wonders of mechanization. His fictional engineer-managers organize a special pageant defending the system they had created and which had created them, a morality play in which a "handsome young engineer" defeats the antitechnology arguments of an "unkempt young radical." The engineer reminds "John Averageman" that automation had made him richer than an emperor of yore and raised American civilization to "the dizziest heights of all time! Thirty-one point seven times as many television sets as all the rest of the world put together! Ninety-three per cent of all the world's electrostatic dust precipitators!"

Such boasting rings hollow to Paul Proteus, who, as son of the country's first national system director, should have been technology's most avid exponent. Looking at a photograph of his family's factory in the nineteenth century, when the humblest floor-sweepers felt "fierce with dignity and pride in their work," Proteus grows disillusioned with a world that uses robots as janitors and leaves humans without hope. Displacement had created psychological torture; in order to stay sane, men had to feel "needed and useful, the foundation of self-respect." Proteus joins the Ghost Shirt Society, a group who proclaims society's power to reject change if it does not truly contribute to human happiness. Ordinary Americans had "changed [their] minds about the divine right of machines, efficiency and organization, just as men of another age changed their minds about the divine right of kings," the rebels declared.[37]

Vonnegut's novel culminates in scenes of the Ghost Shirt revolution, when uncontrollable crowds race to tear down factories, power stations, and the centralized computer system. Shocked at the mindless waste, Proteus mourns the loss of his early innocence, his ability to revel without guilt in the engineering challenge of creating a perfect automaton. He rationalizes that once people have released their pent-up fury in an initial explosion of vandalism, common sense would reassert itself. Americans could "rediscover the two greatest wonders of the world, the human mind and hand," building communities around a healthy respect for labor. As it turns out, the citizens end up having no intention of proving "how well and happily men could live with virtually no machines." Rushing "to recreate the same old nightmare," men begin repairing smashed vending machines and tinkering with ideas for new ones. In ironic ambiguity, Vonnegut emphasized "what thorough believers in mechanization most Americans were, even when their lives had been badly damaged" by it. The national mentality had truly made automation inevitable; mechanical curiosity and an ingrained belief in technological progress would lead society straight back to the same fate of mass unemployment.[38]

Popular discussion brought the relationship between fiction and real life full circle. During the 1955 automation hearings, Vermont Senator Ralph Flanders recommended *Player Piano* to one witness as "a fantastic book which indicates the final development of automation." Vonnegut's description of the future's workerless factory had captured Flanders' engineering heart, while the social ironies apparently had sailed right by.[39]

The 1960s: Displacement as "Major Domestic Challenge"

With unemployment averaging 2 to 4 percent during the early 1950s, it seemed plausible that the country could adjust to automation after all. Yet between 1957 and 1961, joblessness shot up closer to a 5.8 percent national average, making the idea of technological change as a culprit more conceivable. Just as in the 1930s, some of the most troubling evidence appeared after the nation had begun to *recover* from recession, when economists pointed out that the upturn in production did not generate a parallel recovery in employment. At the end of 1958, General Motors announced plans to increase next-quarter production by 25 percent, while expanding the ranks of hourly employees by just 5 percent. The efficiency of automation (plus overtime shifts) would make up the balance, managers said. With a 12 percent rate of joblessness in Detroit, displaced men faced slim prospects. This phenomenon of "productivity unemployment" also seemed to appear in the steel industry, where new equipment had helped raise output per worker by 18.8 percent between 1947–49 and 1958. At the old level of productivity, steel companies would have required 637,000 men to achieve 1958 output, but they actually employed just 536,000. Such trends meant that Pittsburgh, Youngstown, and other steelmaking cities, where unemployment stood over 6 percent, had missed out on one hundred thousand potential jobs. Observers in coal-mining regions, oil-refining centers, and manufacturing towns echoed the fear that the force of automation explained "why jobs are slow to come back" even in good economic times.[40]

In 1962, the New York Transit Authority began testing a subway train that required no human operator or conductor, and the Transport Workers Union promptly threatened a strike to protest the "headless horseman" or "zombie" locomotives. The union ultimately agreed to give automated cars a six-month trial, providing they carried a stand-by motorman. During the same period, New York residents heard that Macy's had adopted the latest type of telephone switchboard, reducing its staff of operators from sixty to sixteen. The store also installed new vending machines, known as "electronic sales girls," allowing customers to serve themselves at stationery, notions, and food counters. New York dairies and brewing companies estimated they would drop a thousand jobs over the next five years (on top of one thousand already lost) with installation of automated production, bottling, canning, and packing equipment. Longshoremen faced radical technological changes in dock work, while office automation threatened to reduce the number of secretaries, clerks, and bookkeepers needed in banks, utilities, and insurance companies. With all this job elimination, New York's mayor judged that the city was fast approaching critical levels of displacement. In language reflecting Cold War emergency awareness, he called for "an early warning system" to detect exactly where jobs would vanish next.[41]

Rising concern about what one union leader called the nation's "first automation recession" played a crucial role in the 1960 presidential campaign. When questioned about science and technology, Richard Nixon spoke about the importance of education; John F. Kennedy chose to emphasize the danger to workers. Kennedy hailed automation as offering "hope of a new prosperity for labor and a new abundance for America" but added that it "carries the dark menace of industrial dislocation, increasing unemployment, and deepening poverty." While "no one—especially labor—is opposed to economic progress," he stressed, too many segments of the population did not enjoy access to the fruits of abundance. The devastated economic and social conditions of West Virginia's mining regions showed at a glance how "steady replacement of men by machines" ended up "menacing the existence of entire communities." Calling up an earlier decade's rhetoric from Robert Wagner, Kennedy said that "workers do want assurance that they will not be tossed on the scrap heap and forgotten like so many obsolete machines."[42]

After the election, Kennedy made it a priority to set the stage for official analysis of automation. Incoming Secretary of Labor Arthur Goldberg vowed that his administration would assume responsibility for the issue, ending years of "indifference to the plight of the displaced workers." In April 1961, the Department of Labor created an Office of Automation and Manpower, whose staff would break down employment statistics by industry and occupation, track and anticipate technological change, and prepare occupational guidance and retraining programs. Goldberg warned that although automation remained "an essential and desirable development" in general, "the second Industrial Revolution" would eliminate 1.8 million manufacturing and agricultural positions over the next year alone. Within a week, Senate leader Everett Dirksen announced that Republicans would set up their own study of labor trends. Kennedy welcomed "all the attention we can get by both parties into what I consider to be a genuine national problem—automation and what happens to the people who are thrown out of work." Despite that show of enthusiasm, partisan rivalries soon began to swirl, as the Republican Policy Committee sniped at Kennedy's proposals for creating a national worker-retraining fund.[43]

Washington's focus on labor displacement upset automation advocates, who disliked being pushed into a "defensive" position. Politicians should stop being obsessed about anyone getting hurt, John Diebold suggested, and encourage business to automate even more "aggressively." Nevertheless, the administration continued to spotlight the issue through a new Presidential Advisory Committee on Labor-Management Policy. Headed by Goldberg, the committee included seven representatives each from business and labor, plus five others—journalists, academics, and members of the public. The committee managed to agree on a statement of purpose which read, "Failure to advance technologically would bring on much more serious unemployment and related social problems than any we now face." The declaration went on to caution, "Achievement of technological progress without sacrifice of human values requires a combination of private and Government action." Beyond vague endorsements of education, retraining, and support for displaced workers, the panel split. AFL-CIO head George Meany and other labor leaders wanted the panel to call for immediate reductions in work hours, while Henry Ford II maintained that no one had yet proven that innovation caused serious job loss. The committee admitted that exact levels of technological unemployment remained a mystery, but the administration hailed mere completion of such a controversial project as a good sign. Kennedy spent more than an hour discussing the report with committee members in what one participant described as "a darned good bull session."[44]

Cold War considerations continued to shape discussion of technological unemployment during the early 1960s. Western analysts remarked that the Soviets had not succeeded in applying automation everywhere; reportedly, some Kremlin offices still performed calculations by abacus. And yet, American experts marveled at the sophistication of Russian technical theory. Seven-year plans put thousands of scientists and engineers to work at national research institutes, and Khrushchev allegedly extolled automation as the way the Communists "shall beat you capitalists." Given such a threat, Senator Jacob Javits told New York unions to accept that the United States must automate or else "slip back to the position of a second-class power" in the "life and death struggle of freedom" against Communism. The International Association of Machinists agreed; though delegates at the 1960 convention expressed grave concern that new electronically controlled machine tools would destroy jobs, the union declared its support for automation as a means to help America face the Soviet military and economic challenge.[45]

As if the United States did not have enough to worry about trying to keep ahead of the Soviet Union, business leaders warned that Western Europe and Japan had also started racing toward automation. The West German government offered companies special tax incentives to adopt the latest production technology, Diebold pointedly said. Reports indicated that Germany's Bremen steel-strip mill, Japan's Honda Motor Company, Sweden's ball-bearing factories, and the Renault automobile plant in France had all installed state-of-the-art machines to triple or quadruple output. American unions' unrealistic wage demands threatened to price the nation's business right out of the global marketplace, the NAM complained. Steelmakers and other industries had no choice but to pursue labor-saving strategies, the NAM maintained, since the surest way to destroy jobs in a technological economy was to have companies fail to modernize fast enough and succumb to foreign competition. The harsh realities of international economics made change inevitable, the group insisted; the only "alternative to automation is economic suicide."[46]

Domestic economic conditions through 1961 seemed to give working-class Americans reason for concern. Commentators credited automation for helping generate a 6 percent jump in national productivity, but they also blamed automation for the fact that unemployment remained stubbornly near 7 percent (twice the level Kennedy considered desirable). With five and a half million people out of work, the highest number in twenty years, some feared that automation had created an "intractable" or "hard-core" form of joblessness. The most vulnerable segments of the population would be hit hardest; due to age discrimination, older men and women confronted special obstacles in retraining and securing new positions. Younger generations faced a phenomenon known as "silent firing"; when businesses adopted labor-saving technology, those just entering the workforce were "not fired, just not hired." Automation permanently cut off employment opportunities for unskilled labor, critics warned, leaving no place in the system for high-school dropouts. Black workers might also suffer disproportionately, since the minority labor force had become concentrated in rapidly automating fields such as mining, longshoring, and factory production. Racial discrimination threatened to compound the misery by limiting black workers' access to alternate jobs, linking the issue of technological displacement to civil rights.[47]

President Kennedy made headlines in February 1962 when he described such displacement as "the major domestic challenge, really, of the Sixties—to maintain full employment at a time when automation, of course, is replacing men." Such trends placed "a major burden upon our economy and on our society," the president told a news conference, since "we have to find over a ten-year period 25,000 new jobs every week to take care of those who are displaced by machines and those who are coming into the labor market." Kennedy had, if anything, understated the difficulty, W. Willard Wirtz, Under Secretary of Labor, later said; Wirtz reported that mechanization forced the job market to produce 35,000 positions each week. Labor Department economists admitted the evidence was inferential but suggested that Kennedy's and Wirtz's numbers might be added together for a total of 60,000: predictions of an annual 1.3 million net increase in the workforce worked out to 25,000 new positions needed each week, while assumptions of a 2.7 percent annual increase in productivity multiplied by a 67.7 million working population equated to another 35,000 jobs lost each week. The business community quickly attacked that estimate, pointing out that gains in productivity would not eliminate jobs on a one-to-one basis. Just as critics had accused Roosevelt thirty years earlier of falling into the Technocrats' pessimistic mindset, Kennedy's remarks brought charges that the president had set a bad example.[48]

Given the possibility of chronic technological unemployment, Kennedy chose to focus on creating a federal initiative to promote labor adjustment. Under the Manpower Development and Training Act of 1962, the Labor Department and the Department of Health, Education, and Welfare prepared programs to retrain displaced workers in steelmaking, mining, railroad industries, and other occupations undergoing rapid change. One cartoonist symbolized that goal by drawing an automated factory on one side of a chasm and a nice suburban house labeled "a new life" on the other; a bridge, labeled "jobless retraining program," closed the gap. In an ideal world, education might offer a perfect solution. One illustrator depicted the retraining program itself as an automatic factory, picking up an old-fashioned blacksmith at one end of a conveyor belt and sending him out smiling at the other end, retooled as a missile-production specialist. In reality, society could not instantaneously remake humans to fit into an updated high-tech workplace, and practical complications doomed the dream of easy retraining.[49]

Workers themselves did not regard retraining as an all-purpose solution, and the issue of declining job security commanded pressing attention in union meetings. Several high-visibility labor disputes of 1963 turned on whether employers should have free rein to cut workers loose after new technology had rendered their functions obsolete. Three thousand members of the New York Typographers Union mounted a bitter 114-day walkout, largely to protest the prospective introduction of machines that could set type twice as fast as human printers. Publishers objected that rising labor costs had already driven seven newspapers out of business, but the union claimed it had legal control over shop rules and a moral justification to protect its members when employers refused to help them.[50]

Similar questions resounded even more powerfully in the railroad business. Carriers complained that unions were forcing them to keep 37,000 firemen on the payroll at an annual cost of millions, even after the switch from steam to diesel freight locomotives had eliminated any need for stoking boilers. Labor felt justified in fighting for jobs, given that automatic equipment now allowed a workforce of 700,000 to handle a level of rail traffic that had formerly required 1,400,000 men. The dispute echoed Depression-era battles over train-crew size, and, in 1963, the issue of "featherbedding" brought the country to the verge of a nationwide rail strike. To head off that catastrophe, Kennedy referred the matter to the Interstate Commerce Commission and further promised to appoint "the ablest men in public and private life" to examine this latest automation crisis. Skeptics doubted whether creating another commission would solve anything, but the episode reinforced Kennedy's assertion that America's economic future revolved around the technology debate.[51]

State governments followed the president's lead in multiplying committees, trying to convene some critical mass of expertise. California created a special Commission on Manpower, Automation, and Technology, leading up to a 1961 Governor's Conference on Automation. The governor signed a bill offering half a year's unemployment compensation to displaced workers who attended retraining programs. In New York, Nelson Rockefeller organized several conferences bringing public officials, labor leaders, business executives, and educators together to discuss the "mixed blessing" of automation. Political watchers interpreted Rockefeller's attention to the issue as signaling his ambition to leave a mark on the national scene. According to one estimate, state and federal agencies, colleges, unions, business groups, and other parties had organized more than three hundred programs since the 1950s to discuss the social and economic implications of automation. Some meetings attempted to bring together labor and management for constructive dialogue, but the inherently controversial nature of the subject often defeated such efforts. In putting together a symposium called "The Educational Implications of Automation," the National Education Association sketched out a set of initial principles recommending that companies shorten work hours and let employees devote the extra time to retraining. Businessmen promptly complained that such steps were both unnecessary and unfeasible, while academics and educators objected that the proposal reduced ideals of an enlightened liberal education to mere vocational prep. The NEA wryly admitted that even if their efforts had not unifed the conference, their starting hypothesis had at least stimulated a response.[52]

The American Foundation on Automation and Unemployment enjoyed a unique opportunity to bridge the labor-management gap. U.S. Industries, a manufacturer of automated equipment, and its primary union, the International Association of Machinists, had jointly created the foundation in 1962 with company funding. The foundation promised to support scholarly research on the economic impact of automation and issues of retraining, with the aim of devising practical measures by which labor and business could cooperate to minimize displacement problems. To stimulate public discussion about the future of work, the foundation brought together 300 representatives of government, unions, and management. That conference at least managed to agree on a starting point, defining automation as *the* serious issue of the day. To express that point symbolically, the foundation coined a special equation, $A^\infty + \Sigma E(md)\,C$: ever-increasing automation (A) and ever-declining employment (E) added up to Kennedy's "major domestic challenge" $([md]C)$. Members admitted that this equation exaggerated the likelihood of machines completely replacing humans and oversimplified complicated causes of job loss, but they insisted that automation remained "second only to the possibility of the hydrogen bomb" in its urgent implications.[53]

As in the Depression, the implications of technological change engaged philosophers and religious thinkers; the immediate spur in the 1960s came from Pope John XXIII's "Pacem in Terris" encyclical, which endorsed citizens' active participation in government. Picking up on that initiative, American religious leaders organized a 1965 meeting in New York at which 250 ministers, priests, rabbis, nuns, and scholars attempted to define a new automation-age theology. Western society valued individuals according to their occupation, speakers observed, but such a one-dimensional attitude could and should end as more and more jobs became obsolete. The National Council of Churches led the call for evolving a modern ethic, one that emphasized service and the healthy use of leisure. Its special "Committee on Human Values in a Changing Technological Society" announced plans to sponsor a two-year series of national conferences in which people could explore how to find "new opportunities for human self-fulfillment" to replace the psychological and social role of employment.[54]

How did Americans actually feel about automation? When asked, most voiced opinions reflecting occupational loyalties and their personal experience. Those in the automation industry and high-tech companies hailed innovation as launching a "whole new world." One plastics worker at IBM called automation "one hell of a fine thing" and doubted it would cause any significant employment problems. "Somebody has to initiate automated procedures, and the somebody who initiates them is people," he commented. Unemployed workers, however, said they had seen firsthand how introduction of machinery took away jobs. One Pittsburgh steelworker said, "New machinery, new methods in the mill did it to me—after nineteen years in the mill. They abolished my department completely because they found a better way of doing it with new machines." Steelworkers in Chicago and Gary, Indiana, eliminated after seventeen or twenty-two years, also blamed automation. Many remained without work for a year or more; a skilled molder who had landed a position as tool-room helper (at a one-third pay cut) considered himself "one of the lucky ones." In Michigan, a gear-maker at Ford observed, "Every time a new machine's put in, two or three jobs are gone. If you don't have seniority to beat it, you're a dead duck." Americans outside the blue-collar world also expressed fears about falling victim to automation. In a letter to the *New York Times*, one lawyer wrote that he was "appalled to learn that there is an 'automatic law clerk' which performs seven man-hours of legal research in a matter of minutes." Another reader described the prevention of automation unemployment as a problem "second in importance only to the prevention of nuclear warfare."[55]

Television helped draw attention to the issue of labor displacement by making it the subject of special investigations. A promotional ad for one program in 1960 asked viewers, "Are Push-Buttons Pushing You out of Your Job?" The teaser continued, "Efficiency—sometimes called Technological Progress—inevitably demands more automation. People want a new job before they lose the old one to some family-less machine. Something, or somebody, has got to give!" In the print media as well, it seemed that virtually every week (as in the 1930s) some publication ran an article offering information or opinion.

Newsweek devoted a seven-page report to "The Challenge of Automation," while *Readers' Digest* invited Americans to decide, "Automation—Friend or Foe?" Where Depression-era reporters had visited the A. O. Smith plant, journalists of the 1960s observed automated oil refineries and offered statistics to illustrate how automobile factories kept turning out greater volume with fewer men. Among Christmas advertisements for cars, cameras, and color television, the *Saturday Evening Post* told the story of how increasing mechanization of meatpacking had forced one man to settle for a night-shift janitor's job. *Life* editors wrote that new production techniques had created a "slag heap" of unnecessary workers, while *Time* described that group as "automation refugees."[56]

A 1963 issue of *McCall's* treated the issue of automation as a woman's concern, declaring that harsh new employment realities placed a grave responsibility on a breadwinner's entire family. The cover teaser promised to help a reader "pick your husband's next job"; the inside title chose a more threatening phrasing, asking, "When Will Your Husband Be Obsolete?" Anxious wives could check the list of thirty-one "dead-end jobs," ranging from appliance-assembly worker to textile worker; if a husband's job appeared on the list, he was "practically certain to be obsolete within the next generation." Instead of fearing or fighting such a trend, a woman should "help [her] husband welcome it" by encouraging him to retrain as an aerospace engineer or plastics worker. The piece, based on advice from John Diebold, told women that by preparing for such "bright future" occupations, a man could protect his family from poverty and simultaneously contribute to national economic strength. Threatened employees should have courage, an upbeat conclusion declared; the "challenge of discovering and developing the latent . . . mental powers" could "keep alive the pioneer spirit that made America."[57]

Newspapers and magazines again had to figure out ways of illustrating the abstract topic of job subtraction. Repeating a technique used in the 1930s, editors frequently ran "before" and "after" photographs; one shot would show a row of operators sitting at a telephone switchboard, while the next showed only banks of equipment, no people in sight. Like the Depression-era cartoons that portrayed mechanization as a force sweeping tiny human figures off a cliff, an editorial cartoon of 1961 showed an automated factory tumbling little men down a disposal chute and into a garbage can marked "jobless workers." The 1963 rail labor dispute spawned a cute cartoon showing a worker waving a hand shovel under the nose of a vicious-looking steam shovel, while announcing defiantly, "Come any further and I'll strike."[58]

For illustrators, the humanoid robot remained an obvious, if still unrealistic, symbol for all the machines that threatened labor. One *Chicago Sun-Times*

Fig. 8. "Profit and Loss." As the large label on the side of the factory in this illustration suggests, the force of automation had seemingly come to define the nature of modern industrial production by the 1960s. The process of discarding labor had itself been streamlined, with a chute incorporated right into the building, dumping employees into the trash can. Symbolically, technological change had reduced human beings to waste, creating a social loss to counter the gain in material output. Illustration by Doyle in *the Philadelphia Daily News*, reprinted in the *New York Times*, April 23, 1961.

cartoon showed a small but game worker falling behind a large robot in a futile footrace. The novel concept of automation gave cartoonists extra scope, though automated equipment could be difficult to draw and harder for readers to identify. A 1964 *New Yorker* cartoon got around that difficulty by showing five men being tossed out of a factory by an unseen force, while bystanders remarked, "I see Fenton's is finally automating." Artists settled on the computer as a convenient visual form to denote all automation, an image that could make a forceful point about labor displacement. One straightforward cartoon from 1962 showed a business-suited man picking up a computer printout reading, "You're fired." A more brutal warning came in a 1963 *Atlanta Journal* cartoon, which showed an enormous mainframe giving an office worker a rough kick in the pants; the caption simply read, "Feedback."[59]

Even pictures accompanying pro-automation articles could carry ambiguous undertones. A rich illustration for a 1965 *New York Times* piece by Peter Drucker represented the Machine as something like a double-headed lion. One half of the monster chisels out a large statue of an idealized laborer, smiling and ready to start work, shirt-sleeves rolled up and tools in hand. The other end of the machine-beast has a terrifying snarl and a clawed grip, which picks up tiny human figures by the seat of the pants to drop them ignominiously in a large scrap heap. The article, entitled "Automation Is Not the Villain," set out to reassure readers that technological change would not create a permanent job crisis. The drawing, however, undercut that message by conveying a deeper tension, literally merging automation's dual promise and menace into a single creature.[60]

As in the Depression, business leaders blamed political radicals and academics for encouraging inflammatory images. A 1965 *Fortune* series, "Technology and the Labor Market," complained that "social scientists engaged in a 'competition in ominousness'" had "wildly and irresponsibly exaggerated" the negatives of automation. Such a stance made it embarrassing when attacks actually came from inside the business community. John Snyder, president of U.S. Industries, told a Senate subcommittee that the Kennedy administration's calculations of the displacement rate represented a "gross underestimate." His own reckoning suggested that forty thousand jobs vanished each week, mostly because of automation. Setting out to puncture "myths" about machines creating work, Snyder urged everyone to admit the "hard truth" that "modern automated equipment requires very little maintenance." After all, he continued, "If it did not, it would not pay to operate it; and if the equivalent number of workers replaced by automation were required to build the machines and systems, there would be no point in automating." Such a traitorous line appalled *Fortune* writers, who counterattacked, saying that automation did not

Fig. 9. "Feedback." Illustrators of the 1960s began substituting the impersonal form of a computer mainframe for the image of a humanoid robot as the visual short-hand for workplace mechanization. Blue-collar men had long worried about change, but the postwar years broadened the question of displacement. Once office computers began to take over more and more data-handling functions, this cartoon and others like it implied, white-collar workers might also get the boot. Drawn by Eric in the *Atlanta Journal*, reprinted in the *New York Times*, April 7, 1963.

come into factories and destroy jobs overnight. U.S. Industries itself so far had sold only eleven of its highly touted TransferRobots, they pointed out, which had apparently displaced a grand total of one worker.[61]

In truth, while businessmen blamed troublemakers for spreading rumors about imminent job loss, the automation industry itself contributed to the impression of trouble. Even as promoters insisted that change would come

Fig. 10. This neat cartoon cleverly captures both the Aladdin and the Frankenstein images of mechanization, literally combining the promise and the threat into a single machine-beast. One half of the creature tosses aside little figures of workers (an illustrative technique popular among Depression-era artists), while the other half builds the new and improved ideal man, sleeves rolled up ready for the new day ahead. Which force was more powerful, destruction or construction? Could society adapt to technological change fast enough? *New York Times Magazine*, January 10, 1965, 26.

slowly enough to allow readjustment, they simultaneously urged employers to realize the joys of labor-saving. When a machine-tool firm bragged that its new machines multiplied production fifty times, it was easy to see why labor might worry. Advocates indicated that machines would always outperform people; a *Business Week* article describing the TransferRobot bore the title, "Tool Does Everything Except Loaf on Job." Advertisements emphasized that companies should seize every opportunity to bypass the human being. A promotion for the Kimball Datatag information handler scolded, "No data processing system is truly automated if your input requires human transcription and multiple handling." Honeywell boasted that its system of "one-man building control" allowed managers to entrust all aspects of security, climate control, and fire prevention to a single employee. Visual images reinforced the impression of labor elimination. Ads for GT&E's Automatic Electric subsidiary pictured a huge hand turning a giant key to start a whole cluster of oil derricks, with the headline, "How to make an entire oil field run itself." Singer's Friden subsidiary ran a photo of a friendly-looking woman over the caption, "Yesterday, Marion Ackerman was a billing clerk. Today, she's a whole department."[62]

Americans would simply have to accept the substitution of technology for people, advertisements indicated. "When you call for reservations" on an airline, National Cash Register told readers, "if a computer answers don't hang up." A 1965 ad from the Warner & Swasey machine tool company informed readers that any intelligent and patriotic worker must recognize how automation added to national wealth. Without technical evolution to keep manufacturing costs down, the ad warned, the United States risked ending up as poor as China, producing too few consumer goods to go around. "Anything such as automation which helps workmen increase their output is the workers' partner; anything restricting production is the workers' opponent which takes rewards rightfully theirs."[63]

Educators and social scientists put forward a multitude of suggestions in the 1960s for helping people accept and adapt to workplace change. Many started from an assumption that as the spread of automation ended more and more jobs, government would have no option but to organize large-scale "first aid" programs for labor casualties. Realists acknowledged some practical objections. Establishing a national system of mass retraining could cost a staggering amount and create bureaucratic nightmares. Adult education entailed some inherent complications and still might not be enough to help displaced workers keep up with the rapidity of technological change. Finally, retraining would ultimately prove futile if automation meant that industry and offices simply did not need as many workers as before.[64]

Realizing that the automation age might not provide enough employment to go around, unions argued that employers must begin to compensate by shortening the workweek. Business representatives disapproved of anything "which reduces the nation's productive capacity at the very time when it ought to be expanded," and warned that cuts in working hours would only give management more incentive to automate. On a more philosophical plane, the idea of a five-day weekend still bothered people intent on preserving a traditional work ethic. Sociologists such as David Riesman worried that most men and women were not psychologically prepared for such overwhelming amounts of leisure time. Television tempted people to waste time, and the passive experience of viewing fostered a mental and physical laziness that might lead to national decay. A future society might have to handle many "productively surplus" individuals, *Esquire* suggested, a challenge foreshadowed in the attitude of Beat Generation youngsters who "refuse to be tricked into behaving as though they were economically . . . necessary, when they are patently not necessary."[65]

How to make an entire oil field run itself

New AE system controls processing from ground to gathering line

Running an oil field is a mammoth job — with thousands of operations to monitor and control.

Even with a large staff, coordination is difficult. It can take 12 or more hours just to find out that a well isn't producing.

Can the job be done better? And faster?

AE says yes. At one big Western oil field, an AE control system will soon perform *all* supervisory and control operations automatically. From one location.

It will read and instantly report liquid levels. Flow rates. Water content. It will control pumps, valves and other equipment.

No one even needs to watch! The system follows the program of its built-in computer, quickly reports any unusual conditions—to an operator if one is there, or to a telephone answering service for relaying.

Benefits? More efficiency, more profit. Actions can be based on timely readings that come in minutes instead of hours.

Are your efficiency and profits as good as they could be? See how an AE control system can help.

Write the Industrial Products Division, Automatic Electric Company, Northlake, Illinois 60164. Or phone 312-562-7100.

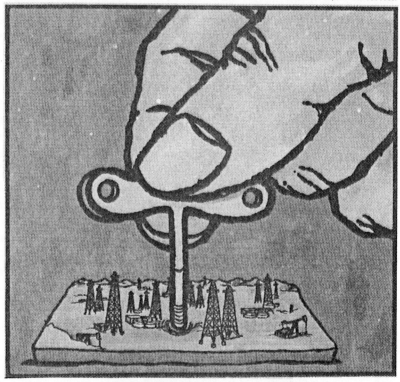

AUTOMATIC ELECTRIC
SUBSIDIARY OF
GENERAL TELEPHONE & ELECTRONICS GT&E

Fig. 11. In their advertising campaigns of the 1960s, the makers of automatic systems often touted the idea that businesses should seek to replace employees. In this case, the illustrator managed to displace most of an entire human being, leaving only one hand as the force to set an automated oil field in motion. "No one even needs to watch!"—the technology would run perfectly without any people around. *Fortune*, April 1965, 165.

If the automation economy did indeed provide employers with the means and incentive to shrink the payroll, society should consider some structural reforms to reduce the number of people seeking work. Measures encouraging more teenagers to attend college would delay their entry into the workforce and also raise their skill level, educators said. Some economists recommended that government and business offer older workers special incentives for early retirement. Politicians might consider creating a federal agency to move displaced workers into regions with jobs available, just as the Agricultural Resettlement Administration had relocated Dust Bowl families. A more extreme proposal called for "dumping" a surplus population overseas by encouraging Americans to emigrate to countries short of labor. The issue of technological unemployment even became entangled with population conditions; radical thinkers suggested that government promote birth control as a way of reducing the automation-age labor supply. One New York State official extended the Depression decade's technotax movement into an idea for solving the country's problems of job loss and civil defense simultaneously. The Internal Revenue Service could tax companies on the labor cost they saved through automation, then use the money by hiring displaced workers to build fallout shelters.[66]

One attempt to create a coherent approach for dealing with technological unemployment came from a group called "The Ad Hoc Committee on the Triple Revolution," whose membership included Michael Harrington, Gunnar Myrdal, Tom Hayden, Todd Gitlin, and Linus Pauling. Recent developments in cybernation, weaponry, and human rights issues had completely altered the basis for global existence, the committee stated, and the displacement of labor especially risked causing "unprecedented economic and social disorder." Retraining could not provide a viable solution, members scoffed, since automation would only proceed to eliminate workers from those new positions. The idea of turning displaced miners into accountants at a time when computers were taking over the office seemed like a ridiculous game—"playing musical chairs without any chairs." Far from advocating rebellion, the Ad Hoc Committee urged Americans to come to terms with automation as destiny, accepting that "the traditional link between jobs and incomes is being broken." Rather than depending on employment for survival, every family should naturally claim the "right" to a guaranteed income.[67] The overtly socialist implications of such radical proposals ensured their rejection, and the country proceeded to address displacement on a case-by-case basis. Union leaders sought stopgap measures to help their own constituents, negotiating with employers who asked, in effect, "How much will it cost us to get a free hand in introducing new labor-saving machinery?"

Since the late 1950s, longshoremen had begun to fear the impact of containerization, the strategy of packaging large amounts of material together rather than shipping it in many separate units. Loading and unloading cargo had always been dangerous, but skilled men had expressed a certain pride in their ability to maneuver difficult loads around a ship. Container technology threatened to render such experience worthless, making it possible for one man operating an overhead crane to pick up a whole 150-ton barge from the dock and transfer it to a hold. Labor worried that "the ultimate goal of the shipowners is to eliminate the longshoremen altogether."[68]

In 1964, the east coast's International Longshoremen's Association negotiated an arrangement that set a guaranteed annual income for senior members, even in the absence of work. This agreement provided some protection for older men, those likely to face the greatest difficulty in retraining and reemployment, but conceded in essence that technological advance left many longshoremen with no future. Figuring that containerization had probably quadrupled shippers' profits, labor argued that the industry could well afford to compensate labor for the human costs of its technological revolution. Containerization had accounted for less than 3 percent of cargo passing through New York in 1966, but that had jumped to 73 percent by 1975. Labor needs shifted accordingly; 48,000 longshoremen had moved 22 million tons of cargo through New York–New Jersey docks in the 1950s, whereas 12,000 workers handled 27 million tons in the 1980s. Thanks to new shipping technology, "a dozen men can accomplish in six hours what . . . once took 100 longshoremen a day," one veteran at the port of Baltimore concluded in 1993. "Automation just killed us."[69]

Troubled by such prospects, west coast longshoremen had convened a special caucus on containerization in the 1950s. The industry's strong commitment to change left labor little choice but to come to terms, union officers concluded. They would simply be "fighting a losing battle" in trying to oppose technological change through strikes, arbitration, or guerrilla resistance. To salvage some "*quid pro quo*, some specific benefits to the longshoremen as our 'share of the machine,'" the International Longshoremen's and Warehousemen's Union reached agreement with the Pacific Maritime Association in 1960. In exchange for accepting containerization, workers wanted a share of the gains. The deal effectively ended restrictive work practices and gave shippers freedom to mechanize, providing they paid a total of 29 million dollars into a fund giving union members a guaranteed income and retirement benefits. While perhaps expedient, the move marked the death of old-style longshoring. Union head Harry Bridges said, "At this rate, by the year 2000 there will be one longshoreman left on the West Coast. But he's going to be the best paid son of a bitch in the United States."[70]

The idea of setting up "automation funds" for endangered workers attracted attention in the meatpacking industry, which had made increasing use of automatic equipment for slicing meat, packaging bacon, and stuffing sausages. The country's second largest packer, Armour and Company, closed eleven old plants between 1956 and 1959, shifting the work to modernized facilities. Total business remained steady, but production and maintenance employment fell 40 percent in those three years, from 25,000 to 15,000. Subsequent contract talks focused on job security, with the union calling for shorter hours and other protection. Armour responded by proposing to create a special automation fund, effectively set at $500,000 over two years. That money, administered by a joint labor-management committee, went toward investigating "problems resulting from the modernization program and making recommendations for their solution." The first half of that mission—investigation—worked well. The fund supported research into meatpacking economics, studies of Armour's employee transfer plans and severance pay, and an analysis of how displaced workers fared in searching for reemployment. The quest for solutions proved more difficult, as the automation committee discovered when it set up an experimental retraining program for individuals displaced by the shutdown of Armour's Oklahoma City factory. Only 60 workers out of 170 met state employment service standards to judge whether they could benefit from retraining. Twenty-six of the forty-seven who completed vocational courses subsequently failed to find new positions, and those who did earned less than they had at Armour. Companies and unions could not address the problem alone, the committee accordingly decided, making retraining and other reemployment programs a matter for national legislation.[71]

Automation funds brought more headaches than relief for the American Federation of Musicians. Following years of agony over the introduction of talking pictures, musicians proceeded to worry about the impact of recorded radio broadcasts, jukeboxes, "piped music," and television. After union members had taken a stand in the early 1940s by refusing to make new recordings, Decca, Columbia, and RCA had agreed to contribute some royalties to a Music Performance Trust Fund. Instead of directly compensating individuals who had lost work, that money paid the musical community at large to put on public concerts and dances. In 1959, the fund spent $5.7 million for more than 42,000 programs, hoping to renew Americans' appreciation for the value of live performance. But, as in the Depression, such educational campaigns failed. "History has taught us that the flood of canned music was not, and could not be, halted or even slowed down by the Music Performance Trust Fund," AFM president Herman Kenin acknowledged in 1961. Furthermore, controversy over strategy dragged the AFM into a series of lawsuits. Los Angeles union members complained that forcing studios to make fund payments only gave them an incentive to import foreign musicians. Discontented parties set up an alternate union to challenge AFM leadership, driving a wedge through the musical world.[72]

From the business perspective, automation funds offered a relatively simple and affordable way to mute opposition to technological change, as well as providing a nice opportunity to demonstrate sympathy for workers. The shipping industry and Armour got significant public relations mileage out of their automation agreements. Labor had reason to accept the terms as the best possible option under existing circumstances, if companies seemed determined to mechanize and unions did not have much leverage or if automating appeared necessary to keep a company afloat. In theory, automation funds could capture, for workers' benefit, a share of the productivity gains resulting from technological improvement.[73]

In practice, as the United Mine Workers found out, members grew frustrated when they sensed that automation funds did not actually address the issue of displacement or offer much relief to affected workers. Starting in 1946, coal companies had contributed five cents per ton (a rate which rose to forty cents per ton by the early 1960s) to a Miners' Fund supporting pensions, medical services, funeral costs, and survivors' benefits. By linking payments to tonnage, planners had hoped to give the union a reason to favor mechanization. Indeed, John L. Lewis emphasized that the UMW "encouraged" the adoption of more powerful continuous-mining machines, new sorting and cleaning equipment, and bigger and faster conveyor systems. Without mechanization, miners would be reduced to starvation wages, he insisted. True, the total amount of mining employment had dropped, but those men still in the field could afford nice houses and send their children to college, "a triumph for the free-enterprise system." Though Lewis maintained that UMW members "understand the need" for mechanization, displaced miners themselves felt they had been sold out. While the Miners' Fund extended pension and health benefits to unemployed members as income permitted, those men resented the lack of more direct assistance. In 1960, after trustees announced new limits on medical coverage for jobless members, unemployed men in Johnstown, Pennsylvania, demanded a reversal. When their ultimatum failed, the protesters shut down local mines for several days, even as union officials encouraged men to cross the picket line.[74]

Such internal disputes obscured a deeper issue for labor, the fear that automation would make unions insignificant. Technology could undermine membership in blue-collar unions such as the United Auto Workers simply by ending the more labor-intensive forms of manufacture. Automation also threatened to neutralize one of labor's best weapons, the strike. When the AFL-

CIO Oil, Chemical, and Atomic Workers Union staged a 1962 walkout over job security, members discovered that Gulf Oil's supervisors and engineers could still keep the refineries running at 65 percent of capacity without them. Telephone unions foresaw that as routine calls became automated, strikes by operators might no longer result in a disruption of service. William Simkin, head of the federal mediation service, warned that automation could radically "change the balance of power" in all future labor relations.[75]

Most union leaders still refrained from blanket condemnations of technical innovation, but in an unusual deviation, George Meany opened the 1963 AFL-CIO convention with a speech denouncing automation as "a curse to society." Even then, the meeting concluded by passing a resolution praising automation as a means of raising living standards and maintaining national greatness. Labor leaders believed they could help at least some displaced workers adapt to new employment realities. The International Brotherhood of Electrical Workers sponsored classes to familiarize members with new technology. Some corporations provided support: AT&T, Bell Telephone, and the Xerox Company funded employee retraining programs, while DuPont, Sperry Gyroscope, and other firms arranged placement services for former employees. The UAW received federal funding under the Manpower Development and Training Act to teach white-collar skills to autoworkers who worried about automation cutting off jobs.[76]

Some such arrangements worked well, or at least seemed better than nothing. Meanwhile, President Lyndon Johnson muted Kennedy's earlier line on the seriousness of technological unemployment. As the 1960s wore on, the escalating war in Vietnam, civil rights, and other issues commanded attention inside and outside Washington, pushing automation out of the headlines. The economy remained strong overall; between the late 1940s and 1972, the typical American's earnings, family income, and per capita consumption all roughly doubled. The system seemed to support a decent rate of economic growth, leaving little fear that automation created an imminent risk of an across-the-board collapse in employment.[77]

A Harris poll in 1965 revealed how much Americans could sympathize with the woes of longshoremen and other endangered occupations. By nearly a five to one margin, those surveyed supported the idea of using public tax dollars to finance government retraining programs. People's consciousness of the displacement issue came through; when asked to identify the most common effects of automation, 51 percent said that it raised unemployment. Only 44 percent linked automation to more efficient production, and just 38 percent believed it would yield better and cheaper consumer goods, an unusually low number considering how strenuously the Machinery and Allied Products Institute, business publications, and automation advocates such as John Diebold had pushed the dream of abundance. Americans' ultimate feelings about new technology seemed to turn on how vulnerable they felt. Only 4 percent of those in managerial or professional positions said they felt threatened by automation; 72 percent of that group said automation would prove on balance a positive development, while only 17 percent considered it negative. Among skilled and unskilled laborers, 14 to 16 percent felt at risk of job loss; 54 percent of unskilled workers judged automation to be a negative, 28 percent positive. Overall, 50 percent of those surveyed believed that having "machines doing jobs that people did before" would bring more good than harm; 32 percent said more harm than good.[78]

Computers, Robots, and a Changing Economy

National economic conditions changed dramatically by 1973, as the oil crisis led to "stagflation," a miserable combination of recession and inflation. The downturn revived concern about displacement, a tendency reinforced by more broad-based criticism of technology. Environmentalists linked modern industry to problems of pollution, resource loss, and other ecological costs, while the 1979 nuclear crisis at Three Mile Island symbolized looming possibilities of technological disaster. In the labor movement, printers grew increasingly militant over the substitution of "cold type" video display terminals for "hot type" linotype machines. In 1975, *Washington Post* contract negotiations came to a standstill over management's desire to save money by running its pressroom with fewer employees. The union not only went on strike, but also vandalized all nine presses (reportedly attacking the night foreman, a fellow union member, who had attempted to stop them). Although episodes of small-scale sabotage had appeared in many workplaces for a variety of reasons, the *Post* incident represented one of the rare occasions when fear of job loss drove workers to comprehensive destruction. But tearing apart machines did not solve anything. Between 1970 and 1983, employment in almost five hundred newspaper composing rooms dropped 53 percent, a loss of 7,600 positions.[79]

The *Post* incident underlined the new centrality of computer technology in the way both blue-collar and white-collar workers regarded job prospects. Office workers of the 1930s had worried about bookkeeping machines, and the postwar introduction of mainframe systems made banking clerks nervous. Though the Labor Department's 1955 study of computerization at one insurance company had reported a virtually painless transition, a later investigation of nineteen insurance firms, banks, and utilities reported that electronic data processing took away five jobs for every one it created. By putting just two accounting operations on computer, one business had crossed 286 of almost 3,200 clerical jobs off its payroll. Growing need for key-punch operators did not automatically absorb all displaced bookkeepers and clerks. Contrary to advocates' promise that technology-intensive jobs would be superior to those eliminated, key-punch work meant low pay, little chance for promotion, and repetitious, stressful assignments. Whereas the original Labor Department study had praised efforts to accommodate workers, this second assessment concluded that too many employers had adopted a "let them eat cake" mentality.[80]

As it turned out, computers did not instantly throw hordes of office workers onto the street. Soaring demand for financial services had spurred a doubling of banking positions and a 50 percent increase in insurance jobs between 1946 and 1962, even with the installation of new mainframes. Nevertheless, economists noted that while total employment in offices was still rising, the *pace* of that gain had slowed down substantially. Upon hearing about the invention of a computer powerful enough to handle the functions of seventy-five older models, one worker commented, "When computers start creating unemployment among computers, it's really time to start worrying."[81]

Fortune called it merely an unfortunate historical coincidence that the "computer happened to come into widespread use in a period of sluggish economic growth and high unemployment," tempting lazy thinkers to posit a causal relationship. Enthusiasts argued that, like the automobile and television before, computers would open up vast employment in entirely new areas. By 1963, computer equipment and services amounted to a $3.5 billion dollar American market, and in the following year, the number of computer programmers in the United States passed forty thousand. Besides, in Peter Drucker's phrase, computers remained a long way from being a perfect "substitute for the brewmaster's nose." Machines could not even match a child at a simple task such as differentiating between people's voices, and so the human species would remain irreplaceable. Still, by the 1980s, development of more flexible and less expensive personal computers such as the IBM PC and the Apple Macintosh held out an unprecedented potential to transform work. Driving that point home, *Time* magazine honored the computer as its "man of the year" in 1983. Letters to the editor regarding that choice were split pro and con, and, significantly, several negative responses focused on the issue of displacement. A Virginia woman wrote, "The title should be shared equally with unemployed Americans. The computer is the big reason why so many Americans are jobless." Another reader commented, "First we have machines taking over jobs. Now one of them is taking over Man of the Year. Was this your idea, or did the computer tell you to do this?"[82]

Meanwhile, the power of the microchip redirected attention to the future of manufacturing labor. Numerical control (NC) machine tools had existed for decades, but their acceptance lagged, partly due to perceptions that the technology remained both expensive and difficult to use. In the 1980s, development of more flexible and sophisticated computer-numerical-control (CNC) equipment spurred new interest. The industrial robot loomed larger than ever as a factor in the evolution of work; Chrysler adopted dozens of spot-welding robots that reportedly did the jobs of 200 men and sped up production 30 percent. General Electric used robots to spray paint and adhesive on refrigerators and dishwashers, displacing one or two workers at each step. Technology companies announced that future generations of "smart robots" would display greater visual capacity and delicate touch control. The philosophy of "flexible automation" promised to extend the economic advantages of robotics to even small production runs.[83]

Where technology advocates of the 1950s had justified the call for intensive automation by pointing to the Communist menace, economists in the 1980s pointed to Japanese competition. Just as Cold War politicians had worried about a missile gap, the new danger seemed to be an automation gap. One consultant referred to foreign robotics as a wake-up call, a "Japanese sputnik." The number of robots in American industry had jumped from 1,300 in 1979 to almost 5,000 units in 1981, but Japanese factories allegedly had up to 10,000 robots in place already, with the encouragement of the country's powerful Ministry of International Trade and Industry. *Fortune* reported on several Japanese plants where robots worked all night under the supervision of a single human. In the ultimate blow, the latest Fujitsu plant would use robots to build one hundred more robots each month. The consequences of falling behind seemed all too clear; Japan had recently averaged productivity growth of over 7 percent a year, while American productivity had reversed, slowing to a negative .9 percent by 1979.[84]

Though Japan might lead in recent productivity growth, experts argued that the United States could yet retake the lead, thanks to its native talent in computer engineering. Promoters reasoned that business would gain an obvious economic advantage from the introduction of robots, considering that union wages had risen to $15 or $20 an hour. Even given an initial price of $40,000 or $70,000 dollars, the "steel-collar worker" averaged a cost of less than $5 an hour when amortized over three shifts a day. One GE executive explained that the company had needed to go through a learning process, but once managers understood the new technology, they would start "attacking where we have most of our people." Such prospects horrified one reader of *Time*'s cover story on robots, who wrote, "The ultimate insult to the blue-collar workers standing in long lines at the unemployment office will be a Civil Service robot electronically reporting, 'Your claim has run out.'"[85]

A 1981 study by Carnegie Mellon University suggested that within a decade, robots might take over one million automaking and other industrial jobs, perhaps two million more after that. Unemployment jumped from 7.5 percent in 1981 to 9.5 percent in 1982 and 1983, drawing further attention to the issue. The International Association of Machinists drafted a "New Technology Bill of Rights," stipulating that all displaced workers deserved to get help with retraining and reemployment. Echoing old technotax ideas, the union argued that "communities, the state and the nation have the right to require employers to pay a replacement tax on all machinery, equipment, robots, and production systems that displace workers." The IAM further asserted that labor organizations should enjoy "an absolute right to participate" in a company's decision whether or not to introduce new technology. By giving workers a seat at the table, such measure could ensure that machines would "be used in a way that creates jobs."[86]

Economists of the early 1980s worried that with the decline of manufacturing, the United States had begun to lose its employment base. While theoretically automation might release assembly-line employees to move up into more rewarding positions, many communities simply did not offer enough opportunities to replace those lost jobs. Robert Kuttner, among others, warned that the disappearance of midlevel manufacturing jobs might force many families into a downward spiral. Such trends could create a segmented or dual labor market, leaving the country economically and socially polarized. Elite professionals with a high-tech educational background could command healthy rewards and job security, while a less-skilled underclass faced unemployment or, at best, low-wage, dead-end work. The mid-1980s brought evidence reinforcing the idea of a declining middle class; 58 percent of all new jobs created between 1979 and 1984 yielded an annual income under $7,012. A changing economy might open up new positions, but if those jobs also paid little, then displaced men and women might end up losing economic ground even once they succeeded in finding new work.[87]

Some employment watchers argued that the elimination of traditional smokestack industries would not matter once the country began to see how many opportunities opened up in computers, robotics, and biotechnology. Critics derided the belief that high-tech development could perform economic miracles, accusing the techno-optimists of inflating expectations. Growth of high-tech business seemed likely to occur relatively slowly compared to the rapidity of job loss in the manufacturing sector. Individuals without advanced

Fig. 12. This cover illustration from *Time* shows the automation ideal in its most extreme form, creating the picture of an all-powerful, multitasking robot simultaneously taking over the jobs of farmers, manufacturers, and even chemists. The huge machine literally dominates the scene, reducing the lone human operator to a miniscule position at the bottom corner—where, ironically, the machine-applied address label pasted onto the magazine almost totally obscures his image. *Time*, December 8, 1980.

technical knowledge or even basic reading and math skills would not find it easy to begin working with computers. Ordinary Americans themselves doubted whether the economic adjustment would come so easily. Even at two high-tech information-processing companies, almost half the workers surveyed disagreed with the assertion that "in the long run, high-tech work will mean more jobs for more workers." It was not that the doubters considered innovation an absolute evil; 75 percent believed that automation would make future jobs cleaner and safer, and all but 4 percent liked the inherent challenge of high-tech work. But when it came to the well-being of labor, almost 65 percent denied the principle that "organizations must always adopt the newest technology even if it means some workers must lose their jobs." Given economic pressures in the business world, such views might seem self-destructive; a firm that fell behind competitors in introducing the latest techniques might collapse, throwing all out of work. At least in the abstract, those polled insisted on the importance of technological choice, giving people the option to delay or reject changes that appeared to violate fundamental human values.[88]

Ongoing concern over job loss made sense considering that through the 1980s, unemployment averaged over 7 percent and sometimes hovered closer to 15 percent in troubled Midwestern states. Ronald Reagan adopted the phrase "Morning in America" to express his faith in supply-side economics and national superiority, while a defense buildup and other government expenditures propelled budget deficits to new heights. Popular culture defined the emergence of a "yuppie" (young urban professional) lifestyle, one that celebrated the free-market glories of moneymaking and conspicuous consumption. Emphasis on material possessions separated society into groups of consumers and excluded alternate measures of well-being, such as community responsibility, racial harmony, moral purpose, and democratic political participation. The 1980s would not be the time for Americans to face tough questions about the relationship between consumerism and progress, between computerization and jobs. That issue would only resonate again during the early 1990s, as national economic stress generated new angst about the implications of workplace technology.

Epilogue

Revisiting the Technological Unemployment Debate

THE ISSUE OF technological unemployment had never vanished; the post-war gospel of automation only raised the stakes of the debate. Engineers made vast strides in developing robotics and computer power, technical achievements that in theory could permanently revolutionize humans' role in the workplace. In practice, the automated future did not arrive overnight. Entry costs and practical complications deterred many factories and offices from instantaneously embracing the latest production equipment. A prevailing sense of economic well-being also helped dampen the urgency of debate, quashing fears that technological unemployment would soar to epidemic levels. And yet, through the 1950s and 1960s, miners, musicians, longshoremen, and railroad men contended that changing technology threatened to destroy at least their specific occupations. An underlying concern about displacement continued to haunt American labor, resurfacing amidst a climate of economic uncertainty.

1990s High-Tech Angst—Insecurity and Inequality?

The early 1990s' downturn sparked alarm over employment trends, especially as the philosophy of "downsizing" gained vogue in business culture. Advocates preferred the term "rightsizing," explaining that too many companies jeopardized their long-term success by overexpanding payrolls. Consultants recommended dramatic action, slashing the ranks at Sears, General Electric, IBM, and other major established companies. Each week's headlines carried news of downsizing at more firms, moves that posed a unique threat to America's upper-middle class. The postwar period had felt like a golden age for salaried professionals, when economic expansion promised to make their skills and credentials increasingly valuable. A cartoonist of the 1960s had shown two assembly-line workers remarking, "Wait till automation reaches management levels—then we won't hear so much about learning to live with it." For years, managers appeared relatively immune; the 1980–81 recession saw three layoffs of blue-collar workers for every one white-collar layoff. The years 1990 and 1991, by contrast, brought one downsizing in management, professional, or clerical staff for every two blue-collar reductions. Wall Street endorsed the mantra of "lean and mean," as investors rewarded firms that announced cutbacks with a leap in stock prices. That mentality transformed job elimination into a corporate virtue. Jane Bryant Quinn warned, "Firms rarely downsize once; no sooner is the blood staunched from the first reduction than management comes back for a second or third blitz."[1]

As the popularity of downsizing forced many Americans to reexamine the stability of employment, it reopened talk about how much technological change contributed to job loss. Suggestive evidence emerged from one American Management Association study, which showed that 13.6 percent of firms downsizing in 1993–94 specifically linked cuts to automation or other new technology. Experts in "reengineering" promised that computers could eliminate many functions of middle management, such as gathering data and monitoring activity. The idea of substituting machines for people might appeal to executives seeking a "quick fix" for corporate doldrums, but it would not help if a firm's true problems lay elsewhere. In an AMA review of five hundred companies with post-1987 layoffs, about half admitted the cuts had not raised profits, and two-thirds reported that downsizing had failed to improve efficiency. Among the most troubling repercussions, over three-quarters of firms experienced declining morale. The experience of watching friends and colleagues terminated could leave the "survivors" worried about their futures and resentful of increased workloads.[2]

Such incidents fueled a growing feeling of vulnerability in America's workforce; writers at *Business Week* commented that only ostriches with their heads in the sand could feel safe anymore. The magazine advised readers to adopt a defensive posture and prepare for the worst. Smart individuals should begin researching alternate careers and building a cushion of savings to keep themselves afloat. Reports suggested that cutbacks could strike virtually anyone, anytime, anywhere; downsizing had taken a toll even in well-off suburban enclaves such as Chicago's North Shore. Economic recovery would not bring back enough jobs for all the "corporate castoffs," *Business Week* declared; 2 million middle-management positions had vanished for good. One study suggested that of 5.6 million individuals displaced since 1987, less than 50 percent had secured new full-time work five years later, and of those, all but 22 percent had had to accept lower pay. Victims of downsizing "face permanent loss of the income, possessions, and status long considered the defining elements of the middle-class life." To classify such Americans, trend-watchers coined the phrase "dumpies," standing for "downwardly mobile professionals."[3]

To the extent that the 1990s economy did generate jobs, they often seemed less secure, less rewarding positions than those that had vanished. Official surveys of unemployment did not acknowledge involuntary part-timing as an indicator of economic stress, but up to 40 percent of people holding part-time jobs expressed a preference for full-time positions. Between 1982 and 1993, contingent work grew at ten times the rate of total employment, making temporary-help business Manpower Inc. the nation's largest private employer. While clerical positions still accounted for the bulk of temp work, companies were hiring increasing numbers of engineers, scientific researchers, computer programmers, accountants, lawyers, and managers as temps. Federal rules allowed the government to keep postal workers and other personnel on a "temporary" basis, without benefits, for as long as four years. IBM, GE, and other manufacturers hired temporary assembly-line labor at hourly wages two to three dollars below the permanent staff working next to them. Some men and women had horror stories of being fired, then hired back the following day through a temp agency, minus insurance and other benefits.[4]

Experts warned that replacing full-timers with temporary workers brought hidden costs; temps possessed little experience with a company's unique needs and less commitment to long-term goals. From a bottom-line perspective, slicing payroll obligations made sense, especially considering the rising cost of health coverage. Observers spoke of a fundamental shift in business principles, creating the age of the "throwaway worker." Businesses that had once cultivated employee loyalty came to prize staff minimalism, following a philosophy of "just-in-time" hiring to match "just-in-time" manufacturing. Firms could develop a two-level employment structure, retaining just a core of vital employees and devolving all other functions to peripheral workers. Extrapolating such trends, business consultants envisioned the rise of a "modular" or "virtual" corporation, consisting of nothing more than an electronically linked network of freelance people, rotated in and out for specific projects.[5]

Back in the 1930s, educators and labor analysts had started warning students that the rapidly changing nature of production technology might force them to change jobs two, three, or more times. Career counselors of the 1990s told college graduates to anticipate a dozen changes in employment, crossing two or three occupational lines. A tradeoff of job security for mobility might appeal to well-educated, entrepreneurial individuals hoping to win pay hikes and advancement by jumping between different forms of work. But less advantageously positioned people might easily fall through the cracks in what *Fortune* called "the new Darwinian workplace." A job market that emphasized survival of the fittest encouraged a "blame the victim" mentality. Henry Ford had insisted that mechanization left plenty of opportunity for intelligent and ambitious workers, while other Depression-era observers called for eugenicists to breed out the nation's misfits. In reality, employment depended on many factors beyond an individual worker's power; the 1990s brought cases in which a factory or an entire industry moved overseas. Nevertheless, labor had to sink or swim; rapid economic changes tested a person's capacity to adapt, and, by definition, those unable to adjust did not belong. The new workplace ideal rested on a premise that humans could and should evolve as fast as computer technology. In order to promote a corporate culture that embraced risk and grew through stress, some employers sent staff on rock-climbing courses and other outdoor adventure programs.[6]

Optimists insisted that flexibility could empower ordinary Americans, but such rhetoric produced only an illusory sense of control for many. Economists such as Frank Levy warned that under recent economic conditions, losers threatened to outnumber winners. Earlier decades had been different; between 1947 and 1969, average family income had virtually doubled, from about $15,000 to $29,000 in constant 1987 dollars. But between 1973 and 1986, average family income had risen just 6 percent, to $30,670. Nonprofessional wages had stagnated, creating the danger of a declining middle class. Even as Washington slashed top income tax rates, more people lost purchasing power thanks to rising social security taxes, state sales taxes, local property taxes, and medical expenses. Younger Americans feared they could never afford to live so well as their parents, whose generation had profited from a lucky postwar coincidence of rising wages and soaring home values. The United States had entered an "age of diminished expectations," Paul Krugman wrote; the 1980s represented "the first decade since the 1930s in which large numbers suffered a serious decline in living standards."[7]

Political commentators cautioned that such economic trends threatened to polarize America's population. Robert Reich foresaw the United States moving toward a two-pronged economy, in which minimum-wage retail and food-service workers would keep losing ground to "symbolic analysts"—the consultants, bankers, lawyers, and scientists who worked in the high-powered information-age sector. The "fortunate fifth" would buy homes in gated communities and send their children to private school, leading a life separated both economically and personally from other citizens. Mickey Kaus emphasized that technology itself would tend to widen the gap, since fortunate youngsters exposed to computers at home and in well-equipped classrooms would get a

head start preparing for high-tech jobs. Computerization would place a premium on advanced education, while also making it less feasible to retrain people without college degrees, who "may simply not be needed anymore."[8]

Bill Clinton's 1992 presidential campaign hammered away at the message
"It's the economy, stupid," as financial uncertainty blew apart George Bush's
post–Gulf War popularity. In a cover story entitled "The Job Drought," *Fortune*
warned, "The Great American Job Machine, which once routinely churned out
millions of high-wage jobs . . . is shifting gears—downward." Media reports
highlighted the fact that employment at Fortune 500 companies had dropped
25 percent between 1981 and 1991, a loss of 3.7 million jobs. True, the period
from 1979 to 1989 had brought 13.6 million new jobs, but Labor Department
numbers showed that more than 4 million of them, 36 percent, came in food
service, retailing, and other low-end sectors, paying wages below the official
family poverty line. Census figures revealed that the proportion of full-time
workers holding low-wage positions had risen from 18.9 percent to 25.7 percent during the decade, pushing down America's median weekly wage from
$409 in 1979 to $399 in 1989 (in constant terms). According to one calculation,
only 15 percent of new jobs created in 1988 offered health coverage; even factoring in benefits, hourly compensation for manufacturing workers fell from
$14.89 in 1980 to $14.31 at decade's end.[9]

Even as business began reviving from the 1990–91 recession, the trend
toward downsizing appeared to escalate. Cutbacks in 1993 ran 13 percent over
1992 and then soared to a record monthly high of 104,000 in January 1994.
Companies such as Compaq Computer, Proctor & Gamble, and General Electric, which all enjoyed strong profits, announced plans to reduce employment.
Downsizing, initially promoted as a short-term emergency measure for troubled firms, had evolved into an all-purpose corporate strategy. Such evidence
prompted talk of a "jobless recovery," terminology that echoed presidential-
level discussions of 1937 and 1962. Both Franklin Delano Roosevelt and John F.
Kennedy had worried that the rise of mechanization had severed the logical
connection between national economic activity and levels of employment.
Once employers had taken advantage of a downturn to install technology
allowing them to operate with fewer workers in basic positions, even an economic rebound would not automatically restore the former number of jobs.
By some accounts, in 1994 the nation remained 3 million jobs short of what
the economic recovery should in principle have created. In a cover story,
"What Ever Happened to the Great American Job?" writers at *Time* proclaimed
that "the rules of the game have changed forever," a situation intensified by
"the relentless and accelerating pace at which technology is changing work as
well as every other aspect of life."[10]

Posing the burning question "What Happened to the Jobs?" writers for *Fortune* answered, "Machines Do It." As one CEO explained, "As business becomes more automated and productive, we need fewer bodies." Given low interest rates and sharp declines in the price of new equipment, electronics companies and automobile makers found it more attractive to add machines rather than people, especially considering the soaring expense of health care and other benefits. Managers at General Electric boasted that automation let them double the output of circuits without putting more men on the factory floor. "Technology has permanently reduced the number of entry-level jobs," *Fortune* warned college classes of '93. Business majors would see insurance firms turning to computers for number-crunching and analysis, while computer-aided design and manufacturing (CAD/CAM) programs threatened to curtail the hiring of young engineers. Even the birth of high-definition television, interactive information technology, and other wonderful new high-tech industries might not restore employment, warned *Harper's*. "New jobs will be created by robotics, but more will be lost," since "robots are getting smarter and more agile all the time."[11]

New Era of Technological Unemployment?

The reenergized question of technological unemployment resounded for workers in some specific occupations, where 1990s fears strikingly paralleled those of the 1930s and 1960s. Reminiscent of how the spread of talking pictures had alarmed Depression-era musicians, the AFM campaigned in 1993 against Broadway's adoption of synthesized music equipment. Producers objected to contracts that obligated major theaters to hire at least twenty-four to twenty-six union members, even for shows such as *A Chorus Line*, which contained parts for just eighteen instruments. Producers had become obsessed with the bottom line, New York musicians complained, seeking "to eliminate first this instrument and then this section until they have done away with all the live musicians and are left with a dead art form." Just as musicians in the Depression had accused Hollywood of cheating moviegoers by substituting cheap recorded sound for the subtle passion of live music, Broadway artists of 1993 called it unfair for producers to "fool" audiences into settling for synthesized sound. Given the ticket prices commanded for seats at hit musicals, the union said, patrons deserved a real orchestra. Demonstrators at an AFM rally in Times Square carried signs reading "Jelly Couldn't Jam Without Live Music!" and "Theatergoers—Crazy for Us!" (referring to the shows *Jelly's Last Jam* and *Crazy for You*).[12]

A theater spokesman denied any intent to end live music, saying that any such move would prove suicidal, "damaging our product and alienating our audiences." But already for one week in Washington, D.C., after the Kennedy Center's orchestra had walked out over stalled contract negotiations, the theater had performed *The Phantom of the Opera* to a digital tape recording handled by the conductor and a single sound engineer. New music technology had become an unavoidable element in labor disputes; reportedly, *Cats* and other Broadway productions had already taped synthesized versions of their scores, ready for use in case of strikes. The New York AFM ultimately backed down, acceding to a clause allowing productions in "special situations" to proceed with fewer musicians than the union minimum. Defenders of electronic music argued that the technology had actually increased show-business employment by providing music for road productions, local theaters, and dance troupes, all of which found a full orchestra prohibitively expensive. Broadway musicians' continued anxiety meant that the battle over synthesized music soon erupted again. The AFM insisted that, for historical accuracy, the 1960s-style revue *Smokey Joe's Cafe* should use a genuine string quartet. Producers won that skirmish, convincing arbitrators that the expense of adding four performers ($1,500 per week each in salary, benefits, and taxes) would push ticket prices to undesirable levels.[13]

The 1990s similarly revived talk about the obsolescence of telephone operators. Computers equipped with advanced voice-recognition software could place collect or card calls, offer directory assistance, handle service requests, and dispatch personnel. Thanks in part to the automation of customer contact, AT&T in 1992 announced plans to shut thirty-one operator centers, one-quarter of its total nationwide. Such a move would displace 3,000 to 6,000 operators, plus 200 to 400 managers. Maintenance staff feared that repair needs would shrink with the spread of fiber optic networks, high-speed digital switching, modular equipment systems, and computerized diagnosis of problems. Such innovations should create positions for systems engineers and experts in satellite communications, while demand for cellular phones might require hiring managers and maintenance staff. Nevertheless, the Bureau of Labor Statistics concluded that, on balance, technical sophistication tended to reduce work opportunity. Industry productivity had risen 5.9 percent per year between 1967 and 1988, largely thanks to mechanization, while job growth had averaged just 0.4 percent per year. After 1979, in fact, telephone-related employment actually had fallen 2.4 percent annually. The shift to high-tech equipment complicated prospects for retraining and reassigning displaced individuals, commented the *New York Times*. Many "workers whose skills are specific to running a plain old phone network may have trouble hitching a ride on the vaunted 'information superhighway.'"[14]

Depression-era Americans would have been familiar with talk of machines displacing telephone operators and musicians, but 1990s technology also seemed to endanger occupations that had felt safe sixty years before. Bank tellers and managers noticed how the introduction of electronic banking systems radically altered the face-to-face nature of their work. Many account-holders embraced the convenience of doing business by machine; by 1993, almost half the customers at ten major banks used ATMs for all cash withdrawals. More than one-third of workers in the private sector had arranged to have paychecks deposited directly into their accounts. Though some older bank customers still preferred interacting with humans when handling money, banks set out to discourage expectations of service. In 1995, First National Bank of Chicago announced it would charge three dollars per transaction for those who insisted on using counter service for withdrawals, deposits, and other simple requests. Banks hoped that inducing customers to patronize ATMs would allow them to consolidate operations; some banks closed as many as one-third of all their branches in the early 1990s as a wave of mergers spread through the industry. The president of Citizens Bank remarked, "The beauty of the ATM is that it's a peopleless activity."[15]

As feelings of economic insecurity spread, Americans across a wide range of occupations began to express dismay about the way that technological change seemed to imperil their jobs. In Chicago, newspaper sellers complained about the proliferation of self-service vending machines. Steelworkers saw manufacture in their industry shifting to sophisticated minimills equipped with computerized furnaces, permanently reducing manpower needs. Postal workers realized only too well how expanding automation had altered the whole process of sorting letters. In the modernized post office, bulk mail preprinted with routing information went directly into barcode scanners, while advanced optical character recognition technology could read and direct handwritten envelopes. Such machines, run by just a few employees, could handle 40,000 pieces of mail per hour. Adding to the unease, the U.S. Postal Service projected that by the year 2000, the popularity of faxes and e-mail might eliminate 25 percent of its business.[16]

Uncertainty about how mechanization might affect jobs exacerbated labor tensions and complicated important contract disputes. The 1993 battle between Caterpillar and the United Auto Workers partly revolved around the company's use of robotic carts and automated assembly technology, which the union blamed for steep cutbacks in employment. Even as timber workers in northwestern states fought against environmental regulations protecting the spotted owl, analysts warned that mechanization posed a far greater threat to the region's labor. According to one Weyerhauser foreman, new sawmills equipped with computers and laser-guided saws could, with half as many employees, surpass the productivity of older facilities. One illustrator captured the controversy of impending technological displacement with a cartoon showing a logger taking out his aggression by attacking a spotted owl, all the while ignoring the automated buzzsaw chasing after him.[17]

Even gas station employees became nervous, the *MacNeil-Lehrer Newshour* reported, after hearing about the development of a self-cleaning washroom. At the touch of a button, the module could spray all walls and fixtures with soapy hot water, then dry them in minutes. With customers already using self-serve fuel pumps and washing their own windshields, a large station needed just one worker to take money—a job itself undermined by installation of pumps with self-service credit-card readers. One station manager admitted that while he loved the prospect of new savings and efficiency, he found invention of self-cleaning bathrooms "kind of worrisome, because eventually they're

Fig. 13. Where did workers' problems really lie? Loggers in the Northwest might blame environmental regulations protecting the spotted owl's habitat, but this illustration accompanying an article on the controversy pointed out that the force of automation had also put jobs in jeopardy. The artist's image of an automated buzzsaw is a terrific symbol, capturing a sense of how difficult it could be to escape from the rapid, inevitable course of workplace technological change. Illustration by Eleanor Mill, for "Timber Troubles," *Washington Post*, April 2, 1993.

going to find . . . a computer to take my job." Economist Robert Gordon ridiculed such notions, declaring that mechanization represented "an unambiguous good." Supposing that innovation doubled productivity growth, the profits realized would provide plenty of funds for job training, and then "it's all a free lunch." Fellow economist Richard Freeman countered, "There's no rule that everybody benefits from productivity growth," and in a rapidly changing technical environment, it's "not a very good deal to be a less skilled worker." Such Americans would have to accept the hard reality that "technology's going to march on" regardless, correspondent Paul Solmon concluded, which made adult education the only way to minimize social costs. "If we don't equip our workers with the skills to keep pace, they could be a drag on economic growth instead of contributing to it," he told PBS viewers, "in which case productivity itself could be a very mixed blessing."[18]

Since the 1930s, observers had often suggested that the inevitability of technical change would force men and women to upgrade their skills. Machines could rivet auto frames and connect phone calls, but they had not taken over the lawyer's chair or the CEO's suite. Optimists predicted that job growth in management and the professions could make up for declines in manufacturing and agriculture. In the 1990s, development of increasingly sophisticated computer applications ripped a hole in the comforting assumption that a specialized education would protect a person from technological unemployment. Client-server computer networking could collect sales figures and prepare financial reports for top executives, allowing them (in theory) to bypass lower managers. According to one projection, 70 percent of jobs in corporate finance and accounting departments might disappear within a decade. The popularity of income tax software for home computers had already started chipping into the business of professional tax preparers, while programs to help people write their own wills could take that business away from lawyers. "Think twice before investing years" developing a knowledge of accounting, law, medicine, travel agentry, financial planning, insurance, or library work, *Forbes* cautioned. "All these professions are beginning to face serious competition from computer programs."[19]

Displacement of professionals on such an all-encompassing scale threatened to exert a real toll on the modern urban economy, the *New York Times* warned. Employers were "making quantum leaps in learning how to produce ever more work with ever fewer people," a front-page story announced. Computerization allowed companies to reengineer document handling and compress information management; Merrill Lynch had closed seven regional service centers as its need for clerical and management workers dropped. Given such trends, the *Times* predicted, hundreds of thousands of New Yorkers might face technological unemployment, and the ensuing drop in consumer spending could spread distress across the manufacturing, retail, and service sectors. Over the long run, computers might produce the efficiency gains necessary to "rejuvenate New York's ailing economy, by increasing its competitiveness in the intensifying battle for global markets." In the meantime, if workplaces from Wall Street to Madison Avenue cut back too far on jobs, city life might temporarily turn disastrous.[20]

Economic pessimism led to the rebirth of a particular species of literature, the popular book on technological unemployment. Just as Stuart Chase, William Ogburn, and many other Depression-era writers had turned out works warning about the power of machines to replace people, so authors of the early 1990s warned that computers would force the United States into a new age of displacement. "We are in big trouble," declared sociologists Stanley Aronowitz and William DiFazio in *The Jobless Future: Sci-Tech and the Dogma of Work*. Manufacturers had used numerical control devices, laser-guided machinery, and robots to build factories where "millions of square feet of space stood relatively empty of living labor." Retail stores had cut down on labor requirements by adopting computerized checkout and inventory systems. General Electric executives hoped that introducing CAD/CAM technology would reduce need for drafters and engineers. Ironically, the latest innovations had even begun cannibalizing high-tech jobs; sophisticated computer programs could automatically construct basic software, driving routine programming out of existence. Automation might bring a small number of people exciting work and high salaries, but those "highly publicized benefits" had been "vastly overblown." Recent economic trends would lead to a day of reckoning, warned Aronowitz and DiFazio, the point at which accelerating "technological change in the midst of sharpened internationalization of production means that there are too many workers for too few jobs."[21]

Aronowitz and DiFazio criticized easy assumptions that retraining could help everyone adapt to the high-tech future and place them in better jobs than those that had been lost. Even if a future computer-age economy managed to create more positions than it destroyed, too many social and practical obstacles complicated prospects for mass retraining. Children of longshoremen had once enjoyed a good chance of following their fathers into waterfront work, but since containerization had foreclosed that option, the world had not opened up many alternative opportunities. Economic disadvantage often left such second-generation members in a permanent second-class status, "unable to accumulate the requisite cultural capital to qualify for employment in one of the knowledge industries." Unless the country wanted to see such casualties increase, Aronowitz and DiFazio concluded, "job-destroying technologies

... should be rigorously *evaluated*" in terms of their impact on both communities and individual workers. Significantly, while the two spoke of a "need to reconsider the pace of technological change," their book did not fundamentally challenge the ultimate inevitability of workplace computerization. The authors continued the tendency to grant innovation an agency of its own, placing it on a historical course forever beyond the control of ordinary Americans.[22]

Where Aronowitz and DiFazio spoke in 1994 of a "jobless future," Jeremy Rifkin in 1995 predicted the advent of "a near-workerless world." *The End of Work* blamed the modern obsession with automation on postwar business managers, who had dreamed of no longer having to deal with human quirks and troublesome unions. "By the mid-decades of the coming century," Rifkin predicted, their wish would be fulfilled, since "the blue collar worker will have passed from history." Engineers would have perfected manufacturing robots, designing self-correcting machines that could even reprogram themselves. That technical ideal would spread into agriculture, where researchers had started developing robots to plow, seed, and spray fields. Retail employment would drop in parallel with the advance of electronic shopping. Even entertainers might find themselves competing against "synthespians" for parts. Digitalization had already brought Humphrey Bogart and Fred Astaire back to life, inserting their images in new commercials. Producers of the 1997 movie *Titanic* had eliminated some need for human extras by using computer graphics to generate crowd scenes.[23]

Like Aronowitz and DiFazio, Rifkin challenged classical and neoclassical economic assumptions that economic growth would naturally create enough jobs to absorb displaced workers. The rise of high tech would not generate a huge number of positions, since fields such as biotechnology were simply not labor intensive. Development of new computer gadgets and other exciting consumer items would not create any boom in manufacturing employment once production systems had become thoroughly automated. Rifkin derided "naive" dreams of retraining millions of displaced manufacturing workers or salesclerks as computer technicians or molecular biologists, pointing to functional illiteracy rates supposedly approaching 33 percent. African American males in urban communities had become the "first casualty of automation," and in coming years, Rifkin forecast, white suburbanites would suffer a similar fate "as the new thinking machines relentlessly make their way up the economic pyramid, absorbing more and more skilled jobs." The unhappiness of economically marginalized populations might build up to a revolt, the technologically unemployed fighting back against the "new cosmopolitan elite of 'symbolic analysts' who control the technologies and the forces of production."[24]

Other commentators in the 1930s and 1960s had linked the stress of displacement to the danger of social unrest, but few went as far as Rifkin in explicitly predicting "open class warfare." For all its cataclysmic rhetoric, Rifkin's picture contained many elements in common with more mainstream opinion. Respected economists, sociologists, political commentators, and workers themselves believed they saw good reason to fear for employment opportunity in the early 1990s. Those people expressed alarm over the possibility of displacement escalating to a point where economic inequality endangered American democracy.

Concern over Technology and Jobs in the Clinton Administration

The question of machines and job loss became a matter for concern within the first term of Clinton's presidency, as it had for Franklin Delano Roosevelt, John F. Kennedy, and other chief executives. Secretary of Labor Robert Reich spearheaded Clinton's approach to technological unemployment, just as David Weintraub had done for Roosevelt and Arthur Goldberg for Kennedy. The Clinton administration internalized an assumption that ongoing computerization and automation would destroy more and more of the country's older sources of employment. The force of world economic competition, combined with the internal momentum of technology itself, had made change the only constant in life. Clinton declared in 1993 that such a process placed constant strain on society. "All the advanced nations are having difficulty creating new jobs even when their economies are growing," making unemployment the "most troubling problem of the new era."[25]

In theory, employment could keep pace with future innovation, Reich indicated, if the country made a successful transition to the new information age. American business would need to hire masters of high tech to maintain a global economic lead. Training enough "symbolic analysts" would take a national commitment, a campaign mobilized from the highest levels of government. Such a call to arms appealed to the president, who launched the theme of education for the twenty-first century by convening a "Conference on the Future of the American Workplace." At that 1993 event, Clinton, Reich, and Commerce Secretary Ron Brown warned that the country's future rested on getting labor and management to join in "reinventing how we work." Once such cooperation had begun to pay off in the creation of new "high performance" jobs, men and women would come to welcome workplace technology as a source of prosperity and progress.[26]

Such a mandate sounded good, but extensive and effective training programs would require substantial investment at a time when politicians shied away from tax increases and when corporate wisdom emphasized downsizing. Meanwhile, the Census Bureau reported that in real terms, 1993 had brought a $300 drop in typical household income. Different yardsticks yielded varying numbers but confirmed the same theme: in constant dollars, compensation for blue-collar workers had fallen from $17.22 per hour in 1987 to $16.50 in 1994, while that of white-collar workers had dropped from $19.95 to $19.76. As economist Lester Thurow explained, even a fall of less than 1 percent a year in real wages "radically alters the distribution of purchasing power" over time. "By the turn of the century the real wages for nonsupervisory workers will be back to where they were at mid-century, fifty years earlier, despite the fact that the real per capita GDP more than doubled over the same period." Such trends took a real psychological toll. Three-quarters of people surveyed in one poll agreed that under the current economy, "middle-class families can't make ends meet." Fifty-seven percent said that with personal debt soaring and job confidence plummeting, the average family had lost its hold on the American Dream of opportunity and wealth.[27]

Several analyses suggested that the pain had not spread evenly, that labor displacement and wage stagnation had widened the economic gulf between America's rich and poor. Seventy-nine percent of all income gains between 1977 and 1990 had gone to the upper 1 percent of all families, a group that monopolized a greater proportion of economic resources in the United States than in any other major industrialized country. Thurow asked, "How far can inequality widen and real wages fall before something snaps in a democracy? No one knows, since it has never before happened." President Clinton seized on that theme for his 1995 Labor Day speech, warning that workplace automation had made economic unevenness a systemic problem of modern society. At present, he declared, "technology is changing so fast that the working people . . . have not gotten their fair share of our prosperity."[28]

Economists at a New York Fed gathering attributed 60 percent of the widening income gap to the way technology had affected the job market (blaming 10 percent on trade factors and 30 percent on miscellaneous reasons such as weak unions). Without giving such a precise breakdown, *U.S. News & World Report* referred to information-age workplace change as "probably the most critical" factor behind the fall in real wages. As auto manufacturing, steelmaking, and other fields had gone high tech, entire categories of employment had evaporated. While "in the short run it may be cheaper to replace workers with technology," business leaders might ultimately find such moves "self-destructive, because there will not be enough purchasing power to grow the economy." In a front-page series on the "battered middle class," the *Chicago Tribune* similarly bemoaned the way robots had replaced assembly workers and textile machine operators, "working-class jobs that once supported a middle-class lifestyle."[29]

Talk about middle-class job loss brought unexpected success in the early 1996 presidential primaries for Pat Buchanan, whose campaign capitalized on economic fear. Buchanan called for placing a protectionist tariff on imports from Mexico and Asia, plus a five-year moratorium on immigration to stop downward pressure on American wages. Though academic economists argued that such steps would not solve problems of wage decline and unemployment, Buchanan's attacks on corporate "executioners" resonated with voters fed up with downsizing. His appeal to economic populism forced fellow Republicans, especially Bob Dole, to follow his lead in denouncing the rush to cut payrolls. Media reports cultivated a popular impression that business leaders' lives had become completely disconnected from the harsh realities facing ordinary Americans. In 1995 there had been a median rise of 31 percent in compensation for CEOs at the nation's largest firms, reaching almost $5 million apiece, more than 100 times that of average workers. Critics charged that the increase in compensation bore little relationship to improved corporate performance, that a rigged system funneled Olympian rewards even to executives with poor track records.[30]

The sense of economic injustice gained strength in January 1996, when AT&T announced plans to eliminate 40,000 positions and company stock promptly jumped more than $2.50 a share. A *Newsweek* cover story on "corporate hit men" noted that AT&T CEO Robert Allen failed to apologize for the human cost of downsizing or make any gesture of sharing employees' pain, such as accepting a temporary reduction in his salary of over $3 million. Anxiety about displacement spawned black humor; people inside the firm joked that downsizing would continue until "AT&T" stood for "Allen & Two Temps." In a belated attempt to rescue its image, AT&T set up a special job bank and ran full-page newspaper advertisements asking other employers to send in help-wanted leads for its former staff. While the company reaffirmed its intent to eliminate 40,000 jobs, spokesmen announced that enough people had accepted retirement offers or transfers so that probably only 18,000 would be laid off. But the harm had been done; the AT&T controversy hit a raw nerve for many Americans, who felt that their well-being no longer counted for anything.[31]

While many people blamed job declines on foreign competition or general flaws in the economic system, 74 percent of the 1,200 men and women surveyed by the *New York Times* said they would place either "a lot" or "some" blame for unemployment on automation and computers. Incidents of technological displacement kept piling up. Since 1990, the shipbuilding company

Tenneco Inc. had adopted new automated equipment and simultaneously dropped 11,000 of its total 29,000 jobs. Cutting out the ribs of a tanker ship, a process formerly involving twenty-one workers, now required just four men using robotics and automatic welding technology. General Motors could manufacture the same number of cars with 315,000 workers in the 1990s that it had done with 500,000 employees in the 1970s. Across the board, the *Times* declared, it seemed that the "progress of technology kept taking tasks from human beings and giving them to machines, undermining the bedrock notion of mass employment."[32]

Democrats emphasized the notion that corporate leaders must accept a responsibility to all "stakeholders," a group that included workers. Congressmen such as Ted Kennedy and Richard Gephardt proposed changing the tax code to remove provisions that encouraged layoffs. They also floated the idea of creating special tax benefits for companies that invested in the workforce by using the profits gained through efficiency to offer wage hikes and fund retraining programs. At a special White House conference focusing on the motto of "good corporate citizenship," Clinton applauded Hewlett-Packard and fifty-nine other firms for "doing well by doing good" to labor. Presidential ceremonies aside, the economic picture for the so-called anxious class had begun to brighten by mid-1996. The Commerce Department announced that average income had risen 2.6 percent over inflation in the past year, the greatest gain in almost a decade. Announcements of new hirings at IBM, Sears, and even that demon downsizer, AT&T, seemed to signal an end to the tidal wave of job loss.[33]

Such good news might seem to show that economic optimists had been correct, that the new efficiency created by downsizing and the new jobs generated through innovation had more than offset any problems resulting from workplace technological change. 1997's unemployment rate dropped to 4.7 percent, the lowest level since 1973. To make matters better, inflation dropped from 3.3 percent in 1996 to 1.7 percent in 1997. That fortunate combination of low inflation and low unemployment seemingly defied economic law: analysts had often warned that falling below the "non-accelerating inflation rate of unemployment" (which conventional wisdom set at 5 or 6 percent) risked setting off an inflationary spiral. Ecstatic Wall Street watchers interpreted the statistics as proof that the United States had leaped forward into a New Economy, a postmodern structure in which deregulation, global competition, and the computer revolution had made old limitations meaningless. In a dizzying swing, financial euphoria superseded talk of the middle-class blues. Newspaper editors stopped commissioning special front-page series on "The

Downsizing of America" and concentrated on reporting the latest record-breaking numbers on the Dow Jones Industrial Average.[34]

Most mainstream economists rejected any idea that the wonders of information-age productivity had rendered the United States invulnerable to recession. Chaos in Asian markets might create a negative ripple effect, while the domestic scene still contained some serious trouble spots. The recent drop in unemployment did not indicate a miraculous economic transformation; half the trend had resulted from an unusual decline in the size of the nation's labor force. Official figures on joblessness did not count the millions of underemployed Americans or factor in those who had become too discouraged to continue searching for work. Furthermore, the United States had not reversed the tendency toward financial inequality; wage gains in 1996 primarily had gone to the professional elite and the top ranks of management, though workers at the bottom got a slight boost from legislation raising the minimum wage. Census Bureau numbers showed that the gap between rich and poor had reached its widest point since the Second World War. About 50 million Americans, including one out of four children, remained below the official poverty line, and a wide economic divide between races persisted. African American households, statistically less likely to own stocks or share in pension funds, held approximately one-tenth as much net worth as white families. Though the rate of joblessness among black males dropped in 1997 to the lowest level in over two decades, it stood nonetheless at double the rate for white men.[35]

For all the technical advances of the late twentieth century, there seemed good reason to wonder how much an increasingly computerized economy had raised the overall standard of living. The promise that workplace mechanization would yield wealth, security, and happiness had not materialized. The downsizing fad of the early 1990s had left lasting scars; even as stock indexes soared, good fortune did not erase workers' feelings of vulnerability. A 1997 survey of 450,000 people in major companies showed that almost 45 percent still often worried about being laid off. A number of Americans sensed that technological growth and other modern trends had permanently undermined old conditions of employment, making the whole job world more stressful than ever. "What's new about the new economy is that it's scary all the time, not just in cycles," a *Time* essayist observed.[36]

The Big Questions Revisited: Technology, Labor, and "Progress"

In the 1990s, as in the 1930s, questions about the impact of workplace technological change turned on a number of explicit and implicit value judgments about the nature of employment, leisure, consumerism, and the ideal economy. During the Depression, union leaders and some social scientists had argued that mechanization must inevitably create mass joblessness unless business slashed working hours to compensate. In an imagined future, industry could satisfy everyone's material desires while requiring people to work just two or three hours a day (or, according to some, a few hours a week). After all, new machinery had already created free time, only it had been badly distributed; what was unemployment but an unwanted excess of leisure? World War II had interrupted the anticipated march toward a leisure society, but postwar observers again proclaimed that automation would naturally shrink the standard workweek to almost nothing. Such an evolution would occur over some unspecified timeline; when labor had pushed for writing a thirty-five-hour week into an immediate agreement, corporate leaders quickly resisted, insisting that foreign business competition and the Soviet menace precluded any cuts at present. Americans of the 1990s still debated the relationship between technological innovation and working hours, this time with the awareness that earlier predictions of universal leisure had not been realized.

In retrospect, former U.S. Senator Eugene McCarthy had regretted that in the 1960s, his Senate Special Committee on Unemployment had given in to warnings that reducing the workweek would choke national economic growth and risk leaving the nation at a Cold War disadvantage. In trading increased leisure for the promise of better living standards, he concluded in the late 1980s, ordinary Americans had ended up losing both. McCarthy hoped to renew the call for shorter hours as a way of both averting unemployment and restoring balance to life. There seemed good reason to reconsider the matter: Some evidence suggested that work time had actually *lengthened* over recent decades; Americans of the 1980s put in a full month's worth of extra hours compared to their 1960s counterparts, two months more than French or German workers. Why did Americans put up with such a system? In part, there seemed little choice. Given the high costs of health care and other benefits, employers preferred to extract more hours from present staff rather than expanding the payroll, citing the pressure of competition as a way of justifying such demands. Meanwhile, organized labor had lost power during the 1980s and also had other battles to fight.[37]

From another perspective, many people may have "chosen" a long week, though perhaps without full cognizance of the trade-offs involved. Technology itself had facilitated the extension of hours; portable phones, laptop computers, e-mail, fax machines, and other new modes of communication added flexibility to where and when people might work, but they also left no excuse for being out of reach, even on vacation. Popular culture induced a consumerist mentality, and the range of available possessions kept expanding to include fancier computers, entertainment equipment, and sport utility vehicles. With the decline in real-term wages squeezing income, many Americans added overtime or began moonlighting to keep up. Slow commutes ate up more hours, while women especially still face a "second shift" of housework and child care. People coped with the "time crunch" by shifting errands into the wee hours; grocery stores, health clubs, and photocopying centers began to stay open later or around the clock, imposing more stress on their own employees. Humans could not keep stretching their natural body schedules without paying a price, physical and psychological studies indicated. Doctors warned that a startling number of men, women, and children were trying to live on a permanent sleep deficit, risking their health, impairing their functional efficiency, and multiplying the risk of accidents.[38]

Some rebelled against such trends, trying to escape the pressures of career and consumerism. Trend-watchers of the 1990s hailed a revival of the 1970s' "voluntary simplicity" movement, driven less by environmental activism than by workplace burnout. Twenty-eight percent of respondents in one survey reported having reevaluated their lives to open up more personal time, reduce stress, and focus on nonmaterial goals. Books such as *Your Money or Your Life* became bibles for those dreaming about "downshifting"; advocates related heartwarming stories about how wonderfully relaxed they became after "dropping out of the rat race." Though analysts spoke of a "new world of work" which empowered professionals to choose slower career tracks or switch to freelancing, not all people enjoyed equal freedom to gain that flexibility. Many employers, especially those in the manufacturing or low-end service sector, would not or could not readily accommodate alternative work patterns to fit some leisured ideal. Moreover, while upscale men and women might voice a wistful urge to trade one or two days' pay for more time off, wage stagnation placed such desires out of reach for their lower-income counterparts.[39]

Even those willing to swap money for leisure might find themselves actually working longer. With threats of downsizing, many workers felt pressured to demonstrate their indispensability by conspicuously putting in extra early-morning and late-night time. In one survey, 82 percent said that if cutbacks appeared imminent, they would increase their visible work hours to raise their chances of surviving. Americans had not lost the drive for leisure; 65 percent favored the principle of starting a new movement to reduce the workweek. But of those, only 58 percent thought the aim was realistic. Americans of the 1930s and 1950s, observing the astounding power of automation, had felt that society must perforce evolve toward shorter hours. By the 1990s, expectations that the production revolution would naturally yield more leisure had vanished.

Newsweek dismissed "the possibility of a society not based on work as we know it" as "a bold and mostly discounted hypothesis" expressed only by radical Europeans. In the United States, acceptance of long hours had won out.[40]

The once-anticipated leisure society had not appeared, but what about the mass consumer wealth promised during the 1930s and 1950s by those who promoted mechanization and automation? Without question, the postwar years had raised consumer expectations and for the most part realized them; in fixed dollars, American consumption per person doubled between 1945 and 1990. Prewar promises of air conditioning and television had been fulfilled, the two- or three-car family had become commonplace, and personal computers, microwave ovens, and VCRs had become "necessities." Advertising had institutionalized a belief in consumption as the key to personal happiness, making a science out of linking a choice of beer brand to success and sex appeal. And yet, psychologists pointed out that the happiness derived from consumerism remained mainly illusory; once the novelty and initial thrill had worn off, most purchases did not provide special pleasure. While Americans in the 1990s owned more automobiles and larger houses than people of the 1950s, the percentage in surveys who described their lives as very happy had not increased. Many families needed to borrow money to maintain their lifestyle; levels of personal debt rose in the early 1990s, matched by a record number of nonbusiness bankruptcies and credit-card delinquencies. Sociologists and psychologists argued that a fixation on consumerism could impair people's mental health, undermine marriages, and cut into time devoted to children and community.[41]

Had problems of the information-age work environment contributed even more directly to loss of community? Membership in religious and political groups, unions and fraternal orders, the Boy Scouts and Parent Teacher Associations had taken a nosedive in recent years, said political scientist Robert Putnam. League bowling had fallen 40 percent during the 1980s, even as the sport's overall popularity had increased; more Americans had started "bowling alone." Lester Thurow, Robert Kuttner, Kevin Phillips, and others linked such observations to economic analysis, suggesting that as modern society promoted technically gifted people into privileged positions, it left many individuals marginalized. People with evaporating employment opportunity lost any sense of personal investment in the technological, economic, and democratic system. Christopher Lasch noted that with the global style of computerized work, high-tech elites felt closer to their professional equivalents overseas than to fellow Americans, whom they disdained as culturally and technologically provincial. Community planners began attempting to recon-

struct civic life through a deliberately nostalgic small-town design. By building houses close together and including porches, architects hoped to renew
interaction between neighbors.[42]

Such arguments raised the possibility that a decline in local involvement
stemmed from the very nature of modern economics, a system that measured
value in terms of personal consumption while excluding any other index of
welfare. For example, standard economic analysis denied the importance of free
time to individual health and community coherence, said Eugene McCarthy,
treating leisure merely as "an empty spot in time devoid of wealth-produc-
ing activities." Counting up the number of automobiles Americans owned
denied how much nonmaterial community conditions contributed to people's
happiness (or lack thereof). Measurements of the gross national product actually turned well-being upside down, a 1995 *Atlantic Monthly* cover story complained. That skewed perspective turned pollution and crime into an
economic good, since they stimulated business for lawyers, security companies, and waste removers. The *Atlantic* authors proposed replacing the GNP
with a different economic barometer, the "Genuine Progress Indicator," which
would add value for community improvement while subtracting it for social
and economic weaknesses such as the growing unevenness of income.[43]

While such a fundamental change in economic measurement remained
unlikely, the argument reflected what those concerned about technological
unemployment had said all along: standard definitions of economic "progress"
discounted significant costs of workplace change. What were the consequences? Certainly nothing like the class warfare Rifkin had predicted; the
1990s had not brought a massive wave of guerrilla attacks on Silicon Valley
CEOs and engineers by those who felt excluded from the best jobs. Yet, reports
on the modern militia movement suggested its leaders had expanded membership by tapping into the vulnerability and anger of economically strapped
Americans, especially those suffering from the decline of blue-collar jobs. The
subject of terrorist resistance to technological development, of course, led to
the Unabomber, whose mail-bomb attacks had primarily targeted professional
figures linked to computers and high-tech business. "Ever since the Industrial Revolution, technology has been creating new problems for society far
more rapidly than . . . solving old ones," the Unabomber wrote in the philosophical manifesto explaining his actions. Mechanization had caused "a disaster for the human race," a situation that demanded "a revolution against the
industrial system."[44]

While the Unabomber's crimes deserve unqualified condemnation, his criticism of technology voiced some familiar points, ones worth consideration.
The Unabomber wrote that "modern man has the sense (largely justified) that

change is IMPOSED on him." Such a sentiment did not seem entirely ridiculous, given how strongly twentieth-century society had embraced the notion that ordinary people must not even try to interfere with the inevitability of automation. While few Americans advocated a return to the nineteenth century, even technophiles such as Clifford Stoll registered misgivings that the recent infatuation with computers had gone too far too fast. Educators who showed children how to use the Internet for everything might neglect to teach them other important lessons, and people caught up in the ease of connecting online to hundreds of faraway people might spend less time communicating with their own families.[45]

Environmentalism and the horrors of Vietnam had spawned a movement of technology criticism in the 1960s and 1970s; during the 1990s, discomfort with computerization coalesced into a "neo-Luddite" sentiment. In many ways, "neo-Luddites" did not deserve this label. Steven Levy called it "a telling fact that the neo-Luddites consist not of blue-collar workers, but elite symbol-shufflers who will never themselves be displaced by computers." Those identified with the philosophy did not lead marches to defend bank tellers replaced by ATMs; when Kirkpatrick Sale took a sledgehammer to a computer, it was merely a token act to impress a lecture audience. But like the original Luddites, neo-Luddites did not issue a blanket condemnation of modernization. While publicizing their skepticism about the link between mechanization and progress, they reserved the right to be "selective about disliking technology."[46]

Images in popular culture reflected persistent attempts to grapple with the question of whether a more computerized workplace represented progress. As in the 1930s, concern about displacement sparked bitter humor; episodes of downsizing in the 1990s inspired references to labor "as roadkill on the 'information superhighway.'" Artists continued to translate workers' fears into visual form, literally drawing on talk about job loss. A 1995 cartoon by Jeff MacNally shows a man announcing to his wife, "Well, the new technology has made it possible for me to work at home full time—I've been laid off." Tom Toles drew a man looking at a newspaper headline, "Still No Jobs," and asking his friend, "Remember how we used to look forward to the day that machines would do all our work for us?" A second Toles cartoon titled "Evolution of the Job" starts with a picture of primitive man hunting, followed by panels showing a man working the assembly line and then just watching a robot put parts together. The final scene shows the fellow walking away from a door marked, "No Help Wanted," with the punchline, "Man discovers why animals never bothered making tools." Tom Tomorrow updated the old story about Walter Reuther asking who would buy cars after machines replaced workers, drawing a cartoon that shows a businessman proudly announcing, "We are replacing *our entire workforce* with *robots.*" When journalists point out that unemployed Americans couldn't keep consumer spending going, the executive answers that he had taken care of that snag: "You see, these are no *ordinary* robots! *These* robots come equipped with their own *Visa cards!*"[47]

Jokes about computers and displacement did not have to be sophisticated to make a point. One cartoonist showed a man telling his boss, "I'm concerned about my future here . . . with all the use of computers and robots to replace workers!" to which the manager replies, "No need for you to worry, Thornapple! I'd be hard-pressed to find a machine that does absolutely nothing!" Though the technology at issue had changed radically over sixty years, cartoonists of the 1990s used images of displaced men which would have struck home with Depression audiences. Sidney Harris drew a panhandler holding a sign, "Replaced by one of those little pocket calculators." Another illustrator showed an outdated computer terminal sitting on the street next to a bum, who tells the machine, "Yeah, I lost my job because of Bill Gates too." Cartoonists referred to real-life controversies, such as the way use of e-mail threatened regular postal service. A 1998 "Blondie" strip shows Dagwood typing e-mail into a laptop and telling Blondie, "Just think! Someday soon this will eliminate letter carriers altogether!"—at which point their friendly neighborhood postman feels a psychic punch in the head. More than anything else, such humor captures a sense of technological change as rapid and unstoppable. Another cartoon shows a teenager gloating, "It's amazing the way computers are able to do more and more work . . . Pretty soon, we won't have to do *anything!*" to which his friend interjects, "Except stand in the unemployment line," drawing the comeback, "No, they can send the checks electronically!"[48]

Awareness of the technological unemployment issue rose and fell with the nation's overall economic fortunes. By mid-1997, as joblessness fell to a twenty-year low, the fear of displacement voiced three years earlier by Aronowitz and DiFazio, two years earlier by Jeremy Rifkin, just the previous year by the *New York Times*, suddenly receded. A soaring stock market, low unemployment, and low inflation once again pushed back concern about the social implications of workplace change. And yet, given the modern power of science and engineering, Americans' continued dependence on employment, and values of consumerism and the traditional work ethic, the issue of displacement continued to lie just below the surface. The next economic crisis seemed likely to reawaken the doubts voiced back in the Depression era, again raising concerns about how Americans should weigh the balance of effects from mechanization.

Fig. 14. Combining retro-style graphics with the latest references to virtual reality, Tom Tomorrow satirized economic trends of the early 1990s in "This Modern World." Recent decades had brought incredible advances in invention, automation, and computerization, and yet somehow it seemed that people had to run faster just to keep up. Did all this really add up to progress? Tom Tomorrow, "This Modern World," *Utne Reader*, November/December 1993, 155.

Throughout the twentieth century, economists and others in the public sphere have had to cope with the fact that the multidimensional relationship between new technology, displacement, reemployment opportunity, and job growth has proved inherently difficult to unravel. Researchers could observe how adoption of computers affected employment in one company, or even how introduction of dial telephones affected the number of switchboard operators needed throughout the industry, but such studies could not complete the picture. A single innovation might cause rippling or even contradictory employment effects. Automatic telephone systems reduced demands on operators but increased the need for engineers and technicians. A business office might stop sending its accounting or information-management operations out to subcontractors after adopting new computer systems, thereby affecting employment in an entirely separate company. Firms have often refused to reveal such information to researchers, out of fear that the details might fan controversy. One journalist investigating computer-related labor displacement in the insurance industry observed that most managers felt "reluctant to specify the impact of technology on payrolls; they fear a Luddite backlash among employees."[49]

Because of the sheer complexity of the subject, experts found it virtually impossible to measure total technological unemployment at any one time. Economists, statisticians, sociologists, and historians mobilized all the tools of their disciplines to guide and legitimate investigations of mechanization. Such studies produced a wealth of statistics and reports, but the results were inherently limited, and, in any case, researchers could quarrel for years about methodology and interpretation. A seemingly straightforward question, What is the number of workers displaced by technical changes in production, became a matter for perpetual and passionate disagreement. Even in the depths of the Depression, no one suggested that mechanization alone accounted for all the millions of unemployed. Beyond that, no consensus emerged. In the absence of objective figures, discussion of technological unemployment became a battle of emotions, assumptions, and vested interests.

Was fear of automation catching? Undoubtedly, when a president such as Roosevelt or Kennedy expressed concern, when each day's newspaper brought stories about new job cutbacks, those events in themselves could make alarm contagious. Such an assessment, however, merely begs the question: Why did fear of technological unemployment seem so very real, so urgent, throughout different decades? It remains difficult today to gauge the precise percentage of Depression-era job loss that stemmed directly from mechanization. It should be safe to assume that at some level, real problems of displacement due to automation did exist. Historians cannot dismiss twentieth-century concerns about mechanization as a simple error or mass hallucination. True, the most extreme forecasts proved way out of line; the Technocrats in the 1930s, Norbert Wiener in the 1950s, and Jeremy Rifkin in the 1990s all falsely predicted catastrophic unemployment and social breakdown. And innovation did create some work; by the 1990s, the personal computer alone had generated hundreds of thousands of jobs in software and hardware development, manufacturing, distribution and sales, repair, and support.

Glib generalizations that technological advance always has and always must produce more opportunity than it destroys, however, tend to overlook harsh human realities. Although computerization had opened many new and exciting dimensions to modern economic life, the transition did not come easily for everyone. Workers on the assembly line and in the office watched as employers brought in automation and talked about how inventors would soon devise still more powerful robots. Musicians and longshoremen suffered at least temporary displacement; some would find new positions relatively soon, but more would experience difficulty. Mechanization aroused unhappiness and uncertainty in the workplace, especially when labor and management already distrusted each other. The ability of robots to perform humanlike work, the enormous capacity of computers to handle information, and the promise of more inevitable change yet to come made worries about technological unemployment credible. Definitions mattered. The word *efficiency* might serve as a technical term referring to a machine's quality, or as an economic measure of a system's effectiveness in production and distribution, or as a social phrase indicating whether a country made optimal use of its human resources. That last meaning made displacement seem like a senseless waste of people's skills, while the other two definitions treated technological unemployment as an unfortunate but ultimately meaningless consequence of economic advance.

In the end, lack of exact numbers on technological unemployment only made the discussion more contentious. Precisely because the impact of machines on jobs remained open to dispute, the question continued to command attention from both expert observers and ordinary Americans for decades. The debate over displacement became entwined with presidential and Congressional politics, with Cold War tensions, and with international economic competition. In thinking about mechanization, Americans had to wrestle with both practical and philosophical questions about the relative value of work and leisure, the importance of material consumption, and whether society ought to provide citizens with some basic job security or safety net. People's nervousness stemmed from all the social tensions and ambiguities inherent in the very process of modernization. Did technical innovation represent a historic force bringing the United States to the zenith of civilization, or did changes in production threaten to break apart all the values on which American life had been grounded? What did the concept of "progress" really mean, and how did the technological revolution of work fit into that picture?

The twentieth-century debate over the effects of mechanization played out under a perspective shaped by the partnership of science, engineering, and corporate public relations. In 1921, Thorstein Veblen's *The Engineers and the Price System* had talked about a fundamental clash of values between the industrialist's bureaucratic mindset and the engineer's commitment to skilled workmanship. Veblen anticipated that in the future, emergence of a "soviet of technicians" might transform the narrow ideology of business into a universally beneficial system of perfect efficiency.[50] By the 1930s, the relationship of professionals had evolved quite differently. Under the pressure created by Depression-era uncertainty, leaders in business, science, and engineering consciously drew on their common interests to form an alliance in defense of mechanization. Dreading a backlash against research, physicists and engineers worried about their popular image more than ever before. Their professional societies redefined annual meetings as public relations exercises, an opportunity to stage speeches, pageants, and ceremonies debunking talk of technological unemployment. Such events frequently drew on the support of nationally prominent business owners and executives, who mobilized their own resources to reinforce faith that technological change always had and always would serve as the primary force driving American progress. Postwar developments only strengthened the mutuality of science, engineering, and business, as promoters made the plan of an automatic factory ever more explicit.

At the same time that advocates of automation dismissed concern about job loss, they made increasingly bold promises about machines' ability to outperform human beings. Dial equipment could work faster than the best switchboard girl, photoelectric eyes could pick up more flaws than the most experienced inspector. Those who touted the dream of production without labor assumed the perfectibility of technology. A. O. Smith, Leaver and Brown, and John Diebold never acknowledged any possibility of mechanical breakdown in their descriptions of the automated plant. In the mindset of both engineers and businessmen, dealing with real workers could get messy; dealing with machines represented the ideal. An economic system built around the principle of product obsolescence could use a similar logic to dismiss labor problems; when technology moved forward, some people might inevitably encounter "occupational obsolescence."

The men who created the A. O. Smith plant of the 1930s, the blueprints for the "automatic factory" in the 1950s, and high-tech robots in the 1980s all spoke of the same goal: manufacture "without men." And yet, those crusaders did not really aim for a completely humanless operation. They envisioned abolishing the lowest ranks of the payroll, the clerks, union members, and assembly-line workers, who got tired, made mistakes, and demanded raises. Mechanization promised to eliminate such undesirable elements while keeping day-to-day command safely in the hands of human executives and engineers. The irony of such notions came across in one Sidney Harris cartoon that showed two managers gazing at a factory floor full of machines and remarking, "Now that we're completely automated, there's no one to yell at."[51]

The ideal of the automatic factory rested on the privilege of expertise. Introduction of computers would render shopfloor labor disposable while making a company increasingly dependent on engineers, researchers, and technicians. Despite advertisements that made mechanization look magically easy, adopting and maintaining new equipment usually proved costly and complicated. Setting up software to handle the functions of accountants and bank tellers might cause countless headaches and require hours of troubleshooting by programmers and systems operators. Automation represented career opportunities for such specialists; far from threatening to displace them, innovation would only generate more demand for their skills. Technical specialists did not come cheap, yet executives generally considered the money well spent. They perceived engineers as sympathetic to management, unlike those troublemakers in the union. Production technologies offered an illusion of control; the very process of mechanization would sort employees into two categories, necessary and redundant.[52]

Optimists applied free-market analysis to suggest that a natural course of economic advance would ultimately absorb all casualties of change. Technological advance might cause short-term harm for certain workers, but, in the long run, such unavoidable incidents became meaningless, a minor note in America's quest for prosperity. Say's Law conveniently told businessmen that they could cure problems of labor displacement by turning *more* production over to machines, speeding up their pursuit of automation in the search for long-term market gain. To those immediately affected, unemployment could never represent an impersonal economic phenomenon. People who had invested years in a job might suddenly find that mechanization had rendered their skills and experience meaningless. Abstract economic theory treated humans like interchangeable units, assuming that the system would slip workers into new slots as old positions closed. In real life, theories of labor fungibility did not hold; the phone company would not hire a displaced switchboard girl as an electrical engineer, and retraining clerks as computer operators took time and money. Even temporary displacement meant financial and emotional pain; as social scientists documented, layoffs brought increased rates of illness, divorce, and suicide. Jobs meant more than just a paycheck; displacement robbed dock workers, musicians, telephone operators, and countless others of an identity and a community.

In diverting attention from specific cases of distress, defenders of mechanization constructed an entire message equating workplace technology with national progress, past, present, and future. Their technocentric narrative defined a progressive history in which engineering and free enterprise carried the United States along a predestined path to global superiority. William Ogburn's cultural lag theory reinforced the assumption that technological advance represented an independent leading variable in modern life, while social institutions trailed behind. Ogburn argued that human curiosity sent invention along a preset course, that if Robert Fulton, say, had died young, ideas already in the air would have inevitably led others to develop the steamboat in the early 1800s. Ogburn and fellow sociologists pushed such ideas hard, incorporating that deterministic outlook into academic literature, government reports, and popular consciousness. In predicting how much the development of cotton-picking machines or photoelectric devices might eliminate employment in future, those forecasters always accepted such advances as a given.[53]

Most leaders in the scientific and engineering world gravitated toward the self-congratulatory idea that new production technology had transformed human life from primitive chaos to an American peak of civilization. The logical corollary to such assumptions suggested that the United States could attain further greatness through and *only* through a national commitment to research and invention. Technology itself would solve the social and economic problems of a technological age; the insignificant labor displacement caused by the introduction of machines would be cured by more intense mechanization. Comparing their critics to Luddites, Karl Compton, Robert Millikan, and Charles Kettering argued that the slightest hesitation in embracing new workplace technology would send twentieth-century civilization back to the Dark Ages. Such accusations did not hold water; no one advocated abandoning all modern medicine, scientific discovery, and technical innovation. Those who worried about unemployment did not sigh naively for some earlier, romantic Eden. Even at the height of alarm about job loss, Americans never really hated technology. Telephone operators who feared displacement did not smash dial equipment or insist that the United States hold a "science holiday," a complete moratorium on research. Longshoremen concerned about containerization did not tear apart hospital operating rooms or throw away their cars and television sets. Nor did economic misery ignite attacks on inventors; the names of Thomas Edison and George Westinghouse retained a powerful aura of heroic genius.

The assumption that history set technological change on an inevitable path steered discussion in certain directions while cutting off others. Some observers compared displacement to an act of God, a natural disaster; the Depression had created a "famine" in jobs or a "sickness" in the community. Those metaphors reflect the deterministic philosophy of mechanization, encapsulated in the 1933 Century of Progress motto, "Science Finds, Industry Applies, Man Conforms." The word "conforms" carried harsh overtones, threatening those who refused or failed to adapt. With such language, twentieth-century Americans effectively gave up any notion that ordinary citizens might choose whether or not to welcome all forms of invention. Government should not attempt to redirect the course of innovation or impose broad conditions on the adoption of new devices. Ogburn's cultural lag theory restricted policy options; if decision makers had no control over the course of technical change, their only hope was to find ways of minimizing the damage. Since discoveries usually appeared on the horizon some time before their implementation became practical, social scientists offered to provide advance warning, giving a wise nation some opportunity to retrain workers and otherwise prepare for the unavoidable.

Assumptions of technological determinism ultimately locked labor into a passive, defeatist mindset. Railroad unions could insist on keeping two-man crews on diesel trains as a temporary attempt to cushion workers from displacement, but featherbedding would not stave off the effects forever. Musicians might try convincing audiences to reject sound movies, but more aggressive attempts to head off change would only prove self-destructive. Union leaders largely went out of their way to emphasize their support for mechanization, accepting the promise that innovations in production would raise the United States to new heights of greatness—as measured in terms of material abundance. Such a stance left few options for rank-and-file workers, many of whom resigned themselves to the force of invention. In the 1990s, the introduction of computer systems that continuously monitored telephone operations seemed to jeopardize jobs of repairmen at Pacific Bell. When a breakdown occurred, central offices could automatically locate and diagnose the problem, preempting any need to send a maintenance expert to investigate. One man who faced the possibility of being discharged twelve years before retirement admired the wonders of such technology and added, "If the result of us being efficient is me being laid off, I hate to say it, but I guess that's progress."[54]

The assumption of a technological imperative was itself a victory for the leaders of science, engineering, and corporate America. Their powerful public relations efforts encouraged people to internalize feelings about the inevitability of automating production along industry-approved lines. It did not have to turn out that way. Especially with one-quarter or more of the population out of work in the Depression, fear of displacement had had the potential to be just as influential. After all, President Franklin Delano Roosevelt had described technological unemployment as a real crisis, and WPA researchers had compiled some compelling statistical evidence to back up such warnings. Several decades later, John F. Kennedy expressed equally grave dismay, referring to machine-related job loss as the country's major domestic challenge. Movies such as *Valleytown*, books such as *The Grapes of Wrath*, and popular cartoonists and illustrators gave the question an emotional urgency, a cultural currency.

At least in theory, such tension over technology contained some seeds for a real revolution. Concern might have instigated substantial changes in government regulation, in social policy, and in economic relationships. Legislators might have passed measures that codified responsibility for ensuring technological change did not unduly burden particular individuals or groups. Sweden, for instance, balanced high productivity with low inflation and low unemployment in part by instituting specific strategies to distribute the costs of mechanization. As Swedish companies moved to modernize, centralized labor-market boards arranged "retraining sabbaticals" for displaced employees.[55] Alternately, the discussion could have inspired a more constructive dialogue between professionals and working-class people, steering technical development toward different avenues. Americans might even have taken an opportunity to reassess personal and social priorities, questioning the consumerist ideal. Everyone would always want and need certain items, and the dream of two cars in every garage and fancier television sets in each living room remained compelling. Nevertheless, people's sense of well-being also rested on nonmaterial factors, such as family strength, community harmony, social justice, and democratic representation in politics, subjects which a narrow focus on automation progress pushed aside.

Obsession with consumerism limited Americans' perspective in significant ways. None of those most concerned about displacement succeeded in articulating alternative values that challenged technological determinism, respected the agenda of people working at the low end of the economic spectrum, and gave them a sense of agency. That failure of vision reinforced the tendency to measure individual happiness and national superiority by mass production and mass consumption. Such a one-dimensional definition of progress imposed certain costs. Back in the 1930s, publicists had boasted that Americans used more coffee and owned more automobiles than any other people. Six decades later, with heightened awareness of environmental problems, limited natural resources, and the stubbornness of global poverty, such statements no longer seemed such an unambiguous statement of superiority. The abundance produced through automation had not brought wealth even to everyone in the United States; a persistent income gap left many families further and further behind, struggling both financially and socially. Optimists had promised that the breakdown of old employment patterns would release people from the psychological constraint of corporate regimentation, allowing them to jump between jobs for freedom and profit. As things turned out, the growing sophistication of information-age technology imposed selective barriers; men and women without computer skills, without a certain education and experience, could not break into new high-tech opportunities. Finally, for all the talk about turning production over to machines, the hope of giving Americans new leisure, more time for family and community life, seemed further off than ever. Consumerism did not come free.

Most seriously, the assumption of technological determinism concealed the fact that technical developments always reflected *some* human choice. Consciously or unconsciously, a decision to pursue certain lines of invention meant dropping alternative avenues. An employer who installed automatic equipment on the assembly line exercised certain value judgments about machines and about people. Managers often argued that they had no option but to pursue mechanization, since unions kept pricing their members out of work by demanding higher wages. Moreover, a company that failed to adopt the latest equipment would be eaten alive by domestic or international competitors. Investing mechanization with such an air of destiny ignored the true complexity of twentieth-century economics. Technology alone would rarely make or break a business; many other factors, from employee loyalty to executive vision, affected operations. Postwar promoters of the automatic factory and the automatic office had assumed that one model fit all cases, even telling engineers to redesign products if necessary to fit production machines. The marketplace in fact had never become so monolithic as to make automation a universal mandate. Installing new technology had its cost, and a small flexible workplace might succeed where larger firms had let the mania for computerization kill off creativity and motivation.

At least in principle, talk about displacement asked whether and how people could reassert a sense of agency, a feeling of control over how technology affected their life. Throughout both the Depression and the postwar period, economists, writers, union leaders, presidents, and ordinary individuals called for acknowledging the tradeoffs inherent in mechanization. While headline writers painted the debate over automation in stark terms—"Aladdin" versus "Frankenstein"—the whole process of change brought both a price and a promise. Critics warned that the drive to modernize production must not overshadow the reality that automation sorted people into winners and losers, often according to race, class, or gender. In the end, faith in technology won the day, pushing social considerations aside. Even in periods of high unemployment, Americans never became disenchanted with the promise of machine-age abundance.

Leaders of science, engineering, and major corporations had actively nurtured the dream of consumerism, using advertising, World's Fair exhibits, and professional authority to equate material wealth with civilization. Residents of Detroit worshipped the automobile even as they worried about robots taking over the assembly line; cars represented excitement, freedom, and simple necessity. Most men and women could not give up driving even if they had wanted to, since cities had literally been built around the assumption that almost everyone owned cars. The twentieth century had similarly made home computers and other personal gadgets into necessities, and those without them were disadvantaged. The majority of people remained fundamentally unwilling, even unable, to envision an alternative world with less emphasis on the material goods turned out by automated production. At century's end, Americans seemed to be racing faster than ever in trying to keep up with changing technology. For all the stress such expectations imposed, for all the trauma expressed in ongoing talk about technological unemployment, the United States had enshrined the gospel of workplace mechanization as progress, automation as destiny.

NOTES

Prologue

1. Harry Braverman, *Labor and Monopoly Capital: The Degradation of Work in the Twentieth Century* (New York: Monthly Review Press, 1974); David Montgomery, *Workers' Control in America: Studies in the History of Work, Technology, and Labor Struggles* (Cambridge: Cambridge University Press, 1979); Richard Edwards, *Contested Terrain: The Transformation of the Workplace in the Twentieth Century* (New York: Basic Books, 1979); David Gordon, Richard Edwards, and Michael Reich, *Segmented Work, Divided Workers: The Historical Transformation of Labor in the United States* (Cambridge: Cambridge University Press, 1982). On the impact of mechanization in specific lines of work, see David Noble, *Forces of Production: A Social History of Industrial Automation* (Oxford: Oxford University Press, 1986); and Andrew Zimbalist, ed., *Case Studies on the Labor Process* (New York: Monthly Review Press, 1979).

2. Patricia Cooper, *Once a Cigar Maker: Men, Women, and Work Culture in American Cigar Factories, 1900–1919* (Urbana: University of Illinois Press, 1987); Irwin Yellowitz, "Skilled Workers and Mechanization: The Lasters in the 1890s," *Labor History* 18, no. 2 (1977): 197–213; Keith Dix, *What's a Coal Miner to Do? The Mechanization of Coal Mining* (Pittsburgh: University of Pittsburgh Press, 1988); and Grace Palladino, "When Militancy Isn't Enough: The Impact of Automation on New York City Building Service Workers, 1934–1970," *Labor History* 28, no. 2 (1987): 196–220.

3. "Danzig Bars New Machinery Except on Official Permit," *New York Times*, March 14, 1933; "Nazis to Curb Machines as Substitutes for Men," *New York Times*, August 6, 1933. By contrast, René Clair's 1931 film, *A Nous la Liberté*, depicted a world in which society had adjusted perfectly to changes in production, suggesting that ideally, mechanization could free workers to enjoy fishing and dancing, rather than spread misery. René Clair, *A Nous la Liberté*, 1931.

Chapter 1: "Economy of a Madhouse"

1. Alan I Marcus and Howard P. Segal, *Technology in America: A Brief History* (Fort Worth: Harcourt Brace Jovanovich College, 1989); James Flink, *The Car Culture* (Cambridge, Mass.: MIT Press, 1975); David Hounshell, *From the American System to Mass Production, 1800–1932* (Baltimore: Johns Hopkins University Press, 1984); John Rae, *The American Automobile Industry* (Boston: Twayne, 1984).

2. Christopher Finch, *Highways to Heaven: The AUTO Biography of America* (New York: Harper Collins, 1992); Clay McShane, *Down the Asphalt Path* (New York: Columbia University Press, 1994); and Robert S. Lynd and Helen Merrell Lynd, *Middletown: A Study in American Culture* (New York: Harcourt, Brace and World, 1929), 251.

3. Joseph J. Corn, *The Winged Gospel: America's Romance with Aviation, 1900–1950* (New York: Oxford University Press, 1983); Roger E. Bilstein, *Flight in America: From the Wrights to the Astronauts* (Baltimore: Johns Hopkins University Press, 1984); Jeffrey L. Meikle, *Twentieth Century Limited: Industrial Design in America, 1925–1939* (Philadelphia: Temple University Press, 1979); Joseph J. Corn, ed., *Imagining Tomorrow: History, Technology and the American Future* (Cambridge, Mass.: MIT Press, 1986); and Ruth Schwartz Cowan, *More Work for Mother: The Ironies of Household Technology* (New York: Basic Books, 1983).

4. Marcus and Segal, *Technology in America* and Leonard Reich, *The Making of American Industrial Research: Science and Business at GE and Bell, 1876–1926* (Cambridge: Cambridge University Press, 1985).

5. "Chances for Inventors," *Literary Digest*, December 13, 1919, 29; "The Role of the Patent System in the United States," *Scientific American*, January 1, 1921; and C. H. Claudy, "Invention as the Foundation of the Nation's Wealth," *Scientific American*, June 14, 1919, 625, 640. See also Don Glassman, "This Inventing World," *Popular Mechanics* 54 (1930): 202–7; and Edward Thomas, "The United States Patent Office as a National Asset," *Scientific American*, June 14, 1919: 642–44. For a skeptical interpretation of the nationalist school of invention, see S. C. Gilfillan, "Who Invented It?" *Scientific Monthly* 25 (1927): 529–34. On Edison, see Albert Shaw, "Mr. Edison as the Typical American," *Review of Reviews* 84 (September 1931): 17–18. The cult of invention frequently compared Edison with another American hero, Abraham Lincoln, crediting the two with a common honesty, optimism, and a desire to serve the nation. Emil Ludwig, "Edison: The Greatest American of the Century," *American Magazine* 112 (December 1931): 66–67, 96, 98; and Albert Shaw, "Mr. Edison's Views of Life and Work," *Review of Reviews* 85 (January 1932): 30–31.

6. Roger W. Babson, "There's Magic in the Air," *Collier's*, April 7, 1928, 8–9, 50–52.

7. Edward Marshall, "If the Answer Is Easy It's Wrong," *Collier's*, December 6, 1924, 5–6; and Waldemar Kaempffert, "The Light of Edison's Lamp," *Survey* 63 (1929–1930): 13.

8. Raymond Francis Yates, "Wall Street and the Research Laboratory," *Scientific American*, November 1929, 382–84.

9. American Social History Project, City University of New York, *Who Built America: Working People and the Nation's Economy, Politics, Culture, and Society*, vol. 2 (New York: Pantheon, 1992).

10. George E. Barnett, *Chapters on Machinery and Labor* (Cambridge, Mass.: Harvard University Press, 1926), v, 32, 67, 71, 132, 137–38.

11. Stuart Chase, *Prosperity: Fact or Myth* (New York: C. Boni, 1929), 173–74, 186–88.

12. J. A. MacDonald, *Unemployment and the Machine* (Chicago: Industrial Workers of the World, 1925), 39–41, 43, 54, 73, 80–89.

13. J. C. Lewis to Herbert Hoover, June 13, 1929, Presidential Papers Subject File "Unemployment Correspondence, 1929, June–July," Herbert Hoover Presidential Library (hereinafter HHPL).

14. Frank Weber to Herbert Hoover, July 10, 1929, Presidential Papers Subject File

"Unemployment Correspondence, 1929, June–July," HHPL; and Arthur Huggins to Herbert Hoover, July 1, 1929, ibid.

15. "Labor Declares War on Machines," *Literary Digest,* December 15, 1928, 12–13; and "Machine Displacement," editorial, *American Federationist* 36, no. 7 (1929): 786.

16. "Report of the Committee on Recent Economic Changes of the President's Conference on Unemployment," *Recent Economic Changes in the United States* (New York: McGraw-Hill, 1929), x, xvii, 514–16; and Dexter Kimball, "Changes in New and Old Industries," ibid., 92–95.

17. Kimball, "Changes in New and Old Industries," and Wesley Mitchell, "A Review," *Recent Economic Changes in the United States,* 877–78.

18. W. M. Kiplinger, "Causes of Our Unemployment: An Economic Puzzle," *New York Times,* August 17, 1930; E. Dana Durand to the Secretary of Commerce, memo, March 6, 1930, Presidential Papers Subject File "Unemployment—President's Emergency Committee on Employment, Correspondence 1930 Feb.–Sept.," HHPL.

19. U.S. Bureau of Labor Statistics, "Digest of Material on Technological Changes, Productivity of Labor, and Labor Displacement," *Monthly Labor Review* 35, no. 5 (1932): 1031–33; "Rail Unions Study Loss in Workers," *New York Times,* November 16, 1930; and "The Case for Shorter Hours of Work," *Railroad Trainman* 49, no. 7 (1932): 388–90. See also Witt Bowden, "Productivity, Hours, and Compensation of Railroad Labor," *Monthly Labor Review* 37, no. 6 (1933): 1275–89. Rail unions claimed that in addition to creating unemployment, recent technological changes had harmed service: the time spent in assembling cars for longer trains delayed shipments, while motorists had to wait longer at crossings for two-mile-long trains to pass. Labor organizations also raised safety issues; longer trains running at higher speeds and controlled by a single operator would have no backup protection if the engineer became incapacitated. Citing those matters of compromised safety and convenience as well as unemployment, unions fought for rules limiting cars per train and requiring railroads to use full crews.

20. T. H. Gerken, "Continuous Sheet Mills—What They Mean to Costs and Competition," *Iron Age,* January 2, 1936, 82–89; D. Eppelsheimer, "The Development of Continuous Strip Mills," *Engineer,* November 4, 1938, 513–14; Douglas A. Fisher, *Steel Serves the Nation, 1901–1951: The Fifty-Year Story of United States Steel* (New York: U.S. Steel Corporation, 1951), 141–44; and F. A. Merrick, "The Machine Myth," *Mining Congress Journal* 16 (January 1930): 8.

21. D. Eppelsheimer, "The Development of Continuous Strip Mills," and John Knox, "Continuous Mills Voracious in Cost, But How They Produce!" *Steel,* October 22, 1934, 18–19, 23. Continuous-strip mills represented only one of several significant technical changes in Depression-era steelmaking. Both blast furnaces and open-hearth furnaces were being built to handle increasing capacities, thereby greatly raising output even as automatic instruments and material-handling machinery (the skip hoist, pig-casting machine, car-dumper, and ore bridge) reduced the number of workers needed. Temporary National Economic Committee (hereinafter TNEC), *Technology in Our Economy,* monograph no. 22 (Washington, D.C.: U.S. Government Printing Office, 1941), 235–37.

22. TNEC Hearings, "Technology and Concentration of Economic Power," part 30, *Investigation of Concentration of Economic Power* (Washington, D.C.: U.S. Government Printing Office, 1940), 16392–94, 16413–14, 16425–29, 16442.

23. Ibid., 16421–22, 16446–47, 16482–84, 16493–94, 16500.

24. Ibid., 16448–51, 16459–60, 16508–9.

25. Ibid., 16448–51, 16459, 16508–9.

26. TNEC, *Technology in Our Economy*.

27. Harold J. Ruttenberg, "85,000 Victims of Progress," *New Republic,* February 16, 1938, 37–38; and Harold Ruttenberg and Stanley Ruttenberg, "War and the Steel Ghost Towns," *Harper's Magazine* 180 (January 1940): 147–55. According to Emil Rieve, president of the CIO's Textile Workers Union, the promise of mechanization had similarly created textile ghost towns. Manufacturers found it easier to construct new factories with the latest equipment than to convert old plants, abandoning towns that had been built around the textile economy. Stranded workers often could not afford to pick up and follow the factories, and, in any case, the new mechanized mills needed fewer employees. Linking his argument to John Steinbeck's powerful depiction in *The Grapes of Wrath* of how farm mechanization hit families, Rieve told Congress that the Joads' story "could well be duplicated a thousand times" by real-life accounts of industrial workers displaced by technological change. TNEC Hearings, "Technology and Concentration of Economic Power," 16847–49.

28. Laura Smith, "Opportunities for Women in the Bell System," *Bell Telephone Quarterly* 11, no. 1 (1932). On the development of the American telephone system, see John Brooks, *Telephone: The First Hundred Years* (New York: Harper and Row, 1975), and Claude Fischer, *America Calling: A Social History of the Telephone to 1940* (Berkeley: University of California Press, 1992).

29. Stephen Norwood, *Labor's Flaming Youth: Telephone Operators and Worker Militancy, 1878–1923* (Urbana: University of Illinois Press, 1990). See also John Schacht, *The Making of Telephone Unionism, 1920–1947* (New Brunswick, N.J.: Rutgers University Press, 1985).

30. Such predictions of future labor shortages rested on an assumption that many other employers would be competing with telephone companies to hire women from the same demographic pool. Karl Waterson, "Change from Manual to Dial Operation," *Bell Telephone Quarterly* 9, no. 3 (1930): 167. As it turned out, AT&T personnel managers had their pick of good applicants during the Depression. Henry LaChance, "Trends in the Training of Telephone Operators," *Bell Telephone Quarterly* 18, no. 1 (1939): 33. On the expansion of the telephone system, see Kirkland Wilson, "The Telephone Problem in the World's Largest Metropolitan Area," *Bell Telephone Quarterly* 13, no. 4 (1934); U.S. Bureau of Labor Statistics, "The Dial Telephone and Unemployment," *Monthly Labor Review* 34, no. 2 (1932): 235–47, and TNEC Hearings, "Technology and Concentration of Economic Power," 16652–56.

31. U.S. Bureau of Labor Statistics, *Handbook of Labor Statistics, 1936* (Washington, D.C.: U.S. Government Printing Office, 1936), 729. On dial system maintenance, see P. C. Schwantes Jr., "Behind the Scenes in a Central Office," *Bell Telephone Quarterly* 15, no. 2 (1936).

32. Ethel Best, "The Change from Manual to Dial Operation in the Telephone Industry," *Bulletin of the Women's Bureau* 110 (Washington, D.C.: U.S. Government Printing Office, 1933).

33. U.S. Bureau of Labor Statistics, "The Dial Telephone and Unemployment," 246–47.

34. TNEC Hearings, "Technology and Concentration of Economic Power," 16653–55.

35. Congressmen who heard Sullivan's testimony about customers who had trouble

learning to dial had experienced similar frustrations firsthand; the Capitol's introduction of new phones in 1930 had aroused vocal protest from some members. One senator, accustomed to making calls by just asking the switchboard, complained that the telephone company was trying to foist its job off on him. While Fiorello LaGuardia snapped that any representative without enough sense to operate a dial phone should find another job, Congress resolved overwhelmingly after just two weeks to restore the human operators. Editors at the *Schenectady Union-Star* commented, "plenty of people wish they could be Senators for a few minutes just to be able to say, as Senator Glass said, 'take these abominable things out.'" "Senate Ouster for Dial Phones," *Literary Digest*, June 7, 1930, 9.

36. Frank Jewett, "Utilizing the Results of Fundamental Research in the Communication Field," *Bell Telephone Quarterly* 11, no. 2 (1932): 160; TNEC Hearings, "Technology and Concentration of Economic Power," 16675–84.

37. Bell Company advertisement, *Survey Graphic* (December 1935).

38. TNEC Hearings, "Technology and Concentration of Economic Power," 16670, 16675–84, 16686. The high number of AT&T's hires in the early 1920s reflected in part the large turnover rate; many women left after marrying or for other personal reasons.

39. Walter Myer and Clay Coss, *The Promise of Tomorrow: The Long, Sure Road to National Stability, Family Security, and Individual Happiness* (Washington, D.C.: Civic Education Service, 1938), 516–18.

40. Jean-Baptiste Say, *A Treatise on Political Economy; or the Production, Distribution, and Consumption of Wealth*, trans. C. R. Prinsep, New American Edition; (Philadelphia, 1844), 85–90. For the history of economic thinking on technological unemployment, see Alexander Gourvitch, *Survey of Economic Theory on Technological Change and Employment*, Work Projects Administration, National Research Project, report no. G-6 (Philadelphia: U.S. Government Printing Office, 1940); and TNEC, *Technology in Our Economy*. See also Ester Fano, "A 'Wastage of Men': Technological Progress and Unemployment in the United States," *Technology and Culture* 32, no. 2 (1991): 264–92.

41. J. C. L. Simonde de Sismondi, *New Principles of Political Economy: Of Wealth in Its Relation to Population*, tran. and ann. Richard Hyse (New Brunswick, N.J.: Transaction, 1991), 555–65.

42. Ibid.

43. David Ricardo, *The Principles of Political Economy and Taxation* (London: J. M. Dent and Sons, 1917), 263–71.

44. J. R. McCulloch, *The Principles of Political Economy* (Edinburgh, [1830] 1864), 142–53; see also Gourvitch, *Survey of Economic Theory*.

45. John Stuart Mill, *Principles of Political Economy with Some of Their Applications to Social Philosophy*, 5th ed. (New York: D. Appleton, 1909); Thomas Robert Malthus, *Principles of Political Economy Considered with a View to Their Practical Application*, 2d ed.; (London, 1836), Karl Marx, *Capital: A Critical Analysis of Capitalist Production*, 3d ed., trans. Samuel Moor and Edward Aveling (New York, 1890), 239, 263, 266–75.

46. Carroll D. Wright, *The Industrial Evolution of the United States* (Meadville, Pa., 1895), 334, 336–42.

47. TNEC, *Technology in Our Economy*.

48. Paul Douglas and Aaron Director, *The Problem of Unemployment* (New York: Macmillan, 1931), 126, 130–32.

49. Ibid., 143.

50. Rexford Tugwell, "The Theory of Occupational Obsolescence," *Political Science Quarterly* 46 (June 1931): 183–85, 194.

51. Ibid., 218.

52. Alvin Hansen, "The Theory of Technological Progress and the Dislocation of Employment," Sumner Slichter, "Lines of Action, Adaptation, and Control," and "Discussion, Session on Technological Change as a Factor in Unemployment," from the 44th annual meeting of the American Economic Association, December 28–30, 1931; all in *American Economic Review* 22, supplement (March 1932). See also Alvin Hansen, "Institutional Frictions and Technological Unemployment," *Quarterly Journal of Economics* 45 (August 1931): 684–97.

53. Elizabeth Faulkner Baker, *Displacement of Men by Machines: Effects of Technological Change in Commercial Printing* (New York: Columbia University Press, 1933), vii, ix, 25, 31.

54. Ibid., 160–61, 184.

55. Ibid., 186, 195–97; and "Unemployment and the Machine," editorial, *New York Times,* September 21, 1933.

56. Elliott Dunlap Smith, *What Are the Psychological Factors of Obsolescence of Workers in Middle Age?* (New York: American Management Association, 1931); and "Raises Spectre of Unemployment, *New York Times,* October 2, 1929.

57. "Report on Age as Related to Employment, 1933," U.S. Bureau of Labor Statistics, *Handbook of Labor Statistics 1936,* 625; "Causes of Discrimination against Older Workers," *Monthly Labor Review* 46, no. 5 (1938): 1138–43; Eugene Fisk, *The Man over Forty: The Relation of Health to his Employment* (New York: American Management Association, 1930); and William Shumway, "Guidance from the Employer's Viewpoint," *Vocational Guidance Magazine* 11, no. 2 (1932): 68.

58. U.S. Bureau of Labor Statistics, "Digest of Material on Technological Changes," 1033–34.

59. Ethel Best, "Technological Changes in Relation to Women's Employment," *Bulletin of the Women's Bureau,* no. 107 (Washington, D.C.: U.S. Government Printing Office, 1935): 32–33.

60. TNEC Hearings, "Technology and Concentration of Economic Power," 16378–79; and "Steel Fingers Twist Pretzels," *Nation's Business* 27 (October 1939): 34.

61. David Weintraub, assisted by Harold Posner, *Unemployment and Increasing Productivity,* National Research Project on Reemployment Opportunities and Recent Changes in Industrial Techniques, report no. G-1 (Philadelphia: U.S. Government Printing Office, 1937), 59.

62. Isador Lubin, *The Absorption of the Unemployed by American Industry* (Washington, D.C.: Brookings Institution, 1929).

63. "Optimism Voiced on Social Trend: Professors Ogburn and Merriam, Hoover Committee Members, Cite Our Resources," *New York Times,* January 3, 1933; and "Reveals Big Shift in Jobs in Ten Years," *New York Times,* July 4, 1932.

64. Paul Douglas, "Technical Changes Affecting Vocational Choice," *Vocational Guidance Magazine* 11, no. 1 (1932): 14.

65. "New Industries," editorial, *New York Times,* December 8, 1938; and Silas Bent, *Machine-Made Man* (New York: Farrar and Rinehart, 1931), 4–6.

66. Mordecai Ezekiel, *Jobs for All Through Industrial Expansion* (New York: Knopf, 1931),

69–71; and Stuart Chase, "New Paths for America, III," *Scribner's Magazine* 91 (March 1932): 139–42. On the impact of the frontier thesis, see Warren Susman, "The Frontier Thesis and the American Intellectual," *Culture as History* (New York: Pantheon, 1984).

67. Douglas and Director, *The Problem of Unemployment:* 131–32.

68. William Green, "Labor versus Machines: An Employment Puzzle," *New York Times,* June 1, 1930.

69. Stuart Chase, "What Hope for the Jobless?" *Current History* 39 (November 1933): 136; and "Keedoozle, Inc.," *Printer's Ink* 173 (December 5, 1935).

70. *The Annals of the American Academy* 154 (March 1931). For the statisticians' arguments over, for example, the selection of data in WPA studies of technological unemployment, see Elmer Bratt, "Did Productivity Increase in the Twenties?" *Journal of the American Statistical Association* 34 (1939): 326–29; and David Weintraub, "A Rejoinder," ibid., 332–34.

71. William Fielding Ogburn, *Social Change with Respect to Culture and Original Nature* (New York: B. W. Huebsch, 1922; New York: Viking Press, 1950), 200–201, 389–91; William Fielding Ogburn, "Technology and Governmental Change," *Journal of Business of the University of Chicago* 9, no. 1 (1936): 1–2, 12–13; and William Fielding Ogburn, "Stationary and Changing Societies," *American Journal of Sociology* 42, no. 1 (1936): 16. See also William Fielding Ogburn, "Trends in Social Science," *Science,* March 23, 1934, 257–62.

72. Arland D. Weeks, "Will There Be an Age of Social Invention?" *Scientific Monthly* 35 (1932): 366–70. As examples of the sociologists' annual lists of inventions, see William F. Ogburn, "Inventions and Discoveries," *American Journal of Sociology* 34 (1928–29): 25–39; William F. Ogburn, "Inventions and Discoveries," ibid., 984–93; Clark Tibbitts, "Inventions and Discoveries," *American Journal of Sociology* 36 (1930–31): 885–93; S. C. Gilfillan, "Inventions and Discoveries," *American Journal of Sociology* 37 (1931–32): 868–75; S. C. Gilfillan, "Inventions and Discoveries," *American Journal of Sociology* 38 (1932): 835–44.

73. "Thomas Edison Talks on Invention in the Life of Today," *Review of Reviews* 83 (January 1931): 38–40; Albert Shaw, "Mr. Edison's Views of Life and Work," *Review of Reviews* 85 (January 1932): 30–31; and "Ford and Edison on the Inventions of the Future," *Literary Digest,* December 7, 1929, 28, 30.

74. Stuart Chase, *Men and Machines* (New York: Macmillan, 1931), 209, 216.

75. Charles Babbage, *On the Economy of Machines and Manufactures,* fourth edition (New York: Augustus M. Kelley, 1971), 336–37, emphasis in the original.

Chapter 2: "Finding Jobs Faster than Invention Can Take Them Away"

1. "Unemployment Laid to Machine Use Here: *London Times* Believes the Problem Will Be Increasing and Chronic from Now On," *New York Times,* March 8, 1930.

2. Charles Beard to Lillian Gilbreth, December 3, 1930, Presidential Papers Subject File "PECE/POUR Records of the Women's Division—Technological Unemployment," Herbert Hoover Presidential Library (hereinafter HHPL).

3. William Green to Herbert Hoover, March 12, 1930, and March 20, 1930, and Hoover to Green, March 13, 1930, Presidential Papers Subject File "Unemployment—President's Emergency Committee on Employment—Correspondence, 1930 Feb.–Sept.," HHPL.

4. Press release, July 29, 1930, Presidential Papers Subject File "Unemployment—Advisory Committee on Employment Statistics, 1930," HHPL; Joseph Willits to Herbert

Hoover, February 9, 1931, Presidential Papers Subject File "Unemployment—Advisory Committee on Employment Statistics, 1931, Feb.," HHPL; and Report of the Advisory Committee on Employment Statistics, February 1931, Presidential Papers Subject File "Unemployment—Advisory Committee on Employment Statistics, 1931, Feb.," 23–27, HHPL.

5. Edward Eyre Hunt to Robert Lamont, memo, March 31, 1930, Presidential Papers Subject File "Unemployment Correspondence—March, 1930," HHPL.

6. Dexter Kimball to Herbert Hoover, May 14, 1930, and November 6, 1930, Presidential Papers Subject File "Kimball, Dexter S.," HHPL.

7. Ethelbert Stewart to the Secretary of Labor, memo, March 20, 1931; Presidential Papers Subject File, "President's Emergency Committee on Employment—Correspondence, 1931 Feb.–Mar.," HHPL.

8. "Hoover Hails Labor's Aid in Unemployment Crisis," New York Times, October 7, 1930.

9. George Nash, The Life of Herbert Hoover: The Engineer, 1874–1914 (New York: W. W. Norton, 1983); David Burner, Herbert Hoover: A Public Life (New York: Atheneum, 1984), 63–64; Joan Hoff Wilson, Herbert Hoover: Forgotten Progressive (Boston: Little, Brown, 1975), 51; and Edwin T. Layton Jr., The Revolt of the Engineers: Social Responsibility and the American Engineering Profession (Baltimore: Johns Hopkins University Press, 1986).

10. Ellis Hawley, "Herbert Hoover and American Corporatism, 1929–1933," and David Burner, "Before the Crash: Hoover's First Eight Months in the Presidency," in Martin Fausold and George Mazuzan, eds., The Hoover Presidency: A Reappraisal (Albany: State University of New York Press, 1974).

11. Herbert Hoover to Arch W. Shaw, February 17, 1933, President's Personal File, "Shaw, Arch W., 1930–1933," HHPL.

12. Ibid.

13. "Barnes Sees Nation on Way to Solution of Jobless Problem," New York Times, April 10, 1930; "Expects Robot Age to Absorb Jobless," New York Times, January 27, 1930; "Denies Machine Age Has Caused Depression: Klein Tells Dairy Convention Rise in Efficiency Has Tended to Increase Wages," New York Times, October 27, 1931; and Julius Klein, "The Challenge of the Machine," CBS radio broadcast, January 30, 1931, reprinted in American Machinis 74 (February 5, 1931): 241–43. During those same months, Secretary of Labor William Doak appointed a committee to assess the relationship between technological change and employment. Like the Commerce representatives, that group favored the most positive interpretation of mechanization, believing that innovation ultimately created enough new jobs to outweigh any problems of displacement. "Doak Picks Board on Unemployment," New York Times, May 20, 1931; and "Finds Labor Unhurt by Technical Gains: Report to Chemical Engineers Says It Is Only a Minor Cause of Unemployment," New York Times, December 11, 1931.

14. President's Organization on Unemployment Relief, "Introductory Statement and Recommendations of the Committee on Employment Plans and Suggestions," October 26 and 27, 1931, Presidential Papers Subject File "Unemployment—President's Organization on Unemployment Relief," HHPL.

15. Albert Romasco, The Poverty of Abundance: Hoover, the Nation, the Depression (Oxford: Oxford University Press, 1965), 147; and Fred Croxton to Theodore Joslin, memo, April 11, 1932, Presidential Papers Subject File "Unemployment—President's Organization on Unemployment Relief," HHPL.

16. Harry Wheeler to Henry Robinson, "Depression and the Shorter Week," October 28, 1931, Presidential Papers Subject File "Unemployment—President's Organization on Unemployment Relief—Correspondence, 1931 Oct.," HHPL; PECE Press Release, "Spreading of Work Becomes a Major Factor Toward Recovery," July 12, 1931, Presidential Papers Subject File "Unemployment—Press Releases of PECE and POUR—Industrial Employment, 1931 July–Aug.," HHPL; and President's Organization on Unemployment Relief, "Introductory Statement and Recommendations of the Committee on Employment Plans and Suggestions," October 26 and 27, 1931, Presidential Papers Subject File "Unemployment—President's Organization on Unemployment Relief," HHPL.

17. "An Open Letter to President Hoover and Congress from the Unemployed Union of Western Queens County, N.Y.," Presidential Papers Subject File "Unemployment—President's Organization on Unemployment Relief, HHPL; and C. H. Christensen to Herbert Hoover, June 15, 1932, Presidential Papers Subject File, "Unemployment—President's Organization on Unemployment Relief. Correspondence 1932, June 1–20," HHPL.

18. "Hoover Sees Peril of Slump Like '29," New York Times, December 3, 1936; and Herbert Hoover Presidential Museum oral history interview with Birge Clark, 11, HHPL. Arthur Schlesinger Jr. called Hoover a man who "decided that wherever he finally dug in constituted the limits of the permissible." He described the president as a "man of high ideals whose intelligence froze into inflexibility." Arthur Schlesinger Jr., The Age of Roosevelt: The Crisis of the Old Order (Boston: Houghton Mifflin, 1957), 246–47.

19. William Ogburn with S. C. Gilfillan, "The Influence of Invention and Discovery," 153–58; Leo Wolman and Gustav Peck, "Labor Groups in the Social Structure," 806–7; Edwin Gay and Leo Wolman, "Trends in Economic Organization," 234; and Ralph Hurlin and Meredith Givens, "Shifting Occupational Patterns," all in Recent Social Trends in the United States (New York: McGraw-Hill, 1933).

20. President's Research Committee on Social Trends, "A Review of Findings," xii, xxvii–xxviii; and William Fielding Ogburn with S. C. Gilfillan, "The Influence of Invention and Discovery," 129, both in Recent Social Trends in the United States.

21. President's Research Committee on Social Trends, "A Review of Findings," lxxi–lxxv. For background on the rise of social science, see Dorothy Ross, "American Social Science and the Idea of Progress," in Thomas L. Haskell, ed., The Authority of Experts (Bloomington, Ind.: Indiana University Press, 1984); and Guy Alchon, The Invisible Hand of Planning: Capitalism, Social Science, and the State in the 1920s (Princeton: Princeton University Press, 1985).

22. For a general assessment of the Roosevelt administration's economic positions and relevant political activity, see Paul Conklin, The New Deal (Arlington Heights, Ill.: Harlan Davidson, 1975); Harvard Sitcoff, ed., Fifty Years Later: The New Deal Evaluated (New York: Knopf, 1985); Ellis Hawley, The New Deal and the Problem of Monopoly: A Study in Economic Ambivalence (Princeton: Princeton University Press, 1966); Arthur Schlesinger Jr., The Age of Roosevelt: The Coming of the New Deal (Boston: Houghton Mifflin, 1958); and The Age of Roosevelt: The Politics of Upheaval (Boston: Houghton Mifflin, 1960). Also see, especially, Carroll Pursell, "Government and Technology in the Great Depression," Technology and Culture 20 (January 1979).

23. "Book by Dr. Ogburn Banned by Fechner: Pamphlet, Picturing Machine as a Monster, Was Called Too Gloomy for CCC Study," New York Times, November 16, 1934.

24. William Ogburn, You and Machines (Chicago: University of Chicago Press, 1934), 29–30, 39–41.

25. Ibid., 5–13, 49.

26. Ibid., 5–13.

27. Ibid., 5–11, 12–14, 20–25, 49–52.

28. Ibid.

29. "Works Relief Disputes Laid Before Roosevelt . . . President Says Recovery to 1929 Basis Would Still Leave 20 Per Cent Unemployed," *New York Times*, September 12, 1935; and "Scrap Heap Exaggerated: Industry and Employment," editorial, *New York Times*, September 14, 1935.

30. Aubrey Williams, "Warning! Permanent Unemployment Ahead," *American Federationist* 43, no. 2 (1936): 137–38.

31. "Permanent Cure Is Sought for Idle," *New York Times*, April 18, 1936; Corrington Gill, *Unemployment and Technological Change*, National Research Project on Reemployment Opportunities and Recent Changes in Industrial Techniques, report no. G-7 (Philadelphia: U.S. Government Printing Office, 1940), 1; Corrington Gill, "WPA Studies 'Foe,' the Swift Machine," *New York Times*, May 31, 1936; and "Job Study by WPA Wins Business Aid," *New York Times*, April 29, 1936.

32. "Philadelphia to be Center for Jobless Survey," *Evening Public Ledger*, December 10, 1935; and "White Collar Jobs Mapped for 24,000," *Philadelphia Inquirer*, December 10, 1935.

33. "A Technological Boondoggle," *Product Engineering* (March 1936).

34. Gill, "WPA Studies 'Foe,' the Swift Machine."

35. David Weintraub to Pierce Williams, November 3, 1937, Box 37, folder marked "Correspondence with Outside Agencies," Records of the Works Progress Administration: Records of the National Research Project 1935–1944, National Archives, Washington D.C. (hereinafter WPA-NRP Records).

36. Irving Kaplan to William Neiswanger, memo on "Machine Building Industries," March 11, 1936, Box 33, "Files of Irving Kaplan," folder marked "Machinery and Equipment Study," WPA-NRP Records.

37. Harry Jerome and J. V. H. Whipple to David Weintraub and Irving Kaplan, memo on "New Products," January 7, 1937; and Irving Kaplan, "Notes on Mr. Whipple's Outline on the Relation of the Machine Building Industry to Labor Displacement and Reemployment Opportunities," all in Box 33, ibid.

38. Weintraub to Gardiner Means, May 10, 1935, Box 60, folder marked "Criticism and Suggestions," WPA-NRP Records.

39. David Weintraub, assisted by Harold Posner, *Unemployment and Increasing Productivity*, National Research Project on Reemployment Opportunities and Recent Changes in Industrial Techniques, report no. G-1 (Philadelphia: U.S. Government Printing Office, 1937), 39–30, 56, 75; and Corrington Gill to Harry Hopkins, transmittal letter, ibid.

40. Ibid.

41. "Jobs and Jobless," *Cleveland Plain Dealer*, March 31, 1937; "WPA Job Report," *Indianapolis News*, April 1, 1937; "The Unemployment Picture," Elizabeth, New Jersey *Journal*, April 1, 1937; all in Box 57, File "Publicity—General," WPA-NRP records

42. "The Future of Unemployment," *New York Times*, March 29, 1937; and "'Guesstimating' Again," *Christian Science Monitor*, April 2, 1937; all in Box 57, File "Publicity—General," WPA-NRP records.

43. "'Guesstimating' Again," David Weintraub to the *Christian Science Monitor*, April

1937; Corrington Gill to Charles Merz of the *New York Times*, April 2, 1937; Corrington Gill to Leonard Kuvin (NICB), March 8, 1937; David Weintraub to Leonard Kuvin, March 26, 1937; and WPA staff notes; all in box 57, Files of E. F. Stone, folder marked "Weintraub's Office, Publicity—Unemployment and Increasing Productivity," WPA-NRP Records.

44. Charles Merz to Corrington Gill, April 20, 1937, Box 57, File "Publicity—General," WPA-NRP Records.

45. Harry Hopkins, speech, May 27, 1937, to the National Conference of Social Workers, Hopkins Papers, Box 10, folder marked "1937 speeches," Franklin Delano Roosevelt Presidential Library (hereinafter FDR Library); "End of Poverty Seen by Hopkins," *New York Times*, May 16, 1937; and *Congressional Record*, 75th Cong., 1st sess. (June 18, 1937) 81, pt. 6: 5965.

46. Boris Stern, *Mechanical Changes in the Woolen and Worsted Industries, 1910 to 1936*, National Research Project on Reemployment Opportunities and Recent Changes in Industrial Techniques, report no. B-3 (Philadelphia: U.S. Government Printing Office, 1938), 1–4; and David Weintraub and Irving Kaplan, *Summary of Findings to Date, March, 1938*, National Research Project on Reemployment Opportunities and Recent Changes in Industrial Techniques, report no. G-3 (Philadelphia: U.S. Government Printing Office, 1938). David Weintraub to Emerson Ross, January 18, 1939, Box 16, File "WPA Statistical Section—General," folder marked "Weintraub," WPA-NRP Records; "Hopkins Says Most Exhaustive Survey Yet Undertaken of Effect of Technological Changes in Industry Ready Soon," *Baltimore Sun*, February 28, 1937; Corrington Gill to Harry Hopkins, March 31, 1938, Box 51, File "E. J. Stone—Correspondence," folder marked "Summary of Findings, March 1938," WPA-NRP Records; and David Weintraub and Harry Magdoff, "The Service Industries in Relation to Employment Trends," ca. 1939, Box 57, folder marked "Services paper," WPA-NRP Records.

47. Weintraub and Kaplan, *Summary of Findings to Date, March 1938*, 138–45.

48. Weintraub and Posner, *Unemployment and Increasing Productivity*, 25–26, 74; Isador Lubin, *The Absorption of the Unemployed by American Industry* (Washington, D.C.: Brookings Institution, 1929); and Weintraub to Emerson Ross, January 18, 1939, Box 16, File "WPA Statistical Section—General," folder marked "Weintraub," WPA-NRP Records.

49. David Weintraub to Gardiner Means, May 10, 1935, Box 60, folder marked "Criticism and Suggestions," WPA-NRP Records; David Weintraub to Emerson Ross, January 18, 1939, Box 16, File "WPA Statistical Section—General," folder marked "Weintraub," WPA-NRP Records; and M. F. Behar, "Uncle Sam Discovers Instrumentation," *Instruments* (December 1938): 287.

50. Corrington Gill to Harry Hopkins, May 27, 1939, Hopkins Papers, Box 88, folder marked "Gill, Corrington," FDR Library.

51. Harold Pinches, "Mechanized Agriculture and Civilization," *Social Science* 12, no. 4 (1937): 450–51; Henry Giese, "The Application of Engineering to the Agricultural Industry," *Science* May 9, 1930, 467–70; S. McKee Rosen and Laura Rosen, *Technology and Society: The Influence of Machines in the United States* (New York: Macmillan, 1941), 78; Fred Henderson, *The Economic Consequences of Power Production* (New York: John Day, 1933), 57; and U.S. Bureau of Labor Statistics, "Mechanization of Agriculture as a Factor in Labor Displacement," *Monthly Labor Review* 33, no. 4 (1931): 749–83.

52. Henry A. Wallace, "The Social Advantages and Disadvantages of the Engineering-Scientific Approach to Civilization," *Science*, January 5, 1934, 2.

53. United States Department of Agriculture, *Is Increased Efficiency in Farming Always a Good Thing?* (Washington, D.C.: Government Printing Office, December 1936), emphasis in original.

54. Lorena Hickok, *One-Third of a Nation: Lorena Hickok Reports on the Great Depression*, Richard Lowitt and Maurine Beasley, eds. (Urbana: University of Illinois Press, 1981), 40–41.

55. Paul Taylor, "Power Farming and Labor Displacement in the Cotton Belt, 1937," *Monthly Labor Review* 46, no. 3 (1938): 595–607.

56. Victor Weybright, "Two Men and Their Machine," *Survey Graphic* (July 1936): 432–33; Isador Lubin to Paul Taylor, November 19, 1937, Isador Lubin Papers, Box 87, File "Taylor, Paul S.," FDR Library; and S. H. McCrory, Roy Hendrickson, and Committee, "Agriculture," in U.S. National Resources Committee, Science Committee, Subcommittee on Technology, *Technological Trends and National Policy, Including the Social Implications of New Inventions* (Washington, D.C.: U.S. Government Printing Office, 1937), 105, 139–43. As it turned out, developing a good cotton picker proved quite tricky; the machine would not really become practical or come into widespread use until after World War II. Gilbert C. Fite, "Mechanization of Cotton Production Since World War II," *Journal of Agricultural History* 54, no. 1 (1980): 190–207.

57. John Hopkins, *Changing Technology and Employment in Agriculture* National Research Project on Reemployment Opportunities and Recent Changes in Industrial Techniques, report no. G-3 (Philadelphia: U.S. Government Printing Office, 1941), 1–17, 56–57, 63, 67–70, 73–74, 176–80.

58. Temporary National Economic Committee (hereinafter TNEC) Hearings, "Technology and Concentration of Economic Power," part 30, *Investigation of Concentration of Economic Power* (Washington, D.C.: U.S. Government Printing Office, 1940), 17031; and Caterpillar Tractor Company, *At Last We Wives Can Have Vacations* (1936), quoted in Corlann Gee Bush, "'He Isn't Half So Cranky as He Used to Be': Agricultural Mechanization, Comparable Worth, and the Changing Farm Family," in Carol Groneman and Mary Beth Norton, eds., *'To Toil the Livelong Day': America's Women at Work, 1780–1980* (Ithaca, N.Y.: Cornell University Press, 1987).

59. TNEC Hearings, *Investigation of Concentration of Economic Power*, 17049–55, 17068; and Taylor, "Power Farming and Labor Displacement in the Cotton Belt, 1937."

60. David Weintraub, "Unemployment and Increasing Productivity," in U.S. National Resources Committee *Technological Trends and National Policy*. For background, see Arlene Inouye and Charles Susskind, "Technological Trends and National Policy,' 1937: The First Modern Technology Assessment," *Technology and Culture* 18, no. 4 (1977): 593–621.

61. "Foreword," vi; William Fielding Ogburn, "National Policy and Technology"; S. C. Gilfillan, "The Social Effects of Inventions," 25, all in U.S. National Resources Committee, *Technological Trends and National Policy*.

62. Bernhard Stern, "Resistances to the Adoption of Technological Innovations," 64, in U.S. National Resources Committee, *Technological Trends and National Policy*.

63. McCrory, Hendrickson, and Committee, "Agriculture," 105, 139–43.

64. "Foreword," vi, viii (emphasis in the original); John Merriam, "The Relation of Science to Technological Trends," 92, and S. C. Gilfillan, "Social Effects of Inventions," 24, all in U.S. National Resources Committee, *Technological Trends and National Policy*.

65. "President Commends the Report as Guide to Nation's Development," and William Laurence, "Inventors' Survey Finds Major Changes Imminent, Urges Labor Safeguard," both in *New York Times*, July 18, 1937

66. "'Permanent Cure' Is Sought for Idle," *New York Times*, April 18, 1936; Isador Lubin, "Radio Address, American School of the Air," November 28, 1938, Isador Lubin Papers, Box 160, "Speeches and Writings," File "Radio Address on Technological Unemployment," FDR Library; and Isador Lubin, "New Frontiers for the American Economy," Christian Science Monitor, March 2, 1940, Isador Lubin Papers, Box 161, "Speeches and Writings," FDR Library.

67. Mordecai Ezekiel, *Jobs for All Through Industrial Expansion* (New York: Knopf, 1939), 42–44, 68–76.

68. Mordecai Ezekiel, "Population and Unemployment," *Annals of the American Academy of Political and Social Science* (November 1936): 1–13; "Advertising and Redirection of Consumption," June 1934, Ezekiel Papers, Box 30, "Speeches and Articles," File "Advertising and Redirection of Consumption," FDR Library; "A Minimum Income of $2500 a Year?" Speech to the Jewish Community Center Forum, January 16, 1938, Ezekiel Papers, Box 33 "Speeches and Articles," File "Minimum Income of $2500 a Year," FDR Library. See also Mordecai Ezekiel, *$2500 a Year: From Scarcity to Abundance* (New York: Harcourt, Brace, 1936).

69. Franklin Delano Roosevelt, State of the Union message, January 3, 1940, in Fred Israel, ed., *The State of the Union Messages of the Presidents, vol. 3, 1905–1966* (New York: Chelsea House, 1967), 2853–54.

70. "New Deal Concept Assailed by Dewey," *New York Times*, January 26, 1940; and "Denies Invention Decreases Work," *New York Times*, January 7, 1940.

71. "'Adequate Tariff' Urged by Wagner, *New York Times*, January 19, 1930; and "Economists Plead for Job Insurance," *New York Times*, December 31, 1930.

72. Hugo Black, "For a Thirty-Hour Work Week," *Congressional Record* 73rd Cong., 1st sess. (April 3, 1933) 77, pt. 1:1114–1128, reprinted in Howard Zinn, ed., *New Deal Thought* (Indianapolis: Bobbs-Merrill, 1966), 265.

73. Robert Wagner, "Problem of Problems—Work," *New York Times*, February 16, 1936; and Robert Wagner, "The Senator Would Center Public Attention on a System of Economic Checks and Balances," *New York Times*, May 9, 1937.

74. Charles Faddis, "Letter to the Editor," *Steel* 104 (April 24, 1939): 9, 87; *Congressional Record*, 75th Cong., 1st sess. (June 18, 1937) 81, pt. 6: 5963–65 (June 18, 1937); and Joshua Bryan Lee, "Technological Unemployment and Relief," in A. Craig Baird, ed., *The Reference Shelf: Representative American Speeches*, vol. 13, no. 3 (New York: H. W. Wilson, 1939), 108–11.

75. TNEC Hearings, "Technology and Concentration of Economic Power," 16248.

76. Jennings Randolph to Franklin D. Roosevelt, June 19, 1936; Jennings Randolph to Franklin D. Roosevelt, November 15, 1937; D. W. Bell, Bureau of the Budget, to Roosevelt, memo, December 6, 1937; and 75th Cong., 1st sess., H.R. 7939 (July 20, 1937); all in Box OF 264, File "Unemployment, 1937," FDR Library.

77. American Technotax Society, *Economic Heirlooms*, pamphlet, n.d. [ca. 1934], Box 37 "Correspondence with Outside Agencies," File "Congress—Data, Comments," WPA-NRP; and "Green Backs Levying on Machine Profits," *New York Times*, November 21, 1933.

78. William H. Byrnes Jr., "The Invisibility of the Obvious," *Kiwanis Magazine*, October 1935, reprint in Box 1, File "Project General: Data, Comments, Correspondence, Etc.," WPA-NRP.

79. *Congressional Record*, 74th Cong., 1st sess. (April 13, 1935) 79, pt. 5: 5589–92; ibid. (April 23, 1935), 79, pt. 6: 6216; ibid. (May 17, 1935) 79, pt. 7: 7745; and ibid. (June 13, 1935) 79, pt. 8: 9246. See also "'Technotax' on Machines Offered as New Deal Ace," *New York Times*, April 14, 1935.

80. Ibid.

81. *Congressional Record*, 76th Cong., 3rd sess. (March 11, 1940) 86, pt. 3: 2649–53.

Chapter 3: "No Power on Earth Can Stop Improved Machinery"

1. Gerald Markowitz and David Rosner, eds., *"Slaves of the Depression": Workers' Letters about Life on the Job* (Ithaca, N.Y.: Cornell University Press, 1987), 98–99; Keith Dix, *What's a Coal Miner to Do? The Mechanization of Coal Mining* (Pittsburgh: University of Pittsburgh Press, 1988), 218; P. J. King, "Rip Van Winkles of Today," *Railroad Trainman* 47, no. 4 (1930): 264; and Temporary National Economic Committee (hereinafter TNEC) Hearings, "Technology and Concentration of Economic Power," part 30, *Investigation of Concentration of Economic Power* (Washington, D.C.: U.S. Government Printing Office, 1940), 16833–36, 16852.

2. Marion Elderton, ed., *Case Studies of Unemployment* (Philadelphia: University of Pennsylvania Press, 1931), 80–81, 163, 90, 225, 281–82; and Harold Ruttenberg, "The Big Morgue," *Survey Graphic* (April 1939): 266–67.

3. Patricia Cooper, *Once a Cigar Maker: Men, Women, and Work Culture in American Cigar Factories, 1900–1919* (Urbana: University of Illinois Press, 1987); and Patricia Cooper, "'What This Country Needs Is a Good Five-Cent Cigar,'" *Technology and Culture* 29, no. 4 (1988).

4. William Green, "Labor versus Machines: An Employment Puzzle," *New York Times*, June 1, 1930.

5. Ibid., and "Figures That Stretch Too Far," *New York Times*, October 4, 1930.

6. "Machinery and Unemployment," and "More Machinery, More Jobless," both in *AFL Weekly News Service*, no. 1135 (January 7, 1933): 1. See also "Reduction of Hours Urged by the Metal Trades Department and American Federation of Labor," *Bulletin of the Metal Trades Department, AFL* 17, no. 11 (1935): 1–2.

7. TNEC Hearings, "Technology and Concentration of Economic Power," 16454, 16626, 16835, 16838, 16847, and 17388. See also "Where Shall Control Be Lodged?" *American Federationist* 43, no. 1 (1936): 20.

8. TNEC Hearings, "Technology and Concentration of Economic Power," 16454. "Green Says Idle Number 8,917,000," *New York Times*, February 8, 1937.

9. William Chafe, *The Paradox of Change: American Women in the Twentieth Century* (Oxford: Oxford University Press, 1991); Leslie Woodcock Tentler, *Wage-Earning Women: Industrial Work and Family Life in the United States, 1900–1930* (Oxford: Oxford University Press, 1979); and "Fact Finding with the Women's Bureau," *Bulletin of the Women's Bureau*, no. 84 (Washington, D.C.: U.S. Government Printing Office, 1931): 4–5.

10. Susan Ware, *Holding Their Own: American Women in the 1930s* (Boston: Twayne,

1982); and Lois Scharf, *To Work and To Wed: Female Employment, Feminism, and the Great Depression* (Westport, Conn.: Greenwood, 1980).

11. T. J. E. Duffy, "Plaint of a Mere Man: Women, He Finds, Are Usurping Too Many Male Prerogatives," Letter to the Editor, *New York Times*, February 20, 1933; *The Unemployed* 1, no. 5: 27, in Alex Baskin and Barbara Baskin, eds., *The Unemployed and the Great Depression* (New York: Archives of Social History, 1975).

12. Mary Pidgeon, "Women in the Economy of the United States of America: A Summary Report," *Bulletin of the Women's Bureau*, no. 155 (Washington, D.C.: U.S. Government Printing Office, 1937): 8; Mary Pidgeon, "Women in Industry," *Bulletin of the Women's Bureau*, no. 91 (Washington, D.C.: U.S. Government Printing Office, 1931): 17–20; and Mary Pidgeon, "Differences in the Earnings of Women and Men," *Bulletin of the Women's Bureau*, no. 152 (Washington, D.C.: U.S. Government Printing Office, 1937): 5, 10–11, 28, 39.

13. Carroll Daugherty, *Labor Problems in American Industry* (Boston: Houghton Mifflin, 1938), 215.

14. Edwin Gabler, *The American Telegrapher: A Social History, 1860–1900* (New Brunswick, N.J.: Rutgers University Press, 1988); and Paul Israel, *From Machine Shop to Industrial Laboratory: Telegraphy and the Changing Context of American Invention, 1830–1920* (Baltimore: Johns Hopkins University Press, 1992).

15. U.S. Bureau of Labor Statistics, "Effects on Employment of the Printer Telegraph for Handling News," *Monthly Labor Review* 34, no. 4 (1932): 756; TNEC Hearings, "Technology and Concentration of Economic Power," 16690–91; and U.S. Bureau of Labor Statistics, "Displacement of Morse Telegraphers in Railroad Systems," *Monthly Labor Review* 37, no. 6 (1933): 276.

16. U.S. Bureau of Labor Statistics, "Displacement of Morse Operators in Commercial Telegraph Offices," *Monthly Labor Review*, v. 34, no. 3 (1932): 514–15; and Ethel Erickson, "The Employment of Women in Offices," *Bulletin of the Women's Bureau*, no. 120 (Washington, D.C.: U.S. Government Printing Office, 1934): 91.

17. Frank Powers, "The Modern Galley Slave—Feminine Gender," *American Federationist* 38, no. 1 (1931): 24–26; "Western Union," *Fortune*, November 1935, 93; J. M. Tuggey Jr., "Modern Business Adopts the Teletypewriter," *Bell Telephone Quarterly* 14, no. 4 (1935): 235, 240, 248; and TNEC Hearings, "Technology and Concentration of Economic Power," 16748.

18. Mary Anderson, "The Woman Office Worker—Her Outlook Today," *American Federationist* 42, no. 3 (1935): 267; Erickson, "The Employment of Women in Offices," 1; Emily Burr, "Girl Victims of the Machine Age," *Survey*, May 15, 1932, 188–89; Mal Stuart, "Robots in the Office," *New Republic*, May 25, 1938, 70–71; Markowitz and Rosner, eds., *Slaves of the Depression*, 174–75; and Walter Myer and Clay Coss, *The Promise of Tomorrow: The Long, Sure Road to National Stability, Family Security, and Individual Happiness* (Washington, D.C.: Civic Education Service, 1938), 306. For background, see Alice Kessler-Harris, *Out to Work: A History of Wage-Earning Women in the United States* (Oxford: Oxford University Press, 1982); Margery Davies, *Woman's Place Is at the Typewriter: Office Work and Office Workers, 1870–1930* (Philadelphia: Temple University Press, 1982); and Sharon Strom, *Beyond the Typewriter: Gender, Class, and the Origins of Modern American Office Work, 1900–1930* (Urbana: University of Illinois Press, 1992).

19. Myer and Coss, *The Promise of Tomorrow*, 306; and TNEC Hearings, "Technology and Concentration of Economic Power," 16764.

20. Erickson, "The Employment of Women in Offices," 34, 52, 108–9.

21. Ibid., 17, 34, 52, 108.

22. Pidgeon, "Women in the Economy of the United States," 29–30. Pidgeon praised dial switchboards for relieving the physical exhaustion, long working hours, and job stress of operators, but she worried about the loss of opportunity for young women. For background on the Women's Bureau, see Mary Anderson, *Woman at Work: The Autobiography of Mary Anderson as Told to Mary N. Winslow* (Minneapolis: University of Minnesota Press, 1951), and J. Stanley Lemons, *The Woman Citizen: Social Feminism in the 1920s* (Urbana: University of Illinois Press, 1973).

23. Ethel Best, "Technological Changes in Relation to Women's Employment," *Bulletin of the Women's Bureau*, no. 107 (Washington, D.C.: U.S. Government Printing Office, 1935): 1, 8–13, 20.

24. Women's Bureau of the U.S. Department of Labor, *Behind the Scenes in the Machine Age* (1931), reel 1.

25. Ibid., reel 3.

26. Harry Ober, *Trade-Union Policy and Technological Change*, Work Projects Administration, National Research Project, report no. L-8 (Philadelphia: U.S. Government Printing Office, April 1940); George E. Barnett, *Chapters on Machinery and Labor* (Cambridge, Mass.: Harvard University Press, 1926); and William G. Haber, "Workers' Rights and the Introduction of Machinery in the Men's Clothing Industry," *Journal of Political Economy* 33, no. 4 (1925).

27. Haber, "Workers' Rights and the Introduction of Machinery in the Men's Clothing Industry," 393; David Weintraub and Harry Ober, "Union Policies Relating to Technological Developments," paper delivered to the 1939 American Economic Association meeting, Box 57, File "Statistical Conference," National Archives, Records of the Works Progress Administration: Records of the National Research Project 1935–1944; and Harry Ober, *Trade-Union Policy and Technological Change.*

28. Alexander Walker, *The Shattered Silents: How the Talkies Came to Stay* (London: Harrap, 1978), 6–7.

29. U.S. Bureau of Labor Statistics, "Effects of Technological Changes upon Unemployment in the Amusement Industry," *Monthly Labor Review* 33, no. 2 (1931): 261–67; and U.S. Bureau of Labor Statistics, "Digest of Material on Technological Changes, Productivity of Labor, and Labor Displacement," *Monthly Labor Review* 35, no. 5 (1932): 1031–37. See also Walker, *The Shattered Silents*, 66; Charles Ross, "Sound Is Welcomed into the Realm of Motion Pictures," *Dun's International Review* 56 (November 1930): 21–23; Joseph Weber, "The Theatrical Situation," *International Musician* 27, no. 4 (1929): 1; and George Kent, "The New Crisis in the Motion Picture Industry," *Current History* (March 1931): 889–90.

30. U.S. Bureau of Labor Statistics, "Effects of Technological Changes upon Unemployment in the Amusement Industry."

31. Walker, *The Shattered Silents*, 76–77, 130; and Scott Eyman, *The Speed of Sound: Hollywood and the Talkie Revolution, 1926–1930* (Baltimore: Johns Hopkins University Press, 1998). In the 1952 MGM film *Singin' in the Rain*, Hollywood itself had fun looking back at studios' conversion to sound. One scene shows a vocal coach leading a silent-

film star (played by Gene Kelly) in ridiculous tongue twisters. Jean Hagen portrayed an actress whose beauty had led to success in silents but whose screechy accent proved disastrous for talkies.

32. U.S. Bureau of Labor Statistics, "Effects of Technological Changes Upon Employment in the Motion-Picture Theaters of Washington, D.C.," *Monthly Labor Review* 33, no. 5 (1931): 1005–18; U.S. Bureau of Labor Statistics, "Effects of Technological Changes upon Unemployment in the Amusement Industry"; Malcolm Willey and Stuart Rice, "The Agencies of Communication," *Recent Social Trends in the United States* (New York: McGraw-Hill, 1933), 208; and "Violinist Lays Woe to 'Canned' Music," *New York Times*, July 23, 1930.

33. "1,404 French Idle as 500 Are Ousted: Musicians Thrown Out of Work by Talkies and American There Face Loss of Jobs," *New York Times*, September 21, 1930.

34. James Petrillo, "What Mechanical Devices Have Done to the Worker," *International Musician* 27, no. 5 (1929): 9.

35. Joseph Weber, "The Art of Music Must Survive," *The International Musician* 27, no. 10 (1930): 1; "President Weber's Annual Report," *International Musician* 28, no. 12 (1931): 9; and James Petrillo, "What Mechanical Devices Have Done to the Worker."

36. Advertisement appearing in *American Federationist* 38, no. 11 (1931): 1313; "President Weber's Annual Report," *International Musician* 28, no. 12 (1931): 9; advertisement in *American Federationist* 38, no. 5 (1931): 519; and advertisement in *American Federationist* 38, no. 1 (1931): 7.

37. "Protest Talkie Music: Many Thousands Write on Attitude to Federation of Musicians," *New York Times*, March 1, 1930; and "President's Report to Boston Convention," *International Musician* 27, no. 12 (1931): 11. See also "Advertising Appeal Rallies Hosts in War on Canned Art," *International Musician* 27, no. 8 (1930): 1.

38. "AFM Charges Attempt to Suppress Advertisements," *International Musician* 27, no. 6 (1929): 21. In eleven cities, the AFM mounted ads on outdoor billboards. "President Weber's Annual Report," *International Musician* 28, no. 12 (1931): 8–9.

39. "President Weber's Annual Report," *International Musician* 27, no. 12 (1931): 9; and Joseph Weber, "Canned Music," *American Federationist*, 38, no. 9 (1931): 1064–70.

40. American Federation of Musicians Local 802 (New York), Executive Board Meeting Minutes, October 31, 1930, and January 13, 1936, microfilm rolls 5260 and 5335, New York University Wagner Labor Archives; "Official Proceedings of the Thirty-Sixth Annual Convention of the American Federation of Musicians," *International Musician* 29, no. 1 (1931): 14; and "One Hundred Cities Hold Living Music Days," *International Musician* 30, no. 6 (1932): 1.

41. George Kent, "The New Crisis in the Motion Picture Industry," 890; and National Resources Committee, *Technological Trends and National Policy* (Washington, D.C.: U.S. Government Printing Office, 1937): 228–29; and "Resources Committee Erects Signposts for the March of Television," *New York Times*, July 25, 1937. Weber continued to hope that audiences would reject new sound technology. Attempts to play recorded music on television would prove such a failure that stations would be forced to broadcast live concerts, he predicted, ultimately helping musicians.

42. Patricia Cooper, "'What This Country Needs Is a Good Five-Cent Cigar,'" 799–800.

43. "Deprecates 'Jazz' in Modern Industry," *New York Times*, March 9, 1930; TNEC Hearings, "Technology and Concentration of Economic Power," 16862–63.

44. "Murray Demands Labor Use Science," *New York Times*, February 4, 1940; and S. J. Woolf, "Lewis Challenges Labor's Old Order," *New York Times*, March 15, 1936. On the CIO's creation, see Walter Galenson, *The CIO Challenge to the AFL: A History of the American Labor Movement, 1935–1941* (Cambridge, Mass.: Harvard University Press, 1960); and Christopher Tomlins, "AFL Unions in the 1930s: Their Performance in Historical Perspective," *Journal of American History* 65, no. 4 (1979): 1021–42.

45. "Labor Council Asks Union Drive in South," *New York Times*, October 7, 1929; 27:1; and Woolf, "Lewis Challenges Labor's Old Order."

46. William Green, "Mechanization," n.d. [ca. 1932]; Green Papers, Box 14, Folder 15, "Speeches," George Meany Memorial Archives, Silver Spring, Md.; TNEC Hearings, "Technology and Concentration of Economic Power," 16457; and Fannia Cohn, "Three Papers on Economic Security: Social Responsibility," *Bulletin of the Taylor Society* 18, no. 3 (1933): 68.

47. TNEC Hearings, "Technology and Concentration of Economic Power," 16377, 16866.

48. I. M. Ornburn, "Men before Machines," *American Federationist* 41, no. 2 (1934): 141–47.

49. Harry Calkins and Frank Finney, "Why 22 Million Are on Relief," *American Federationist* 42, no. 10 (1935): 1052–61.

50. Green, "Labor versus Machines"; and TNEC Hearings, "Technology and Concentration of Economic Power," 16377, 16385, 16734–35.

51. "Reduction of Hours Urged by the Metal Trades Department and American Federation of Labor," 1–2; William Green, "Labor's Plan for Recovery: A Five-Day Week for Workers," *New York Times*, July 17, 1932; and "Green Says Goal Is Thirty-Hour Week," *New York Times*, September 3, 1935. Debate over working hours continued through the 1930s. Most NRA codes did not mandate cuts in hours, and after the Supreme Court's 1935 decision invalidating the NRA, more industries kept stretching the workweek beyond forty hours. David Roediger and Philip Foner, *Our Own Time: A History of American Labor and the Working Day* (New York: Greenwood, 1989). See also Benjamin Hunnicutt, *Work without End: Abandoning Shorter Hours for the Right to Work* (Philadelphia: Temple University Press, 1988).

52. "The Case for Shorter Hours of Work," *Railroad Trainman* 49, no. 7 (1932): 389; "Greater Mass Purchasing Power Should Be Our Immediate Objective," *Railroad Trainman* 52, no. 2 (1935): 67; Brotherhood of Railroad Trainmen, *Shorter Workday: A Plea in the Public Interest, 1937* (Cleveland, Ohio: Brotherhood of Railroad Trainmen, 1937) 36; "CIO Is Going On, Lewis Tells AFL," *New York Times*, December 17, 1937; and "Shorter Hours" editorials, *American Federationist* 38, no. 1 (1931): 22–23 and 38, no. 2 (1931): 145–46. On working hours, see also Weintraub and Ober, "Union Policies Relating to Technological Developments"; and "Excerpt from Report of Chairman John L. Lewis" and "Resolution No. 58," CIO Convention, Box 37 "Correspondence with Outside Agencies," File "Congress—Legislation on Technological Unemployment," all in National Archives, Records of the Works Progress Administration: Records of the National Research Project 1935–1944.

53. "Five-Hour Work Day Is Urged for Labor," *New York Times*, October 2, 1930; William Green, "The Five-Day Week," *Harvard Business Review* 9, no. 3 (1931): 270–73; "Shortening the Worker's Task," *American Federationist* 38, no. 4 (1931): 456–62; and William Green, "Labor's Plan for Recovery."

54. "Wage Cuts Are the Backward Road," *American Federationist* 38, no. 6 (1931): 725–30; and Matthew Woll, "More and Faster Increases in Wages Seen as a Remedy," Letter to the *New York Times,* June 22, 1930.

55. Ober, *Trade-Union Policy and Technological Change;* Elderton, ed., *Case Studies of Unemployment:* 80–81, 163, 90, 225, 281–82; and "Trolley Men Run Green Line Buses," *New York Times,* June 23, 1935.

56. TNEC Hearings, "Technology and Concentration of Economic Power," 16750–56, 16855–65; and Sumner Slichter, *Union Policies and Industrial Management* (Washington, D.C.: Brookings Institution, 1941): 213–14, 227.

57. Max Danish, "Taxing Machines to Relieve Jobless," *American Federationist* 39, no. 10 (1932): 1104–6; TNEC Hearings, "Technology and Concentration of Economic Power," 16506; and John Frey, "Wrong Use of Scientific Developments Barrier to Progress of Mankind toward Better World," *Bulletin of the Metal Trades Department, AFL* 17, no. 2 (1933): 1–3.

58. Danish, "Taxing Machines to Relieve Jobless"; "Machines Must Pay Men They Made Idle," *New York Times,* August 17, 1932; and "Machines and Pressers," editorial, *New York Times,* August 31, 1932.

59. Ober, *Trade-Union Policy and Technological Change.*

60. John Riegel, *Management, Labor, and Technological Change* (Ann Arbor: University of Michigan Press, 1942), 26–27.

61. Edward Brady, "Machinery: What Are We Going to Do about It?" letter to the editor, *Railroad Trainman* 51, no. 3 (1934): 164, 169.

62. *Proceedings of the Thirty-Third Constitutional Convention of the United Mine Workers of America,* January 23–31, 1934, Vol. 1: 189–91, 196–99, 203–6.

63. Ibid., and "Miners Ask Share in Machine Profits," *New York Times,* January 25, 1934.

64. *Proceedings of the Thirty-Fourth Constitutional Convention of the United Mine Workers of America,* January 28–February 7, 1936 (Indianapolis: Cheltenham Press, 1936), Vol. 4: 61, 146; and *Proceedings of the Thirty-Sixth Constitutional Convention of the United Mine Workers of America,* January 23–February 1, 1940 (Indianapolis: Cheltenham Press, 1940), Vol. 2: 5–6, 21.

65. Riegel, *Management, Labor, and Technological Change,* 32–34.

66. Slichter, *Union Policies and Industrial Management,* 180, 183.

67. Lorena Hickok, *One-Third of a Nation: Lorena Hickok Reports on the Great Depression,* Richard Lowitt and Maurine Beasley, eds. (Urbana: University of Illinois Press, 1981), 115.

68. Emil Rieve, "The Hosiery Workers Union," *Labor Information Bulletin* 6, no. 2 (1939).

69. Historians have identified many reasons for labor's weakness. David Montgomery has pointed to the 1920–22 economic downturn as one force undermining union strength. The Red Scare helped eliminate militancy, as mainstream labor leaders purged their ranks of Communists and other radicals. Alternately, David Brody has identified the decade's prosperity as a force generating "confidence in America's bountiful mass-production system," leading to worker "satisfaction . . . with the status quo." The promise of welfare capitalism also may have undercut the appeal of union activity. Employers' antiunion tactics (including the establishment of company unions), antilabor legal decisions, and the inadequacies of craft-based organization also made labor weak, according to Irving Bernstein. See David Montgomery, *The Fall of the House of Labor: The Workplace, the State,*

and American Labor Activism, 1865–1925 (Cambridge: Cambridge University Press, 1987); David Brody, *Workers in Industrial America: Essays on the Twentieth Century Struggle* (Oxford: Oxford University Press, 1980), 61–62; and Irving Bernstein, *The Lean Years: A History of the American Worker 1920–1933* (Boston: Houghton Mifflin, 1960).

70. Craig Phelan has argued that while many top AFL men accommodated employers because they felt the union did not have power to win any concessions, Green's faith in a nineteenth-century style social gospel led him to believe in the dream of cooperative labor relations. Craig Phelan, *William Green: Biography of a Labor Leader* (Albany: State University of New York Press, 1989). See also William Green to J. E. Smith, June 30, 1933, Green Papers, Box 6, Folder 14, George Meany Memorial Archives. On Lewis, see Curtis Seltzer, *Fire in the Hole: Miners and Managers in the American Coal Industry* (Lexington, Ky.: University Press of Kentucky, 1985), 45; and Melvyn Dubofsky and Warren Van Tine, *John L. Lewis: A Biography* (New York: Quadrangle, 1977).

71. Stanley Vittoz, *New Deal Labor Policy and the American Industrial Economy* (Chapel Hill: University of North Carolina Press, 1987); Irving Bernstein, *The Turbulent Years: A History of the American Worker 1933–1941* (Boston: Houghton Mifflin, 1969); Hugh Aitken, *Taylorism at Watertown Arsenal: Scientific Management in Action, 1908–1915* (Cambridge, Mass.: Harvard University Press, 1960); David Montgomery, *Workers' Control in America: Studies in the History of Work, Technology, and Labor Struggles* (Cambridge: Cambridge University Press, 1979), 145; and John Bodnar, *Workers' World: Kinship, Community, and Protest in an Industrial Society, 1900–1940* (Baltimore: Johns Hopkins University Press, 1982), 159. For background, see Robert Zieger, *American Workers, American Unions, 1920–1985* (Baltimore: Johns Hopkins University Press, 1986).

Chapter 4: "Machinery Don't Eat"

1. Marion Elderton, ed., *Case Studies of Unemployment* (Philadelphia: University of Pennsylvania Press, 1931), xxxix; and Gerald Markowitz and David Rosner, eds., *"Slaves of the Depression": Workers' Letters about Life on the Job* (Ithaca, N.Y.: Cornell University Press, 1987), 70–71.

2. Robert S. McElvaine, ed., *Down and Out in the Great Depression: Letters from the "Forgotten Man,"* (Chapel Hill: University of North Carolina Press, 1983), 41, 177, 193.

3. Lorena Hickok, *One-Third of a Nation: Lorena Hickok Reports on the Great Depression,* Richard Lowitt and Maurine Beasley, eds. (Urbana: University of Illinois Press, 1981).

4. "Labor Plan Urged by Episcopalians," *New York Times,* January 27, 1930; and "Benefits of Machines Should Extend to All," *AFL Weekly News Service,* August 27, 1932, 1.

5. As examples, see Waldemar Kaempffert, "The Man over Forty: A Machine-Age Dilemma," *New York Times Magazine,* March 6, 1938; L. J. Nations, "The Old South Facing the Machine," *Current History* 33 (October 1930); Emily Burr, "Girl Victims of the Machine Age," *Survey,* May 15, 1932; Louis Stark, "Cars and the Men," *Survey Graphic* (April 1935); "Invention and Unemployment," *New Republic,* March 11, 1940; Charles N. Edge, "The Price of Recovery," *Living Age* 345 (November 1933); Ralph Aiken, "More Machines and Less Men," *North American Review* 231 (May 1931); "The Apex of Mass Production," *Literary Digest,* June 28, 1930; "Machines and Unemployment," *Review of Reviews* 83 (March 1931); D. Marshall, "What about the Machine?" *Commonweal,* June 21, 1935; Josiah Stamp, "Do We Need Birth Control for Ideas?" *Rotarian* 44 (April 1934).

6. Peter Van Dresser, "New Tools for Democracy," *Harper's Magazine* 178 (March 1939): 397–403.

7. Ruth Crawford, "The Jersey Devil Came," *Scribner's Magazine* 93 (April 1933): 239–44.

8. Frazier Hunt, "Will the Goblins Finally Get Us? No!" *Good Housekeeping*, April 1933, 42–43.

9. Stuart Chase, *Men and Machines* (New York: Macmillan, 1931), 178, 198, 205, 208–12, 216, 340.

10. Ibid., 209, 216.

11. "See Problem for NRA: Stuart Chase and Prof. Mills Cite Technological Unemployment," *New York Times*, May 13, 1934; and Stuart Chase, "What Hope for the Jobless?" *Current History* 39 (November 1933): 131–32.

12. "Industrial Growth of Nation Is Traced," *New York Times*, August 6, 1932; "Declares Machines Add to Unemployed: Research Director Finds Great Technical Gains in the Past Three Years," *New York Times*, August 21, 1932; and Wayne Parrish, "What Is Technocracy?" *New Outlook* 160 (November 1932): 13–14.

13. Frank Arkwright, *The ABC of Technocracy* (New York: Harper and Bros., 1933); Howard Scott, *Introduction to Technocracy* (New York: John Day, 1933); and "Sees Price System Doomed in Industry," *New York Times*, June 16, 1932.

14. "Finds Coolie's Pay Higher Than Ours: Survey Shows Even Chinese Get More for What They Produce Than Americans," *New York Times*, September 17, 1932; "Recovery Remedies Scorned as Futile: Dr. Howard Scott Declares That Only Dropping Price System of Control Will Do Any Good," *New York Times*, September 3, 1932; Wayne Parrish, "Technocracy's Challenge," *New Outlook* 161 (January 1933): 13–14; Arkwright, *The ABC of Technocracy*, 60–65, 70; and Scott, *Introduction to Technocracy*, 48.

15. "Ridicules Technocracy," *New York Times*, January 15, 1933; "Technocracy Held Paralyzing Force," *New York Times*, January 10, 1933; and "A Panicky Theory: Technocracy Attacked by Lee J. Eastman in Advertising Talk," *New York Times*, January 24, 1933.

16. "Scan Technocracy, Hold It Is in Error; Three Experts Say Idleness of Machines, Not a Mechanized Era, Wipes Out Jobs," *New York Times*, January 23, 1933; "King Replies to Technocrats," *New York Times*, December 30, 1932; "Dr. Fisher Scoffs at Technocrats," *New York Times*, January 15, 1933; and "'A Dangerous Sophistry': Professor Fisher Attacks Technocracy as Campaign of Fear," *New York Times*, January 25, 1933.

17. Stuart Chase, *Technocracy: An Interpretation* (New York: John Day, 1933), 12–15, 28, 30–32; and "Dr. Butler Disavows Link to Technocracy; Says Columbia Only Houses Engineer Group," *New York Times*, January 18, 1933.

18. Arkwright, *The ABC of Technocracy*; Scott, *Introduction to Technocracy*; and "Sees Price System Doomed in Industry," *New York Times*, June 16, 1932.

19. David Peeler, *Hope Among Us Yet: Social Criticism and Social Solace in Depression America* (Athens, Ga.: University of Georgia Press, 1987); and Richard Pells, *Radical Visions and American Dreams: Culture and Social Thought in the Depression Years* (New York: Harper and Row, 1973), 202.

20. John Steinbeck, *The Grapes of Wrath* (New York: Penguin, 1977), 42–43, 50, 160, 385, 556.

21. Upton Sinclair, *Little Steel* (New York: Farrar and Rinehart, 1938), 174–75.

22. Theodore Dreiser, *Tragic America* (London: Constable, 1931), 16–17, 22–27, 55, 68–69, 106, 182–88, 357, 388–91.

23. Sherwood Anderson, *Perhaps Women* (New York: Horace Liveright, 1931).

24. Frances Lockridge and Richard Lockridge, *Death on the Aisle* (New York: Pocket Books, 1942), 5.

25. Hugo Gernsback, "Wonders of the Machine Age," *Wonder Stories* 3, no. 2 (1931): 151, 284–86. My thanks to John Cheng for this source.

26. Hugo Gernsback, "Wonders of the Machine Age."

27. Karel Capek, *R.U.R. (Rossum's Universal Robots)*, reprinted in *Toward the Radical Center: A Karel Capek Reader* (Highland Park, N.J.: Catbird, 1990), 49, 51–52.

28. John Campbell, "Twilight" (1934), reprinted in Robert Silverberg, ed., *Science Fiction Hall of Fame* (New York: Avon, 1970).

29. See Paul Carter, *The Creation of Tomorrow: Fifty Years of Magazine Science Fiction* (New York: Columbia University Press, 1977).

30. Granville Hicks, with Richard M. Bennett, *The First to Awaken* (New York: Modern Age, 1940; reprint, New York: Times Books, 1971), 68–70, 121, 165–66, 174, 271–71, 277–80.

31. Virginia Lee Burton, *Mike Mulligan and His Steam Shovel* (Boston: Houghton Mifflin, 1939), 3, 13–16, 20.

32. Ibid., 3, 13–16, 20.

33. Harold Rugg, *A History of American Civilization: Economic and Social* (Boston: Ginn, 1930), 475, 542–46, 553–56, 593–608, and 613–15.

34. Harold Rugg, *An Introduction to Problems of American Culture* (Boston: Ginn, 1931), 6–9; Frances Fitzgerald, *America Revised: History Schoolbooks in the Twentieth Century* (Boston: Little, Brown, 1979); and Gary B. Nash, Charlotte Crabtree, and Ross E. Dunn, *History on Trial: Culture Wars and the Teaching of the Past* (New York: Knopf, 1997).

35. Helen Noll Crowell, "Cotton Pickers," *Survey Graphic* (August 1937): 427.

36. "Mechanization Pace Quickens," *Coal Age* 42, no. 6 (1937): 260–61; and George Korson, *Coal Dust on the Fiddle* (Philadelphia: University of Pennsylvania Press, 1943), 138–39, 141; quoted in Keith Dix, *What's a Coal Miner to Do? The Mechanization of Coal Mining* (Pittsburgh: University of Pittsburgh Press, 1988), 85–86.

37. Gertrude Munter, "The Iron Man," *International Musician* 27, no. 6 (1930): 10.

38. *International Musician* 27, no. 10 (1930): 1, and 27, no. 11 (1930): 3.

39. Clarence Day, "Animals in a Machine Age," *Harper's Magazine* 163 (July 1931): 219, 223; and *The Unemployed*, in Alex Baskin and Barbara Baskin, eds., *The Unemployed and the Great Depression* (New York: Archives of Social History, 1975).

40. "Mr. Rogers Reviews Our Status and Cites the Reasons Therefore," *New York Times*, December 31, 1930.

41. Charles Coughlin, "The National Union for Social Justice," radio address, November 11, 1934; "Share the Profits with Labor," radio address, December 2, 1934; and "The Declaration of Independence," radio address, March 10, 1935, all in *A Series of Lectures on Social Justice* (Royal Oak, Mich.: Radio League of the Little Flower, 1935), 11–12, 48–51, 208.

42. Ibid., 12, 49, 208. See also Alan Brinkley, *Voices of Protest: Huey Long, Father Coughlin and the Great Depression* (New York: Vintage, 1982).

43. Pare Lorentz, *The Plow That Broke the Plains* (1936) and *The River* (1938), scripts reprinted in Pare Lorentz, *FDR's Moviemaker: Memoirs and Scripts* (Reno, Nev.: University of Nevada Press, 1992), 44–50.

44. Pare Lorentz, *Ecce Homo! Behold the Man* (1938), ibid., 99, 103–4, 120–23.

45. Ibid., 80–81, 120–23. See also Robert L. Snyder, *Pare Lorentz and the Documentary Film* (Norman, Okla.: University of Oklahoma Press, 1968).

46. Willard Van Dyke, *Valleytown* (1940); and Amalie Rothschild, *Conversations with Willard Van Dyke*, Anomaly Films (1980).

47. Van Dyke, *Valleytown*, and Rothschild, *Conversations with Willard Van Dyke*.

48. George Cukor, *Dinner at Eight*, MGM (1933).

49. George Basalla, "Keaton and Chaplin: The Silent Film's Response to Technology," in Carroll W. Pursell Jr., ed., *Technology in America: A History of Individuals and Ideas* (Cambridge, Mass.: MIT Press, 1981), 197, 201.

50. Rugg, *An Introduction to Problems of American Culture*, 8; David Cushman Coyle, "—And Jobs," *Survey Graphic* (April 1936): 209; and Raymond Yates, *Machines over Men* (New York: Frederick A. Stokes, 1939), 138–39.

51. Terry Smith, *Making the Modern: Industry, Art, and Design in America* (Chicago: University of Chicago Press, 1993), 305–6.

52. Theodore Knappen, "The Machine Turns on Its Master," *Magazine of Wall Street*, May 3, 1930, 11; and John Van Deventer, "Among the Robots," *Iron Age*, March 12, 1931, 846.

53. AFM advertisements, *American Federationist* 38, no. 2 (1931): 135, and 38, no. 6 (1931): 663.

54. William Ogburn, *You and Machines* (Chicago: University of Chicago Press, 1934).

55. Ibid., 7.

56. Harry Laidler, *Unemployment and Its Remedies* (New York: League for Industrial Democracy, 1931); Baskin and Baskin, eds., *The Unemployed and the Great Depression*, 18; Rugg, *An Introduction to Problems of American Culture*, 7; and *New York Times*, February 25, 1940.

57. "Robot to Speak in Texas for Labor Bureau: He Apologizes, Then Defends Machinery," *New York Times*, May 31, 1936.

Chapter 5: "The Machine Has Been Libeled"

1. Justin Macklin, "Labor-Saving Machines Make Jobs," *Nation's Business* 28 (January 1940): 21; and Benjamin M. Anderson Jr., "Technological Progress, The Stability of Business, and the Interests of Labor," *Chase Economic Bulletin* 17 (April 13, 1937): 18–21; F. A. Merrick, "The Machine Myth," *Mining Congress Journal* 16 (January 1930): 7–10. For background on the Depression-era business community, see Louis Galambos and Joseph Pratt, *The Rise of the Corporate Commonwealth: U.S. Business and Public Policy in the Twentieth Century* (New York: Basic Books, 1988); Robert Collins, *The Business Response to Keynes, 1929–1964* (New York: Columbia University Press, 1981); and Thomas Cochran, *American Business in the Twentieth Century* (Cambridge, Mass.: Harvard University Press, 1972).

2. Macklin, "Labor-Saving Machines Make Jobs," 86–89; and John Van Deventer, "The Machine Has Been Libeled," *Iron Age* December 24, 1931, 1605–6.

3. "Kettering Scouts Economic Despair," *New York Times*, May 29, 1932"; and "Denies Machine Age Is Displacing Man," *New York Times*, January 18, 1933.

4. Anderson, "Technological Progress, the Stability of Business, and the Interests of Labor"; and Macklin, "Labor-Saving Machines Make Jobs," 86–89.

5. Temporary National Economic Committee (hereinafter TNEC) Hearings, "Technology and Concentration of Economic Power," part 30, *Investigation of Concentration of Economic Power* (Washington, D.C.: U.S. Government Printing Office, 1940): 16391–92; and John Van Deventer, "The Machine Has Been Libeled," 1605.

6. "Call Unemployment the Price of Progress," *New York Times*, February 25, 1931; and Harold Ruttenberg, "The Big Morgue," *Survey Graphic* (April 1939): 268.

7. Stuart Chase, *Men and Machines* (New York: Macmillan, 1929), 289, 296–300.

8. David Noble, *America by Design: Science, Technology, and the Rise of Corporate Capitalism* (New York: Oxford University Press, 1977).

9. Philip Bliss, "New Machines Needed to Create Profits under the New Order," *Rand McNally Bankers Monthly* (December 1933): 660–62, 682; E. F. Du Brul, "Re-adjustments of the Machine Age," *Mechanical Engineering* 52, no. 6 (1930).

10. Advertisement for Acme Visible Records, *Nation's Business* 19 (February 1931): 77, emphasis in original; James Harbord, "The Present Problem of the Machine," *Mechanical Engineering* 55, no. 1 (1933); and Bliss, "New Machines Needed to Create Profits under the New Order."

11. Chase, *Men and Machines*, 186–87; Macklin, "Labor-Saving Machines Make Jobs"; and "Machinery's Contribution to the Economy of Plenty," *Automotive Industries* 72 (June 15, 1935).

12. U.S. Bureau of Labor Statistics, "A Digest of Material on Technological Changes, Productivity of Labor, and Labor Displacement," *Monthly Labor Review* (1932): 1031–57; Louis Stark, "Cars and the Men," *Survey Graphic* (April 1935): 187–91; Corrington Gill, *Wasted Manpower: The Challenge of Unemployment* (New York: W. W. Norton, 1939), 73–74; "Inspecting Push Rods Automatically," *American Machinist* 81 (September 22, 1937): 833; and Charles Herb, "The Electric Eye in Industry," *Machinery* 45, no. 9 (1939): 610–11.

13. "The Manufacturers Reply," *Survey Graphic* (April 1935): 190; "Sloan Warns Jobs Depend on Buying . . . Holds Technology Is Key," *New York Times*, January 10, 1938; "The Creation of Jobs," *New York Times*, April 14, 1937; and "Handwork Costly," *New York Times*, March 20, 1938.

14. W. J. Cameron, "Machines Create Employment," quoted in John Younger, "A Cure for Unemployment," *Engineering Experiment Station News*, 10, no. 5 (1938): 13–10.

15. Stephen DuBrul, "Technology and Competition," in *Employment and the Engineer's Relation to It: Addresses and Digest of Discussion* (Washington, D.C.: American Engineering Council, 1938), 33–34.

16. "The Automatic Manufacture of Automobile Frames, Part 1 and 2," *Engineer*, January 10 23, 1931; L. R. Smith, "We Build a Plant to Run Without Men," *Magazine of Business* 55, no. 2 (1929): 135–37; and "New Building for A. O. Smith Corporation Engineers," *Factory and Industrial Management* 80 (August 1930): 313.

17. Stuart Chase, "The Iron Bouncer," *Journal of Adult Education* 3, no. 1 (1931): 12–14; A. W. Redlin, "Handling of Materials and Frame Manufacturing Process Are Automatic at Smith Plant," *Automotive Industries* 62 (March 22, 1930): 468; and Dexter Kimball, "The Social Effects of Mass Production," in Charles F. Roos, ed., *Stabilization of Employment* (Bloomington, Ind.: Principia, 1933), 52.

18. F. B. Fletcher, "New Research Building Completed," *Steel* 89 (December 21, 1931); and Frazier Hunt, "Will the Goblins Finally Get Us? No!" *Good Housekeeping* (April 1933): 42–43.

19. "Expects Robot Age to Absorb Jobless," *New York Times*, January 27, 1930; and Stanley Koon, "10,000 Automobile Frames a Day," *Iron Age*, June 5, 1930, 1728.

20. "Kettering Scouts Economic Despair"; S. McKee Rosen and Laura Rosen, *Technology and Society: The Influence of Machines in the United States* (New York: Macmillan,

1941), 175–76; and Edward A. Filene, *Successful Living in This Machine Age* (New York: Simon and Schuster, 1932).

21. Merrick, "The Machine Myth," 7.

22. Van Deventer, "The Machine Has Been Libeled"; Hearings, TNEC, "Technology and Concentration of Economic Power"; John Van Deventer, "Jobs in the World of Tomorrow and . . . a Job for the 'World of Tomorrow,'" *Iron Age*, February 9, 1939, 25; and "Again the Tax on Machines!" *Iron Age*, February 27, 1936, 62.

23. Van Deventer, "Jobs in the World of Tomorrow," 24; and Van Deventer, "The Machine Has Been Libeled," 1605–6. For background on the history of public relations, see Richard Tedlow, *Keeping the Corporate Image: Public Relations and Business, 1900–1950* (Greenwich, Conn.: JAI Press, 1979).

24. Gary B. Nash, Charlotte Crabtree and Ross E. Dunn, *History on Trial: Culture Wars and the Teaching of the Past* (New York: Knopf, 1997).

25. Filene, *Successful Living in This Machine Age*, 53.

26. "Big Spending Cycle Ahead, Study Finds; J. M. Mathes Predicts Research Will Bring About Solution of Unemployment," *New York Times*, May 12, 1936; "Press Urged to Tell 'Real' Business News," *New York Times*, February 19, 1938.

27. General Electric advertisement, *Survey Graphic* (November 1937): 550. For background on Depression-era advertising and consumerism, see Roland Marchand, *Advertising the American Dream: Making Way for Modernity, 1920–1940* (Berkeley: University of California Press, 1985); Stuart Ewen, *All Consuming Images: The Politics of Style in Contemporary Culture* (New York: Basic Books, 1988); and Susan Strasser, *Satisfaction Guaranteed: The Making of the American Mass Market* (New York: Pantheon, 1989).

28. General Electric advertisement, *Survey Graphic* (June 1937): 357; and *Survey Graphic* (September 1937): 454.

29. "Previews of Industrial Progress in the Next Century," and Alfred Sloan to Franklin D. Roosevelt, May 24, 1934, both in Box OF264, File "GM Conference, Unemployment—Special Folder 1934," Franklin Delano Roosevelt Presidential Library.

30. "Previews of Industrial Progress in the Next Century."

31. "Machines and Jobs," *Printers' Ink* 184 (August 4, 1938): 70–71; and "'Machines and Men': Wherein, Banged around a Bit by Ancient-Subscriber Baggs, We Speak Our Mind about Technology," *Printers' Ink* 184 (September 15, 1938): 13–16.

32. Lionel Stern, "The Outcome of the Machine Age and Unemployment," *Iron Age*, June 7, 1934, 16.

33. Macklin, "Labor-Saving Machines Make Jobs," 19, 21.

34. "Says New Machines Do Not Reduce Jobs: International Chamber of Commerce Body Reports on Technological Unemployment," *New York Times*, April 13, 1931; National Industrial Conference Board, *Machinery, Employment, and Purchasing Power* (New York: NICB, 1935); and *Outlines of Eleven Talks on Timely Questions Affecting the American Free Enterprise System* (Washington, D.C.: Chamber of Commerce of the U.S., n.d. [ca. 1940]), 5–6.

35. *Outlines of Eleven Talks*, 13–18.

36. "Machinery Called Stabilizer of Jobs: H. H. Lind Cites Depression as Defense of Mass Production in Industry"; and George Paull Torrence, "Modern Material Handling," *American Machinist* 76 (April 28, 1932): 549–50.

37. Machinery and Allied Products Institute, *Machine-Made Jobs* (Chicago: MAPI,

1936), 1, 4, 11; and Machinery and Allied Products Institute, *Machinery and the American Standard of Living* (Chicago: MAPI, 1939).

38. Machinery and Allied Products Institute, *Machinery and the American Standard of Living;* Machinery and Allied Products Institute, *Technology and the American Consumer* (Chicago: MAPI, 1937); and Machinery and Allied Products Institute, *Capital Goods and the American Enterprise System* (Chicago: MAPI, 1939), 24–25.

39. James McGraw Jr., "An Editorial Service to Meet Industry's Major Problem," "A Good Insurance Policy," "What Machines Mean to Bill Smith in Transportation," "What the Records Show," and "Why a Public Relations Program," all in "Public Relations for Industry" section, *American Machinist* 82 (September 21, 1938).

40. "And There Are 8,000,000 Bill Smiths," ibid.

41. Chamber of Commerce of the United States, *The American Economic System Compared with Collectivism and Dictatorship* (Washington, D.C.: Chamber of Commerce of the U.S., 1936), 35–36; and Hartley Barclay, *Labor's Stake in the American Way* (Boston: Atlantic, 1938).

42. Barclay, *Labor's Stake in the American Way;* Machinery and Allied Products Institute, *Capital Goods and the American Enterprise System;* and National Industrial Conference Board, *Machinery, Employment, and Purchasing Power.*

43. "Previews of Industrial Progress in the Next Century."

44. Virgil Jordan, "Economic Realism," manuscript, speech to the Associated Industries of Cleveland and the Cleveland Chamber of Commerce, October 14, 1936, Conference Board Records, Series 5, File "Jordan, Virgil 1936," Hagley Library.

45. Macklin, "Labor-Saving Machines Make Jobs," 91; Van Deventer, "The Machine Has Been Libeled," 1605–6; "Thirty-Hour Week Will Lower Living Standards," *Iron Age,* November 29, 1934, 65–66; "Slump Laid to Cut in Work Hours; 'Balance' Now Gone, Moulton Says: Wages Far ahead of Technological Growth, on Which They Depend, Brookings Head Tells Engineers—For Less Rapid Pace," *New York Times,* January 27, 1938; and "Wolman Criticizes Labor 'Advances,'" *New York Times,* February 15, 1938.

46. William Z. Ripley, "Railroads Now Face Their Gravest Crisis," *New York Times,* December 13, 1931; and "Sharing With Machines," editorial, *New York Times,* April 13, 1933.

47. Franklyn Hobbs, "The Machine Age and Its Consequences," *American Machinist* 74 (January 29, 1931): 193–95.

48. Cyrus Ching, "Retraining and Adjustment in Occupation," *Vocational Guidance Magazine* 11, no. 1 (1932): 16–17.

49. "Ford Defends Life in Industrial Age," *New York Times,* July 31, 1930; and "Shortage of Labor and Housing Found," *New York Times,* May 10, 1936.

50. "Denies Machine Age Is Displacing Man," *New York Times,* January 18, 1933, 26:1.

Chapter 6: "Innocence or Guilt of Science"

1. "Is Scientific Advance Impeding Human Welfare?" *Literary Digest,* October 1, 1927, 13–14. See Carroll Pursell, "'A Savage Struck by Lightning,': The Idea of a Research Moratorium," *Lex et Scientia* (October–December 1974): 148.

2. On the Depression-era scientific community and professional organizations, see Peter Kuznick, *Beyond the Laboratory* (Chicago: University of Chicago Press, 1987); and Marcel LaFollette, *Making Science Our Own* (Chicago: University of Chicago Press, 1990).

3. Josiah Stamp, "Do We Need Birth Control for Ideas? A 'Technique of Accommodation," and Charles Kettering, "Do We Need Birth Control for Ideas? Inventors Don't Invent Enough," both in *Rotarian* 44 (April 1934): 6–9; "Science and the State," *New York Times*, September 27, 1936.

4. Robert Budington, "The Innocence and Guilt of Science," *Ohio Journal of Science* 33, no. 4 (1933): 259–70; and Alfred Stock, "The Present State of the Natural Sciences," *Science*, April 1, 1932, 349.

5. "New Social Sense Seen in Election," *New York Times*, December 2, 1936; and "Green Says Science Is Not Foe of Labor," *New York Times*, October 4, 1931.

6. Franklin D. Roosevelt to Carl Byoir (President, Society of Arts and Sciences), May 20, 1936, File PPF 700 "Science," Franklin Delano Roosevelt Presidential Library (hereinafter FDR Library).

7. Frederick Winslow Taylor, *The Principles of Scientific Management* (New York: Harper and Bros., 1911); Samuel Haber, *Efficiency and Uplift: Scientific Management in the Progressive Era, 1890–1920* (Chicago: University of Chicago Press, 1964); and Samuel Hays, *Conservation and the Gospel of Efficiency* (New York: Atheneum, 1980).

8. William Ogburn, "National Policy and Technology," in National Resources Committee, *Technological Trends and National Policy, Including the Social Implications of New Inventions* (Washington, D.C.: U.S. Government Printing Office, 1937), 13.

9. Budington, "The Innocence and Guilt of Science," 262–64; Karl Compton, "Engineering in an American Program for Social Progress," in Jesse Thornton, ed., *Science and Social Change* (Washington, D.C.: Brookings Institution, 1939), 496.

10. Michael Pupin, "The Immortal Cosmic Harmony as a Scientist Conceives It," *New York Times*, February 7, 1932; "Romance of the Machine," *Scribner's Magazine* 87 (February 1930): 136; and Charles Kettering, "The Present Day View-Point of Science," in Charles Ross, ed., *Stabilization of Employment: Papers Presented at the Atlantic City Meeting of the AAAS* (Bloomington, Ind: Principia, 1933), 24–25. See also J. D. Bernal, *The Social Function of Science* (New York: Macmillan, 1939).

11. "Aladdin or Frankenstein?" *Banking* 31 (November 1938): 126–27; Karl Compton, "Place of the Machine," *New York Times*, December 25, 1932; Compton, "Engineering in an American Program for Social Progress," 509–10.

12. Karl Compton, *The Social Implications of Scientific Discovery* (Philadelphia: American Philosophical Society, 1938), 6, 12–14. For background, see Robert Kargon and Elizabeth Hodes, "Karl Compton, Isaiah Bowman, and the Politics of Science in the Great Depression," *Isis* 76 (1985): 301–18.

13. Compton, "Engineering in an American Program for Social Progress."

14. Compton, *The Social Implications of Scientific Discovery*; and Karl Compton, "Science and Prosperity," *Science*, November 2, 1934, 387–94.

15. Karl Compton, "Science in an American Program for Social Progress" transcript, speech to the National Industrial Conference Board, NBC radio, May 28, 1936, Series 5, "Addresses and Speeches," File "Conference Board," Hagley Library.

16. Karl Compton, "Put Science to Work" (manuscript) and Franklin D. Roosevelt to Karl Compton, November 13, 1934, File PPF 1975, "Compton, Dr. Karl," FDR Library. Minutes, first meeting, Science Advisory Board, August 21 and 23, 1933, Millikan Papers, Folder 9.9 "Science Advisory Board"; and Resolutions adopted by the Council of the AAAS, December 29, 1934; Barton collection, Box 3, Folder 5 "Correspondence, Karl

Compton," American Institute of Physics (hereinafter AIP), The Center for History of Physics, Niels Bohr Library, College Park, Md. On the Science Advisory Board, see A. Hunter Dupree, *Science in the Federal Government: A History of Policies and Activities,* rev. ed. (Baltimore: Johns Hopkins University Press, 1986); Ronald Tobey, *The American Ideology of National Science, 1919–1930* (Pittsburgh: University of Pittsburgh Press, 1971); Lewis Auerbach, "Scientists in the New Deal," *Minerva* 3, no. 4 (1965): 457–82; and Joel Genuth, "Groping toward Science Policy in the United States in the 1930s," *Minerva* 25 (1987): 238–68.

17. Karl Compton to the Science Advisory Board, December 24, 1934, Millikan Papers, Folder 9.11, "Science Advisory Board"; Frank Jewett to Karl Compton, December 6 and 20, 1934, and December 9, 1935, Millikan Papers, Folders 9.11 and 9.13, "Science Advisory Board."

18. Robert Millikan, "Science—The Nation's Inexhaustible Reserve," manuscript, speech to the Association of Life Insurance Presidents, December 12, 1930, Millikan Papers, Folder 59.25; Robert Millikan, "A Physicist's Dream" manuscript, speech at University of Kansas Commencement ceremony, June 12, 1933, Millikan Papers, Folder 59.41; Robert Millikan, "The New Deal in Science," manuscript, 1934, Millikan Papers, Folder 60.8; Robert Millikan, "Whither Scientific Research in Industry?" manuscript, speech to the U.S. Institute for Textile Research, November 14, 1935, Millikan Papers, Folder 60.17; and Robert Millikan, "Whither Civilization" manuscript, speech to the California State Bar, September 20, 1935, Millikan Papers, Folder 60.14.

19. Charles Kettering, "The Present Day View-Point of Science," in Charles Ross, ed., *Stabilization of Employment: Papers Presented at the Atlantic City Meeting of the AAAS* (Bloomington, Ind.: Principia, 1933); and Dugald Jackson, "Machinery and Unemployment," ibid., 50.

20. Dexter Kimball, "The Social Effects of Mass Production," *Science* 77, no. 1984 (1933): 1–7, emphasis in the original.

21. Henry Barton to Karl Compton, September 20 and October 24, 1933; and Karl Compton to Henry Barton, October 27, 1933; Box 3, Folder 2A, "Correspondence, Compton, 1933," AIP.

22. Henry Barton to Karl Compton, October 24, 1933; Karl Compton to Henry Barton, October 27, 1933; and Henry Barton to Karl Compton, January 8, 1934, ibid.

23. Henry Barton to Karl Compton, December 14, 1933; and Karl Compton to Henry Barton, February 17, 1934, ibid. Henry Barton to Albert Einstein, February 19, 1934; and Albert Einstein to Henry Barton, February 21, 1934, Box 1, Folder 1, "Correspondence from the Director's Office," AIP. Karl Compton to Franklin D. Roosevelt, January 16, 1934, Box 3, Folder 2A, "Correspondence, Compton, 1933," AIP. Henry Barton to Louis Howe, February 5, 1934; and Franklin D. Roosevelt to Karl Compton, February 12, 1934, File PPF 1280, "American Institute of Physics," FDR Library.

24. Memo, Joint meeting set-up, January 19, 1934, and "Science Makes More Jobs" (press release), Barton Collection, Box 9, File 1, "'S' Correspondence, 1931–1941," AIP; and Karl Compton to Robert Millikan, January 17, 1934, Box 3, Folder 2A, "Correspondence, Compton, 1933," AIP.

25. "Science Makes More Jobs: How Fundamental Inventions Contributed to Employment," Museum of Science and Industry, February, 1934, Barton collection, Box 9, File 1 "'S' Correspondence, 1931–1941," AIP. See also Charles R. Richards, "Statistical Charts

Regarding Employment Exhibited at the New York Museum of Science and Industry," *Scientific Monthly* 38 (1934): 390–94.

26. Karl Compton, "Science Makes Jobs" in "The Contributions of Science to Increased Employment," *Scientific Monthly* 38 (April 1934): 297, 299.

27. Robert A. Millikan, "The Service of Science," in "The Contributions of Science to Increased Employment," *Scientific Monthly* 38 (April 1934): 306, 302, 309.

28. Frank Jewett, "Science and Industry," and W. D. Coolidge, "Scientific Discoveries and their Application," both in "The Contributions of Science to Increased Employment," *Scientific Monthly* 38 (April 1934): 306, 302, 309.

29. Owen Young, "The Contributions of Science to Increased Employment," *Scientific Monthly* 38 (April 1934): 309. Henry Barton to Karl Compton, March 8, 1934; Box 3, Folder 2A, "Correspondence, Compton, 1933," AIP. "Science and Jobs," *New York Times*, February 24, 1934.

30. Henry Barton, "The American Institute of Physics," *Scientific Monthly* 34 (March 1932): 287–90; AIP Report of the Director for 1936 and "What the Science of Physics Is Today," n.d. [ca. 1936], Barton Collection, Box 1, Folder 6, AIP; and AIP report of the Governing Board, February 26, 1938; Barton Collection, Box 1, Folder 5, AIP.

31. "Report of Second Meeting, New York, October 28, 1936," Barton collection, Box 1, Folder 7, AIP; George Harrison to Henry Barton, December 10, 1937, Barton Collection, Box 2, Folder 9, "Atoms in Action by G.R. Harrison," AIP; and George Russell Harrison, *Atoms in Action: The World of Creative Physics* (New York: William Morrow, 1939).

32. Harrison, *Atoms in Action*, 63–64, 70–73.

33. Ibid., 294, 333, back cover.

34. William McClellan, "The Engineer a Creator of Leisure," *Electrical Engineering* 53 (May 1934): 778. See also G. W. Starr, "Present Day Popular Economic Thought," *Mechanical Engineering* 62, no. 3 (1939): 209–12.

35. Bancroft Gherardi, "Engineers and Progress," *Bell Telephone Quarterly* 12, no. 1 (1933): 7–9; and C. C. Furnas, *The Next Hundred Years: The Unfinished Business of Science* (New York: Reynal and Hitchcock, 1936), 201, 341–42.

36. Dexter Kimball, "The Century's Great Inventions," *Mechanical Engineering* 59, no. 7 (1937): 507–10; and "Lauds Machine Age, Assails 'Idealists': Mehren Tells Engineers That They Are Dispelling Victorian Gloom While Critics Scoff," *New York Times*, April 6, 1930.

37. McClellan, "The Engineer a Creator of Leisure," 777; and "Publicity for Engineering," *Mechanical Engineering* 60, no. 8 (1938): 594.

38. McClellan, "The Engineer a Creator of Leisure," 778; and Elliott Dunlap Smith, "Engineering Encounters Human Nature," *Mechanical Engineering* 53, no. 1 (1931): 1–4, emphasis in original.

39. Ralph Flanders, "New Pioneers on a New Frontier," *Iron Age*, December 5, 1935, 20–21, 110, 116. See also Ralph Flanders, "A Still Shorter Work Week . . . Is This the Road to Recovery?" *American Machinist* 78 (February 28, 1934): 181–83; Ralph Flanders, *Taming Our Machines: The Attainment of Human Values in a Mechanized Society* (New York: Richard R. Smith, Inc., 1931), 12, 14–15, 77; and Ralph Flanders, "The Engineer in a Changing World," in Jesse Thornton, ed., *Science and Social Change* (Washington, D.C.: Brookings Institution, 1939), 217.

40. Dugald Jackson, "Engineering's Part in the Development of Civilization," Part 1,

Mechanical Engineering 60, no. 7 (1938): 529–34; and ibid., Part 2, *Mechanical Engineering* 60, no. 8 (1938): 619–24.

41. Ibid., Part 3, *Mechanical Engineering* 60, no. 9 (1938): 703–8; and ibid., Part 4, *Mechanical Engineering* 60, no. 10 (1938): 745–50.

42. Ibid., Part 4, *Mechanical Engineering* 60, no. 10 (1938): 745–50; ibid., Part 5, *Mechanical Engineering* 60, no. 11 (1938): 839–43; and ibid., Part 6, *Mechanical Engineering* 60, no. 12 (1938): 949–53.

43. "History of a Culture," *Mechanical Engineering* 60, no. 7 (1938): 527.

44. C. F. Hirshfeld, "The Engineer of Today and Tomorrow," *Mechanical Engineering* 58, no. 8 (1936): 475–78.

45. "Control," *Mechanical Engineering* 52, no. 5 (1930): 512; and "Lauds Machine Age, Assails 'Idealists': Mehren Tells Engineers That They Are Dispelling Victorian Gloom While Critics Scoff," *New York Times*, April 6, 1930.

46. Charles F. Scott, "Epilogue," *Mechanical Engineering* 59, no. 4 (1937): 279; and "Research Parade" invitation, File PPF 4143, "National Committee on Centennial Celebration of the American Patent System," FDR Library.

47. "The Balancing of Economic Forces: Suggesting Lines of Attack on the Interrelated Problems of Consumption, Production, and Distribution," *Mechanical Engineering* 54, no. 6 (1932): 415–23; and "The Balancing of Economic Forces," *Mechanical Engineering* 55, no. 4 (1933): 211–24.

48. C. A. Norman, "Production and Purchasing Power," *Mechanical Engineering* 55, no. 9 (1933): 549–55; "Developments and Trends in Mechanical Engineering," *Mechanical Engineering* 54, no. 12 (1932): 857–58; and "Econometrics and Engineering," *Mechanical Engineering* 55, no. 7 (1933): 489–91.

49. W. D. Coolidge, "Scientific Discoveries and Their Application"; "The Hell Gate Plant of the United Electric Light and Power Co., NY.," *Mechanical Engineering* 54, no. 12 (1932): 820; and Knox Powell, "Engineers Must Service Their Technology," *Mechanical Engineering* 55, no. 5 (1933): 326–27.

50. "Developments and Trends in Mechanical Engineering," *Mechanical Engineering* 54, no. 12 (1932): 857–858; R. E. Kinsman, "Engineers and Economics," *Mechanical Engineering* 54, no. 11 (1932): 803; and O. N. Bryant, "Will Economic Stability Suffice?" *Mechanical Engineering* 54, no. 12 (1932): 879. Ralph Flanders, "Engineering, Economics, and the Problem of Social Well-being: A Statement of the Problem and a Discussion by Representatives of These Two Professions of the Steps That Must Be Taken in Its Solution—The Engineer's View," *Mechanical Engineering* 53, no. 2 (1931): 99–104.

51. Dexter Kimball, "Engineers in Public Life," *Mechanical Engineering* 55, no. 1 (1933); R. E. Kinsman, "Engineers and Economics," *Mechanical Engineering* 54, no. 11 (1932): 803; A. E. Kittredge, "Engineers and Economics," *Mechanical Engineering* 55, no. 3 (1933): 200.

52. Bryant, "Will Economic Stability Suffice?" 879; and Charles Comstock, "Training the Engineer to Think," *Mechanical Engineering* 55, no. 6 (1933): 388–89.

53. "Developments and Trends in Mechanical Engineering," *Mechanical Engineering* 54, no. 12 (1932): 857–58; Flanders, "Engineering, Economics, and the Problem of Social Well-being," 99–104; Dexter Kimball, "Engineers in Public Life," *Mechanical Engineering* 55, no. 1 (1933); and "Econometrics and Engineering," *Mechanical Engineering* 55, no. 7 (1933): 489–91.

54. Frazier Hunt, "Will the Goblins Finally Get Us? No!" *Good Housekeeping*, April 1933,

200; and Roy Wright, "The Engineer Militant: Presidential Address at ASME Annual Meeting," *Mechanical Engineering* 54, no. 1 (1932): 20–22.

55. Bryant, "Will Economic Stability Suffice?" 879; and Knox Powell, "Engineers Must Service Their Technology."

56. C. F. Hirshfeld, "Whose Fault?" *American Scientist* 20, no. 1 (1932): 8–20; "What It's All About," *Mechanical Engineering* 54, no. 3 (1932): 170–71; Charles Jay Seibert, "Principal or Accessory to the Crime? A Reply to 'Whose Fault?'" *Mechanical Engineering* 54, no. 9 (1932); Eugene Szepesi, "Management Essentials for Recovery," *Mechanical Engineering* 55, no. 4 (1933); L. P. Alford, "An Appraisal of Economic Forces, *Mechanical Engineering* 55, no. 6 (1933): 345; and George Eaton, "Unemployment and Patent Monopoly," *Mechanical Engineering* 55, no. 6 (1933): 387.

57. McClellan, "The Engineer a Creator of Leisure": 779; and Charles Kettering, "Research and Social Progress: An Optimistic View of the Future of Engineering for Those Who Face Their Opportunities Rightly," *Mechanical Engineering* 58, no. 4 (1936): 211–14.

58. "Depicts Machine Power: Prof. Roe Says Our Output Equals That of 12 Billion Slaves," *New York Times*, September 14, 1930; and Theodore Knappen, "The Machine Turns On Its Master," *The Magazine of Wall Street*, May 3, 1930, 11.

59. Frederick Allner, "Plan and Purpose of Forum," in *Employment and the Engineer's Relation to It: Addresses and Digest of Discussion*, supplement to AEC Bulletin for July 1938 (Washington, D.C.: American Engineering Council), 4–5.

60. Leo Wolman, "Labor Policy and Employment," Leonard Fletcher, "Contributions of Technology," and "Discussion," all in *Employment and the Engineer's Relation to It*, 9–12, 13–16, 22–23.

61. "400 Leaders Hear Technocracy Plea: Howard Scott Leaves Bankers and Industrialists Skeptical on New Solution," *New York Times*, January 14, 1933; and "Engineers Assail Technocracy Aims," *New York Times*, January 15, and February 5, 1933.

62. Leon Cammen, "A Chance for Engineering Progress," *Mechanical Engineering* 54, no. 12 (1932): 859.

63. Charles Kettering, "Research and Social Progress: An Optimistic View of the Future of Engineering for Those Who Face Their Opportunities Rightly," *Mechanical Engineering* 58, no. 4 (1936): 211–14.

64. Roy Wright, "The Engineer Militant"; and Elliott Dunlap Smith, "Engineering Encounters Human Nature," *Mechanical Engineering* 53, no. 1 (1931): 1–4.

65. "Cultural Emphasis in Engineering Urged: Kaempffert Tells Carnegie Tech Group the Profession Must Study Social Implications," *New York Times*, December 7, 1932; and Henry A. Wallace, "The Social Advantages and Disadvantages of the Engineering-Scientific Approach to Civilization," *Science*, January 5, 1934, 3–4.

66. Edison Bowers and R. Henry Rowntree, *Economics for Engineers* (New York: McGraw-Hill, 1931), 413.

67. Ibid., 13, 412, 416.

68. Review, *Engineering Experiment Station News* 10, no. 5 (1938); and Karl Compton, "American Adventure in Education: Experiment at MIT to Prove the Cultural Value of Sound Technological Training Enters Its Second Year," *New York Times*, September 15, 1935.

69. Presidential press release, October 21, 1936, Box OF 2430, File "'Little Waters' Report, 1936–41," FDR Library.

70. "Engineer Schools Dispute 'Balance,'" *New York Times*, November 1, 1936; and "Engineers Make Reply," *New York Times*, October 25, 1936.

71. "MIT Head Fears 'Relief Palliative' Hampers Science," *New York Times*, October 25, 1936; and "Lehigh Head Hits 'Political Gunning,'" *New York Times*, October 28, 1936. Investigating the balance between technical and nontechnical requirements for engineering majors at seven institutions, the *New York Times* reported a wide variation in curricula. Engineering students at Purdue devoted the most energy to liberal arts studies, 50 percent of their work, while those at the Polytechnic Institute of Brooklyn spent just 18 percent of their time on liberal arts. "Engineer Schools Dispute 'Balance,'" *New York Times*.

72. "Lehigh Head Hits 'Political Gunning,'" *New York Times*; Robert Allen, "Science versus Unemployment," *Science*, May 26, 1939, 474–76, 478; and "Norwich President Replies," *New York Times*, October 25, 1936.

73. "Unbalanced Efficiency," *Bulletin of the Metal Trades Department, American Federation of Labor*, 13, no. 1 (1932): 7–8.

74. Leo Wolman, "The Bearing of Industrial Equilibrium on Regularity of Operations and of Employment," and Stuart Chase, "Discussion," at December 4, 1929, meeting of the Taylor Society, printed in *Bulletin of the Taylor Society*, 15, no. 1 (1929): 6–7, 9–11.

75. Paul Douglas, "Technological Unemployment: Measurement of Elasticity of Demand as a Basis for Prediction of Labor Displacement," at December 5, 1930 meeting of the Taylor Society, printed in *Bulletin of the Taylor Society* 15, no. 6 (1930): 254–56.

76. Sumner Slichter, "Discussion," at December 5, 1930, meeting of the Taylor Society, ibid., 264–66.

77. Henry Kendall, "Discussion," at December 5, 1930, meeting of the Taylor Society, ibid., 269–70; "Comment," ibid., 253; and Harlow Person, "The New Challenge to Scientific Management," at December 5, 1930 meeting of the Taylor Society, printed in *Bulletin of the Taylor Society* 16, no. 2 (1931): 62.

78. Harlow Person, "The New Challenge to Scientific Management" and Robert McFall, "Planning Industry: A Plea for Balance Between Labor Saving and Labor Utilization," at April 20, 1931, meeting of the Taylor Society, printed in *Bulletin of the Taylor Society* 16, no. 3 (1931): 91–92, 95–96.

79. Harlow Person, "Technology and Big Business," in Felix Morley, ed., *Aspects of the Depression* (Freeport, N.Y.: Books for Libraries Press, 1968), 263–265, 266, 269; and Harlow Person, "Technology, Business and Depressions," *Journal of Business of the University of Chicago* 5 (October 1932): 102–3, 107–9.

80. Michael Mok, "Will You Lose Your Job Because of a New Machine?" *Popular Science Monthly*, March 1931, 20–21. See also "What Science Is Doing for You," *Popular Mechanics* 55 (March 1931): 386; Wilfred Owen, "Machine-Made Jobs," *Science News Letter* 31, no. 830 (1937): 150–54; and "Our Point of View: Technological Unemployment," *Scientific American*, 162, no. 2 (1940): 68.

81. Karl Compton, "Science in an American Program for Social Progress" manuscript, speech to the National Industrial Conference Board, NBC radio, May 28, 1936; Series 5, "Addresses and Speeches," File "Conference Board," Hagley Library.

82. Edwin T. Layton Jr., *The Revolt of the Engineers: Social Responsibility and the American Engineering Profession* (Baltimore: Johns Hopkins University Press, 1986).

Chapter 7: "What Will the Smug Machine Age Do?"

1. Silas Bent, *Machine-Made Man* (New York: Farrar and Rinehart, 1931), 4.

2. Charles Beard, "The Idea of Progress," *A Century of Progress* (New York: Harper and Bros., 1933), 3–4, 8, 16.

3. Roger Burlingame, *Engines of Democracy* (New York: Charles Scribner's Sons, 1940), 116–17, 124–25, 534–36.

4. Ibid., 520, 536–39, 540–41.

5. Harold Rugg, *The Great Technology: Social Chaos and the Public Mind* (New York: John Day, 1933), 78, 171, 289.

6. Lewis Mumford, "The Drama of the Machines," *Scribner's Magazine* 98 (August 1930): 150, 153, 159–60.

7. Lewis Mumford, *Technics and Civilization* (1934; New York: Harcourt Brace Jovanovich, 1963), 182–85, 228, 266, 317–18, 325, 393.

8. Ibid., 279, 366, 405, 413.

9. Ralph Borsodi, *Flight from the City* (New York: Harper and Bros., 1933), xiii.

10. Ralph Flanders, *Taming Our Machines: The Attainment of Human Values in a Mechanized Society* (New York: Richard Smith, 1931), 13–14.

11. "Times Good, Not Bad, Ford Says; Sees the Dawn of a Bright Future," *New York Times*, February 1, 1933; Dexter Kimball, "The Social Effects of Mass Production," *Science* January 6, 1933; and "Machine Age Slowing: Dean Kimball Sees Natural Curb on Mass Production," *New York Times*, December 30, 1932.

12. Sumner Boyer Ely, "The Effect of the Machine Age on Labor," *Scientific Monthly* 37 (September 1933): 261–62.

13. C. A. Prosser, "Vocational Advisement in a Changing Economic World," *Vocational Guidance Magazine* 11 (January 1932): 52, 64; and John Garvey, "The Function of the Personnel Department in Technological Unemployment," in *Personnel Administration and Technological Change* (New York: American Management Association, 1935), 14–20.

14. R. B. Cunliffe, "Vocational Guidance That Functions," *Vocational Guidance Magazine* 11 (January 1933), 162; and Spencer Miller, "The Need of Education for Labor," *Journal of Educational Sociology* 5, no. 8 (1932): 509. See also Verne Fryklund, "Training and Changing Technology," *Industrial Arts and Vocational Education* 22 (December 1933) and Morse Cartwright, *Unemployment and Adult Education: A Symposium* (New York: American Association for Adult Education, 1931).

15. William Green, "The Five-Day Week," *Harvard Business Review* 9, no. 3 (1931): 270–73; and "Labor's Plan for Recovery: A Five-Day Week for Workers," *New York Times*, July 17, 1932. See also "Unemployment or Leisure?" *American Federationist* (March 1931): 320–27.

16. C. C. Furnas, *America's Tomorrow: An Informal Excursion into the Era of the Two-Hour Working Day* (New York: Funk and Wagnalls, 1932), 18, 222–25, 238–43, 269, 277, 282.

17. "Deny Machine Age Stifles Culture: Four Speakers Assert It Gives Leisure and Means for Real Development," *New York Times*, February 16, 1930; "Technocratic By-Products," letter to the editor, *New York Times*, January 20, 1933; and "Technocracy Seen as Leading to Sin," *New York Times*, January 30, 1933.

18. "The Leisure Problem," editorial, *New York Times*, October 1, 1932; and L. P. Jacks, "Leisure as a Slavery," *New York Times*, May 10, 1931.

19. H. A. Overstreet, *A Guide to Civilized Loafing* (New York: W. W. Norton, 1934), 27–29. Harold Bowden, bicycle manufacturer and former chairman of the British Olympic Association, suggested in 1935 that society would soon create "paid leisure as well as paid labour." Starting from the premise that mechanization had created an unparalleled economic dilemma, he called for "a scientific adjustment of working hours and leisure" to satisfy the new conditions. Once such executive arrangements had been completed, he promised, "the word 'unemployment' will have disappeared from our economic vocabulary." Harold Bowden, "The Future of Leisure," in Hubert Williams, ed., *Man and the Machine* (London: George Routledge and Sons, 1935). See also Cecil Delisle Burns, *Leisure in the Modern World* (New York: The Century Co., 1932).

20. Arthur Newton Pack, *The Challenge of Leisure* (Washington, D.C.: National Recreation and Park Association, 1934), 4, 64–68, 211; and Paul Frankl, *Machine-Made Leisure* (New York: Harper and Bros., 1932), 180.

21. Mabel Hermans and Margaret Hannon, *Using Leisure Time* (New York: Harcourt, Brace, 1938), iii–iv, 6–7, 10–12, 35, 73–75.

22. C. Gilbert Wrenn and D. L. Harley, *Time on Their Hands: A Report on Leisure, Recreation, and Young People* (Washington, D.C.: American Council on Education, 1941), 46–49.

23. L. P. Alford, "Progress in Manufacturing," *Mechanical Engineering* 52, no. 4 (1930): 401–3; and Ethel Best, "A Study of a Change from 8 to 6 Hours of Work," *Bulletin of the Women's Bureau*, no. 105 (Washington, D.C.: U.S. Government Printing Office, 1933): 1–9. The Women's Bureau emphasized that employees' reaction to a reduced workday depended largely on financial conditions. Although one employer had raised wages 25 percent to compensate for shorter shifts (rationalizing that the switch from an eight-hour day to a six-hour day had raised productivity), other businesses did not adjust pay. When female workers faced such a sharp income reduction, far fewer approved of the change. Significantly, though, many women said they would accept shorter hours as a Depression emergency measure to help spread employment. Ethel Best, "A Study of a Change from One Shift of 9 Hours to Two Shifts of 6 Hours Each," *Bulletin of the Women's Bureau*, no. 116; (Washington, D.C.: U.S. Government Printing Office, 1934): 4–6. On the relationship between domestic technology and housework, see Ruth Schwartz Cowan, *More Work for Mother;* (New York: Basic Books, 1983).

24. "Denies Machine Age Adds to Happiness: Dr. C. E. K. Mees, in Engineering Symposium, Says Life Was Sweeter 4,000 Years Ago," *New York Times*, July 12, 1931; and "Ancient and Modern Cities," editorial, *New York Times*, July 19, 1931.

25. Virgil Jordan, "Economic Realism" manuscript, speech to the Associated Industries of Cleveland and the Cleveland Chamber of Commerce, October 14, 1936, Conference Board Records, Series 5, File "Jordan, Virgil 1936," Hagley Library.

26. Stuart Chase, *Men and Machines* (New York: Macmillan, 1929), 188–89.

27. Robert Littell, "The Price of Comfort," *American Mercury* 34 (April 1935): 430–35; and Borsodi, *Flight from the City*, 122, 146, 154.

28. Michael Hare, "Why Have a Fair?" December 22, 1936, quoted in Joseph Cusker, "The World of Tomorrow: Science, Culture, and Community at the New York World's Fair," in Helen A. Harrison, guest curator, *Dawn of a New Day: The New York World's Fair, 1939/40* (New York: New York University Press, 1980), 6.

29. Ralph Steiner and Willard Van Dyke, *The City* (1939).

30. *Official Guide Book—New York World's Fair, 1939* (New York: Exposition Publications, 1939), 26–27.

31. Henry Ford, "Machines as Ministers to Man," *New York Times*, World's Fair Section, March 5, 1939, 70.

32. Warren Susman, "The People's Fair: Cultural Contradictions of a Consumer Society," and Cusker, "The World of Tomorrow". See also Jeffrey Meikle, *Twentieth-Century Limited* (Philadelphia: Temple University Press, 1989), which explains the role of industrial designers at the fair in molding the world into a streamlined machine image, cutting through all obstacles to future consumption.

33. John Van Deventer, "Jobs in the World of Tomorrow and a Job for the 'World of Tomorrow,'" *Iron Age*, February 9, 1939, 24–28.

34. Ibid.

35. Ford, "Machines as Ministers to Man," 10, 70.

36. Harrison, *Dawn of a New Day*, 94–95; and advertisement for Sheffield Farms, *New York Times*, World's Fair Section, March 5, 1939, 70.

37. AT&T used similar displays of dial technology at other 1930s fairs, including the ones in Chicago and San Francisco. John Mills, "The Bell System Exhibit at the 'Century of Progress' Exposition," *Bell Telephone Quarterly* 12, no. 1 (1933): 1, 51; and Thomas Williams, "Our Exhibits at Two Fairs: At the New York World's Fair," *Bell Telephone Quarterly* 19, no. 1 (1940): 69.

38. *The Middletons Visit the New York World's Fair* (Westinghouse, 1939), excerpted in *The World of Tomorrow*, documentary for *The American Experience* series, PBS, 1988.

39. *All's Fair at the Fair*, 1939 cartoon excerpted in *The World of Tomorrow* series, PBS, 1988.

40. "World of Tomorrow," *Mechanical Engineering* 61, no. 6 (1939): 466; and "New York World's Fair," *Mechanical Engineering* 61, no. 7 (1939): 540–41:

41. Hearings, Temporary National Economic Committee, "Technology and Concentration of Economic Power," part 30, *Investigation of Concentration of Economic Power* (Washington, D.C.: U.S. Government Printing Office, 1940), 16208, 16304.

42. Ibid., 16246, 16256, 16295–97.

43. Ibid., 16335–36.

44. Ibid.: 16360–91, 16379–80.

45. Ibid., 16520, 16541, 16597–99, 16626–32, 16906–15.

46. Ibid., 16456, 16505–6.

47. Ibid., 17228.

48. Lewis Lorwin and John Blair, *Technology in Our Economy*, TNEC Monograph No. 22 (Washington, D.C.: U.S. Government Printing Office, 1941), 188, 219–20.

49. John Scoville and Noel Sargent, eds., *Fact and Fancy in the TNEC Monographs* (New York: National Association of Manufacturers, 1942), ix–xi, 90, 96, 331–37.

50. Henry R. Luce, "The American Century," 1941, reprinted in Warren Sussman, ed., *Culture and Commitment, 1929–1945* (New York: George Braziller, 1973), 323.

51. "Machines That Serve Consumers Can Also Serve Soldiers," *Nation's Business* 28 (September 1940): 28–29.

52. Robert L. Snyder, *Pare Lorentz and the Documentary Film* (Norman, Okla.: University of Oklahoma Press, 1968).

53. Willard Van Dyke, *Valleytown* (1940); and Amalie Rothschild, *Conversations with Willard Van Dyke* (1980).

Chapter 8: "Automation Just Killed Us"

1. John Kenneth Galbraith, *The Affluent Society* (Boston: Houghton Mifflin, 1958). On wartime consumerism, see Frank W. Fox, *Madison Avenue Goes to War: The Strange Military Career of American Advertising, 1941–45* (Provo, Utah: Brigham Young University Press, 1975); on the 1950s, see David Halberstam, *The Fifties* (New York: Villard Books, 1993).

2. Vannevar Bush, *Science, the Endless Frontier: A Report to the President,* Office of Scientific Research and Development (Washington, D.C.: U.S. Government Printing Office, 1945). See also Nathan Reingold, "Vannevar Bush's New Deal for Research: or The Triumph of the Old Order," *Historical Studies in the Physical Sciences,* 17, no. 2 (1987): 299–344. On science in the Cold War, see Stuart Leslie, *The Cold War and American Science: The Military-Industrial-Academic Complex at MIT and Stanford* (New York: Columbia University Press, 1993); Paul Forman, "Beyond Quantum Electronics: National Security as the Basis for Physical Research in the United States, 1940–1960," *Historical Studies in the Physical Sciences* 18, no. 1 (1987): 149–229; and Peter Galison and Bruce Hevly, eds., *Big Science: The Growth of Large-Scale Research* (Stanford: Stanford University Press, 1992). On the "kitchen debate," see Robert H. Haddow, *Pavilions of Plenty: Exhibiting American Culture Abroad in the 1950s* (Washington, D.C.: Smithsonian Institution Press, 1997).

3. *Automation and Technological Change,* hearings before the Subcommittee on Economic Stabilization of the Joint Committee on the Economic Report, Congress of the United States (Washington, D.C.: U.S. Government Printing Office, 1955), 245, 246, 253–62.

4. "The Automatic Factory" and Eric W. Leaver and J. J. Brown, "Machines without Men," *Fortune,* November 1946, 165–66, 192–204.

5. Leaver and Brown, "Machines without Men."

6. Ibid.

7. Norbert Wiener, *Cybernetics: or Control and Communication in the Animal and the Machine* (New York: John Wiley and Sons, 1948), 37–38; and Norbert Wiener, *The Human Use of Human Beings: Cybernetics and Society* (Boston: Houghton Mifflin, 1950), 186, 188–89.

8. "Automation: A Factory Runs Itself," *Business Week,* March 29, 1952, 146–50; "Automatic Factories," *Time,* January 18, 1954; "Coming Industrial Era: The Wholly Automatic Factory," *Business Week,* April 5, 1952, 96–99; David Noble, *Forces of Production: A Social History of Industrial Automation* (New York: Oxford University Press, 1986); and "Automation," *Business Week,* October 1, 1955, 74–102.

9. "Coming Industrial Era: The Wholly Automatic Factory," *Business Week;* "How a Robot Factory Would Work," *Business Week,* July 21, 1951, 56–60; James Rogers, "Automation's Pushbutton Techniques Spread," *Nation's Business* 42 (December 1954): 30–31.

10. Rogers, "Automation's Pushbutton Techniques Spread."

11. "Automation," *Business Week,* October 1, 1955, 74–102; and Council for Technological Advancement, *Trends in Technology and Employment* (Chicago: Council for Technological Advancement–Machinery and Allied Products Institute, 1954) and Council for Technological Advancement, *Automation and Job Trends* (Chicago: CTA-MAPI, 1954).

12. Lawrence E. Davies, "Banking Machine 'Reads' Checks," *New York Times*, September 22, 1955, 46.

13. Edmund L. Van Deusen, "The Coming Victory over Paper," *Fortune*, October 1955, 130–32, 196, 199; and *Automation and Technological Change*, 279–89.

14. *Automation and Technological Change*, 243–45, 255–58; and "Adjustments to Automation in Two Firms," *Monthly Labor Review* 79, no. 1 (1956): 15–19.

15. *Automation and Technological Change*, 263–64, 290–300; and "Adjustments to Automation in Two Firms," *Monthly Labor Review* 79, no. 1 (1956): 15–19.

16. *Automation and Technological Change*, 1, 2, 603.

17. Ibid., 53, 65, 199, 220.

18. Ibid., 65, 169, 203, 247.

19. Ibid., 57–58, 204, 432, 433. Emphasis added.

20. Ibid., 25–26, 246, 384–85, 402–3.

21. Ibid., 24–29, 63, 170, 246, 374–75, 384–85, 431, 447, 498.

22. Ibid., 171–72, 430.

23. Ibid., 181, 401–2, 433, 590, 594–95.

24. Ibid., 101–2, 107, 120, 126. Emphasis added.

25. Ibid., 103, 108–9, 125, 130.

26. Ibid., 455–60.

27. Ibid., 335, 516–25, 532, 534, 538. Emphasis added.

28. Ibid., 336, 337, 340–46.

29. Ibid., 347.

30. Ibid., 117–19, 564–65, 596–98, 615–17.

31. Ibid., 110, 347.

32. Chic Young, "Blondie," *Des Moines Register*, October 14, 1955.

33. "Push Button Plant: It's Here—Machines Do the Work and a Man Looks On," *U.S. News & World Report*, December 4, 1953, 41–44.

34. Walter Lang, *Desk Set* (Twentieth-Century Fox Film Company, 1957).

35. Kurt Vonnegut, *Player Piano* (New York: Dell, 1952), 1–2, 4, 18, 23, 27, 61–63, 111, 179–80. For comment, see "Kurt Vonnegut's Player Piano: An Ambiguous Technological Dystopia," in Howard Segal, *Future Imperfect* (Amherst: University of Massachusetts Press, 1994).

36. Vonnegut, *Player Piano*, 1–4, 13, 18, 27, 42, 61–63, 102, 110–11, 142–44, 179–80.

37. Ibid., 7, 151, 183–91, 260–62, 285, 287, 291.

38. Ibid., 218–19, 286–88, 291–93.

39. *Automation and Technological Change*, 242.

40. "Automation Unemployment," *New York Times*, May 1, 1961; "The Jobs That Are Gone Forever," *Business Week*, December 20, 1958, 39–42; "Why Jobs Are Slow to Come Back," *U.S. News & World Report*, February 13, 1959, 88–91.

41. Ralph Katz, "Shuttle to Begin Automated Runs," *New York Times*, January 4, 1962; John I. Snyder Jr., "How to Develop an Early Warning System for Automation" and "Epilogue: The Implications of Automation," in Charles Markham, ed., *Jobs, Men and Machines: Problems of Automation* (New York: Frederick A. Praeger, 1964), 120–26, 152–54. See Stanley Levey, "Mayor Seeks Plan to Avert Strikes over Automation," *New York Times*, May 2, 1963.

42. "Candidates on Science," *Science News Letter*, 78, no. 6 (1960), 83, 93.

43. Alfred R. Zipser, "New Administration May Study Automation's Effects on Nation," *New York Times*, November 16, 1960; A. H. Raskin, "Goldberg Studies Automation Move," *New York Times*, January 15, 1961; Interview with Secretary of Labor Arthur J. Goldberg, "Changes Coming in Labor Policies," *U.S. News & World Report*, February 27, 1961, 60–65; Stanley Levey, "Goldberg Sets Up Office to Cope with Automation," *New York Times*, April 20, 1961; "A Cure for Automation Woes?" *Business Week*, April 29, 1961, 81; and Arthur J. Goldberg, "The Challenge of 'Industrial Revolution II,'" *New York Times Magazine*, April 2, 1961. On the Republicans, see "G.O.P. Will Form Panel on Jobless," *New York Times*, April 21, 1961; "Transcript of the President's News Conference," *New York Times*, April 22, 1961; and "G.O.P. Panel Scores Automation Bills," *New York Times*, August 14, 1961.

44. Arthur Burns, president of the National Bureau of Economic Research, further complained that asking employers to contribute to retraining would cut profit margins, limit economic growth, and harm industry's ability to compete overseas; he also pointed out that government funding for such programs would raise budget deficits. "Urgent Need: All-Out Automation," *Nation's Business* (July 1961): 29; "Automation Problem Tackled," *Business Week*, May 6, 1961, 28; Peter Braestrup, "Automation Buffer Is Favored by Panel," *New York Times*, January 12, 1962. "Two Sides Agree on ABCs of Automation," *Business Week*, January 20, 1962, 26; and "What Machines Are Doing to Jobs: How Experts Differ," *U.S. News & World Report*, January 22, 1962, 71.

45. "Russians Drive for Automation," *Business Week*, November 19, 1960, 59–68; "When Machines Replace Men . . . How Can Automation Create More New Jobs?" *Newsweek*, June 19, 1961, 78–80; "The Challenge of Automation," *Newsweek*, January 25, 1965, 73–74; Ralph Katz, "Javits Bids Labor Temper Demands," *New York Times*, October 22, 1961; Donald Janson, "Automation Stirs Machinists' Union," *New York Times*, October 9, 1960; and "Union Meetings Air Job Security Issue," *Business Week*, September 17, 1960, 136.

46. Will Lissner, "Foreign Rivalry Rouses Industry," *New York Times*, December 10, 1960; and "The Automation Race," *Time*, September 1, 1961, 60–61.

47. "Automation Unemployment," *New York Times*, May 1, 1961; and "Business in 1961: Automation Speeds Recovery, Boosts Productivity, Pares Jobs," *Time*, December 29, 1961, 50–54.

48. "Transcript of President's News Conference on Foreign and Domestic Affairs," *New York Times*, February 15, 1962; Bernard D. Nossiter, "New Numbers Game Touched Off on Job Needs in Automation Era," *Washington Post*, February 15, 1962; and "Rx: More Automation for the U.S., Not Less," *Business Week*, February 24, 1962, 148.

49. "When Men Lose Jobs to Machines—Who Gets Aid," *U.S. News & World Report*, March 26, 1962, 48; John D. Pomfret, "Federal Efforts Get a Good Start," *New York Times*, November 18, 1962; and Peter F. Drucker, "Automation Is Not the Villain," *New York Times Magazine*, January 10, 1965, 82.

50. Joseph A. Loftus, "Security Is Key in Many Strikes," *New York Times*, January 7, 1963; A. H. Raskin, "Automation Looming Large in Labor Picture," *New York Times*, April 7, 1963; Stanley Levey, "Automation Gains Seat at Bargaining Table," *New York Times*, April 15, 1963.

51. The issue of railroad jobs and new technology dragged on for years, a persistent flash point in labor negotiations. John D. Pomfret, "Rail Crisis Points Up Impact of Automation on Labor," *New York Times*, July 14, 1963; John D. Pomfret, "Kennedy Asks

I.C.C. Power to Referee Rail Dispute in 2-Year No-Strike Plan," *New York Times*, July 23, 1963; and "Automation Held Deep Administration Concern," *New York Times*, July 24, 1963.

52. Stanley Levey, "Rockefeller Urges Economic Growth," *New York Times*, June 2, 1960; A. J. Jaffe and Joseph Froomkin, *Technology and Jobs: Automation in Perspective* (New York: Frederick A. Praeger, 1968); and Luther H. Evans and George E. Arnstein, eds., *Automation and the Challenge to Education* (Washington, D.C.: National Education Association, 1962).

53. "A Formula and an Introduction," in Markham, ed., *Jobs, Men and Machines*, 1–3.

54. Paul L. Montgomery, "Scholars Seeking a New Theology," *New York Times*, March 2, 1965; Paul L. Montgomery, "Pope John Letter Gains New Weight," *New York Times*, March 7, 1965; and "Church Looking at Machine Age," *New York Times*, August 7, 1965.

55. "Among the Jobless—How They Live, What They Say," *U.S. News & World Report*, February 27, 1961, 66–68; "Button Pushers Are Worried, Too," *New York Times*, April 7, 1961; "When Machines Replace Men . . . How Can Automation Create More New Jobs?" *Newsweek*, June 19, 1961, 78–80; Henry Mayer, "To Deal with Automation," Letter to the Editor, *New York Times*, September 19, 1961; and Albert Craig, "Automation," Letter to the Editor, *New York Times Magazine*, April 19, 1964, 48, 51.

56. Advertisement for "Featherbedding," *New York Times*, December 9, 1960; Lester Velie, "Automation: Friend or Foe?" *Readers' Digest*, October 1962, 101–6; Ben Bagdikian, "'I'm Out of a Job, I'm All Through,'" *Saturday Evening Post*, December 18, 1965, 32–46; "Automation: Its Benefits and Its 'Slag Heap,'" *Life*, January 19, 1964, 4; "Business in 1961: Automation Speeds Recovery, Boosts Productivity, Pares Jobs," *Time*, December 29, 1961, 50–54; and "The Challenge of Automation," *Newsweek*, January 25, 1965, 73–80; See also Ben Seligman, "Man, Work, and the Automated Feast," *Commentary* 34 (December 1962): 9–10.

57. John Diebold, as told to Patricia Lovelady Cahn, "When Will Your Husband Be Obsolete?" *McCall's*, July 1963, 64–65, 118–19.

58. Arthur J. Goldberg, "Challenge of 'Industrial Revolution II,'" *New York Times Magazine*, April 2, 1961, 11, 36–37; cartoon by Doyle for *Philadelphia Daily News*, reprinted in *New York Times*, April 23, 1961; and cartoon by Eric for *Atlanta Journal*, reprinted in *New York Times*, July 14, 1963.

59. Cartoon by Burck for *Chicago Sun-Times*, reprinted in *New York Times*, December 31, 1962; *New Yorker* cartoon, reprinted in *Newsweek*, January 25, 1965, 77; *Punch* cartoon reprinted in *New York Times Magazine*, March 25, 1962; and cartoon by Eric for *Atlanta Journal*, reprinted in *New York Times*, April 7, 1963.

60. Drucker, "Automation Is Not the Villain," 26.

61. Charles E. Silberman, "The Real News about Automation," *Fortune*, January, 1965, 124–27, 220–28; Snyder, "Epilogue" in Markham, ed., *Jobs, Men and Machines*, 153; John D. Pomfret, "Automation Called Major Cause in Loss of 40,000 Jobs a Week," *New York Times*, October 4, 1963; John I. Snyder Jr., "Automation: Threat and Promise," *New York Times Magazine*, March 22, 1964, 16, 62–66; and "John Snyder's Marvelous Mythmaking Machine," *Fortune*, January 1965, 127.

62. "Tool Does Everything Except Loaf on Job," *Business Week*, May 28, 1960, 110–12; Advertisement for Friden, *Fortune*, April 1965, 81; advertisement for Friden, *Newsweek*, October 14, 1963, 103; advertisement for Honeywell, *Fortune*, March 1965, 33; advertise-

ment for Kimball, *Fortune*, June 1965, 18; advertisement for Automatic Electric, *Fortune*, April 1965, 165.

63. Advertisement for National Cash Register, *Fortune*, February 1965, 76; advertisement for Warner & Swasey, *Newsweek*, January 25, 1965, 1. Significantly, the Warner & Swasey ad appeared right inside *Newsweek*'s cover, which that week highlighted a story on "The Challenge of Automation." The ad would be the first thing readers saw upon opening the magazine, letting machine-makers get in their pro-automation message.

64. Donald N. Michael, "Cybernation: The Silent Conquest," in Morris Philipson, ed., *Automation: Implications for the Future* (New York: Vintage Books, 1962), 99, 118, 120; Jerry M. Rosenberg, *Automation, Manpower, and Education* (New York: Random House, 1966).

65. "Automation: Shorter Week?" *Newsweek*, July 3, 1961, 57–58; "Teaching the Jobless New Tricks," *Business Week*, April 15, 1961, 135–38; David Riesman, "Work and Leisure: Fusion or Polarity?," David Riesman with Warner Bloomberg Jr., "Leisure and Work in Postindustrial Society," and David Riesman with Robert S. Weiss, "Some Issues in the Future of Leisure," all in *Abundance for What? and Other Essays* (Garden City, N.Y.: Doubleday, 1964); and Paul Goodman, "The Mass Leisure Class," *Esquire*, July 1959.

66. Michael, "Cybernation," 103–4; and "Labor Aide Offers Plan for Jobless," *New York Times*, November 5, 1961.

67. John D. Pomfret, "Guaranteed Income Asked for All, Employed or Not," *New York Times*, March 23, 1964; and "The Triple Revolution," in Robert MacBride, *The Automated State: Computer Systems as a New Force in Society* (Philadelphia: Chilton Book Company, 1967), 191–207.

68. Everett M. Kassalow, "Labor Relations and Employment Aspects After Ten Years," in Philipson, ed., *Automation: Implications for the Future*, 328; and William DiFazio, *Longshoremen: Community and Resistance on the Brooklyn Waterfront* (South Hadley, Mass.: Bergin and Garvey, 1985), 31, 33, 51, 64, 68.

69. DiFazio, *Longshoremen*, 31, 33, 51, 64, 68; Thomas Kennedy, *Automation Funds and Displaced Workers* (Cambridge, Mass.: Harvard University, 1962); and Suzanne Wooton, "Waiting for Work," *Baltimore Sun*, June 27, 1993, F1–F2.

70. Herb Mills, "The San Francisco Waterfront: "The Social Consequences of Industrial Modernization," in Andrew Zimbalist, ed., *Case Studies on the Labor Process* (New York: Monthly Review Press, 1979), 127–55; "Longshoring and Meatpacking Automation Settlements," *Monthly Labor Review* 82, no. 10 (1959): 1109–10; DiFazio, *Longshoremen*, 31, 33, 51, 64, 68; Robert Kuttner, *The Economic Illusion: False Choices between Prosperity and Social Justice* (Boston: Houghton Mifflin, 1984), 181–82; and Kennedy, *Automation Funds and Displaced Workers*, 76–77.

71. "Coping with Mechanization," *Business Week*, September 19, 1959, 56–58; "Longshoring and Meatpacking Automation Settlements," *Monthly Labor Review* 82, no. 10 (1959): 1109–10; and Kennedy, *Automation Funds and Displaced Workers*, 143. Labor agreements at four smaller meatpacking companies subsequently arranged for automation funds, though none of them experienced displacement on the same scale as Armour.

72. Kennedy, *Automation Funds and Displaced Workers*, 68; and James P. Kraft, *Stage to Studio: Musicians and the Sound Revolution, 1890–1950* (Baltimore: Johns Hopkins University Press, 1996).

73. George W. Taylor, "Collective Bargaining," in *Automation and Technological Change* (Englewood Cliffs, N.J.: Prentice-Hall, 1962).

74. Interview with John L. Lewis, "More Machines, Fewer Men—A Union That's Happy about It," *U.S. News & World Report*, November 9, 1959, 60–64; and Kennedy, *Automation Funds and Displaced Workers*.

75. "Office Automation Hits UAW," *Business Week*, April 9, 1960, 58; William E. Blundell, "Oil Union Finds Strikes Often Are Ineffective at Automated Plants," *Wall Street Journal*, January 19, 1962; William J. Eaton, "Mediator Sees Strike Becoming Obsolescent," *Washington Post*, May 24, 1965; and A. H. Raskin, "Automation Has Made Strikes Senseless," *New York Times Magazine*, October 31, 1965, 45.

76. Damon Stetson, "Meany Declares Automation Evil," *New York Times*, November 11, 1963; "Labor and Automation," *New York Times*, November 21, 1963; and "Learning New Jobs Before Old Ones Fade," *Business Week*, April 11, 1964, 114–16.

77. David R. Jones, "Federal Panel Discounts Job Peril in Automation," *New York Times*, December 23, 1965; and "Automation Panel Splits on Report," *Business Week*, December 25, 1965, 66.

78. Louis Harris, "U.S. Sees Good in Automation, 5 to 3; 90 Percent Feel No Personal Threat," *Washington Post*, May 24, 1965.

79. Wassily Leontief and Faye Duchin, *The Future Impact of Automation on Workers* (New York: Oxford University Press, 1986), 31; Michael Wallace, "Brave New Workplace: Technology and Work in the New Economy," *Work and Occupations* 16, no. 4 (1989): 363–92; David F. Noble, "Present Tense Technology," Part 3, *Democracy* 3, no. 4 (1983): 71–93; Robert G. Kaiser, "*Post* Publishes Curtailed Edition Despite Strike," *Washington Post*, October 3, 1975. On the broader topic of worker sabotage, see Craig A. Zabala, "Sabotage in an Automobile Assembly Plant: Worker Voice on the Shopfloor," in Robert Asher, ed., *Autowork* (Albany: State University of New York Press, 1995), 209–25.

80. Ida Russakoff Hoos, "When the Computer Takes over the Office," *Harvard Business Review* 38, no. 4 (1960): 102–12.

81. Drucker, "Automation Is Not the Villain," 75; Phil Hirsch, "Automation, Clerks, Bookkeepers," *New Republic*, May 7, 1962, 13–14; A. H. Raskin, "Fears about Automation Overshadowing Its Boons," *New York Times*, April 7, 1961.

82. Charles E. Silberman, "The Real News about Automation," *Fortune*, January 1965, 124–26 and 220–28; Otto Friedrich, "The Computer Moves In," *Time*, January 3, 1983, 14–24; Janice Moglen-Dietrich and Judy Pohlman, "Man of the Year," *Time*, January 24, 1983, 4, 6.

83. "Robots Join the Labor Force," *Business Week*, June 9, 1980, 62–76; Gene Bylinsky, "The Race to the Automatic Factory," *Fortune*, February 21, 1983, 52–64; Joann S. Lublin, "Steel-Collar Jobs," *Wall Street Journal*, October 26, 1981; and Otto Friedrich, "The Robot Revolution," *Time*, December 8, 1980, 72–83.

84. "Robots Join the Labor Force," *Business Week*, June 9, 1980, 62–76; Bylinsky, "The Race to the Automatic Factory"; Lublin, "Steel-Collar Jobs"; and Friedrich, "The Robot Revolution."

85. James M. Kahn, "The Age of Robots," Letters to the Editor, *Time*, December 29, 1980, 2. See also "The Speedup in Automation," *Business Week*, August 3, 1981, 58–67.

86. "Workers' Technology Bill of Rights," *democracy* 3, no. 2 (1983): 25–27.

87. Kuttner, *The Economic Illusion*, 4, 151–52, 168–82; Bob Kuttner, "The Declining Middle," *Atlantic Monthly*, July 1983, 60–72; James Fallows, "America's Changing Economic Landscape," *Atlantic Monthly*, March 1985, 47–68; and Bennett Harrison and Barry Blue-

stone, *The Great U-Turn: Corporate Restructuring and the Polarizing of America* (New York: Basic Books, 1988), ix. See also Michael Piore, *Dualism and Discontinuity in Industrial Societies* (New York: Cambridge, 1980). The 1980s fear of America's manufacturing decline connected with an entire set of economic and social issues beyond automation, especially the pattern of firms moving operations overseas. See Barry Bluestone and Bennett Harrison, *The Deindustrialization of America: Plant Closings, Community Abandonment, and the Dismantling of Basic Industry* (New York: Basic Books, 1982).

88. Lenny Siegel and John Markoff, *The High Cost of High Tech: The Dark Side of the Chip* (New York: Harper and Row, 1985); Tom Forrester, *High-Tech Society: The Story of the Information Technology Revolution* (Cambridge, Mass.: MIT Press, 1987); Leontief and Duchin, *The Future Impact of Automation on Workers*, 57, 59; and Wallace, "Brave New Workplace," 363–92.

Epilogue: Revisiting the Technological Unemployment Debate

1. Robert M. Tomasko, *Downsizing: Reshaping the Corporation for the Future* (New York: AMACOM, American Management Association, 1990); Al Ehrbar, "'Re-Engineering' Gives Firms New Efficiency, Workers the Pink Slip," *Wall Street Journal*, March 16, 1993, A1, A11; Lester C. Thurow, *The Future of Capitalism: How Today's Economic Forces Shape Tomorrow's World* (New York: William Morrow and Company, 1996); *Punch* cartoon, reprinted in *New York Times Magazine*, March 25, 1962, 83; and Jane Bryant Quinn, "The Good-Job Market: R.I.P.," *Newsweek*, November 30, 1992.

2. R. C. Longworth and Sharman Stein, "Temp Jobs Gaining Permanence," *Chicago Tribune*, August 21, 1995; Tomasko, *Downsizing*, xvi, 2, 34; and Richard Baumohl, "When Downsizing Becomes 'Dumbsizing,'" *Time*, March 15, 1993, 55.

3. Bruce Nussbaum, "Downward Mobility," *Business Week*, March 23, 1992, 56–63.

4. Chris Tilly, "Short Hours, Short Shrift: The Causes and Consequences of Part-Time Employment" and Eileen Appelbaum, "Structural Change and the Growth of Part-Time and Temporary Employment," both in Virginia L. duRivage, ed., *New Policies for the Part-Time and Contingent Workforce* (Armonk, N.Y.: M.E. Sharpe, 1992). Bennett Harrison, *Lean and Mean: The Changing Landscape of Corporate Power in the Age of Flexibility* (New York: Basic Books, 1994), 201–2, 205; Louis Uchitelle, "Temporary Workers Are on the Increase in the Nation's Factories," *New York Times*, September 6, 1993; Longworth and Stein, "Temp Jobs Gaining Permanence." See also Kevin D. Henson, *Just a Temp* (Philadelphia: Temple University Press, 1996); and Jeffrey Pfeffer and James N. Baron, "Taking the Workers Back Out: Recent Trends in the Structuring of Employment," in *Research in Organizational Behavior* 10 (1988): 257–303. The 1997 United Parcel Service strike drew attention to complaints of part-time workers who claimed they had long been promised promotion to full-time. Ironically, UPS treated part-timers better than many companies, offering health coverage and some pension benefits. Polls suggested, though, that over 50 percent of the public sympathized with the Teamsters, and in settling the strike, UPS agreed to create 10,000 full-time positions. Matthew Miller, "Packaging the Strike," *U.S. News & World Report*, September 1, 1997, 44–46; and Jeff Madrick, "The UPS Strike Delivers a Message," *Washington Post National Weekly*, August 18, 1997, 21.

5. Laura L. Nash, "The Virtual Job," *Wilson Quarterly* (Autumn 1994): 72–81.

6. Stratford Sherman, "The New Darwinian Workplace," *Fortune*, January 25, 1993,

51–56; and Emily Martin, *Flexible Bodies: The Role of Immunity in American Culture from the Days of Polio to the Age of AIDS* (Boston: Beacon Press, 1994).

7. Paul Krugman, *The Age of Diminished Expectations: U.S. Economic Policy in the 1990s* (Cambridge, Mass.: MIT Press, 1990), ix, 22; Frank S. Levy and Richard C. Michel, *The Economic Future of American Families: Income and Wealth Trends* (Washington, D.C.: The Urban Institute Press, 1991), 9; and Katherine S. Newman, *Declining Fortunes: The Withering of the American Dream* (New York: Basic Books, 1993): 198; and Kevin Phillips, *The Politics of Rich and Poor: Wealth and the American Electorate in the Reagan Aftermath* (New York: Random House, 1990), xviii, 8. See also Paul Krugman, *Peddling Prosperity: Economic Sense and Nonsense in the Age of Diminished Expectations* (New York: W. W. Norton, 1994).

8. Robert B. Reich, *The Work of Nations: Preparing Ourselves for 21st Century Capitalism* (New York: Knopf, 1991), 216–17; Mickey Kaus, *The End of Equality* (New York: Basic Books, 1992), 29, 38–39, 47–48, 72–74, 219; and Paul Blumberg, *Inequality in an Age of Decline* (New York: Oxford University Press, 1980).

9. Brian O'Reilly, "The Job Drought," *Fortune*, August 24, 1992, 62–74; and George J. Church, "Jobs in an Age of Insecurity," *Time*, November 22, 1993, 32–39.

10. Gene Koretz, "Big Companies Are Still Axing Jobs with a Vengeance," *Business Week*, May 10, 1993, 16; Robert B. Reich, "Companies Are Cutting Their Hearts Out," *New York Times Magazine*, December 19, 1993, 54–55; Church, "Jobs in an Age of Insecurity," 32–39; Thurow, *The Future of Capitalism*, and Nussbaum, "Downward Mobility."

11. Jaclyn Fierman, "When Will You Get a Raise," and "What Happened to the Jobs?"; Louis S. Richman, "CEOs to Workers: Help Not Wanted," and Alan Farnham, "Out of College, What's Next?" *Fortune*, July 12, 1993, 34–45, 58–64; and Richard J. Barnet, "The End of Jobs," *Harper's Magazine* 287 (September 1993): 47–52; Lucinda Harper, "Businesses Prefer Buying Equipment to Hiring New Staff," *Wall Street Journal*, September 3, 1993. See also Kenneth Labich, "The New Unemployed," *Fortune*, March 8, 1993, 40–49.

12. Glenn Collins, "Broadway Musicians Rally to Defend Their Sound," and "'Phantom' in Washington with Tape," *New York Times*, September 9, 1993.

13. Donald G. McNeil Jr., "New Show Is First Not to Have to Pay Idle Musicians," *New York Times*, February 8, 1995. Robert Epstein, "Strike Out the Band?" *Los Angeles Times*, November 28, 1991.

14. U.S. Bureau of Labor Statistics, *Outlook for Technology and Labor in Telephone Communications*, Bulletin 2357 (July 1990): 4, 5, 7; John J. Keller, "AT&T to Replace as Many as One-Third of Its Operators with Computer Systems," *Wall Street Journal*, March 4, 1992; Adam Bryant, "Along Digital Path, Dead-End Jobs," *New York Times*, February 15, 1994; and Steven Prokesch, "Service Jobs Fall as Business Gains," *New York Times*, April 18, 1993.

15. Theresa L. Morisi, "Commercial Banking Transformed by Computer Technology," *Monthly Labor Review* 119, no. 8 (1996): 30–36; Jonathan D. Glater and Frank Swoboda, "Next Window, Please, to the ATM," *Washington Post National Weekly*, July 24–30, 1995, 21–22; Saul Hansell, "Banks Shutting Local Branches to Trim Costs," *New York Times*, October 23, 1994; Saul Hansell, "Banks Go Interactive to Beat the Rush of Services," *New York Times*, October 19, 1994; Kerry Hannon, "The Teller in Your PC," *U.S. News & World Report*, October 16, 1995, 91, 92, 94; and Connie Guglielmo, "Here Come the Super-ATMs," *Fortune*, October 14, 1996, 232, 234.

16. Joel Achenbach, "Why Things Are," *Washington Post*, April 23, 1993; Marvin Balousek

Sr., "Take Honor Boxes," letter to the editor, *Chicago Tribune*, July 1, 1992; and John Green-wald, "Zapping the Post Office," *Time*, January 19, 1998, 46–47. See also Gene Koretz, "Technology Is Fueling Retail Productivity, but Slowing Job Gains," *Business Week*, May 10, 1993.

17. "Union Foes: Weak Demand, Automation," *New York Times*, June 13, 1993; and Jim Pissot, "Timber Troubles," *Washington Post*, April 2, 1993.

18. Paul Solmon, report for *MacNeil-Lehrer Newshour*, PBS, June 18, 1993.

19. Maria Mallory, Susan Brink, Katia Hetter and David Fischer, "Professionals Feel the Heat," *U.S. News & World Report*, April 1, 1996, 44–48, and Philip E. Ross, "Software as Career Threat," *Forbes*, May 22, 1995, 240–46. *Forbes* pointed out that in many cases computers did not completely eliminate jobs but displaced professionals by shifting work down to lower-paid, lower-status employees. Insurance companies and mutual funds used telemarketers rather than agents to sell products, while physicians' assistants performed doctors' services at half the price.

20. Steven Prokesch, "Service Jobs Fall as Business Gains" and "Short Cuts in One Firm's Paper Trail," *New York Times*, April 18, 1993.

21. Stanley Aronowitz and William DiFazio, *The Jobless Future: Sci-Tech and the Dogma of Work* (Minneapolis: University of Minnesota Press, 1994), 1–2, 6, 15, 107–8, 312–17.

22. Ibid., 312–17, 326–27, 343, emphasis in original.

23. Jeremy Rifkin, *The End of Work: The Decline of the Global Labor Force and the Dawn of the Post-Market Era* (New York: G. P. Putnam, 1995), xv, 68, 134, 161; and Bruce Handy, "Have Gigabites, Will Act," *Time*, September 1, 1997, 72.

24. Rifkin, *The End of Work*, xvii, 56, 77, 80, 88, 290. For detailed discussion of race, unemployment, and urban decline, see William J. Wilson, *Truly Disadvantaged: The Inner City, the Underclass, and Public Policy* (Chicago: University of Chicago Press, 1987), and William J. Wilson *When Work Disappears: The World of the New Urban Poor* (New York: Knopf, 1996).

25. Steven Pearlstein and Frank Swoboda, "Conflicting Goals on Jobs, Free Trade," *Washington Post*, July 10, 1993; "Jobs," editorial, *Washington Post*, July 7, 1993; and Hobart Rowen, "Slow Growth," *Washington Post National Weekly*, November 1–7, 1993: 5.

26. Frank Swoboda, "Conference Tries to Define 'High-Performance' Jobs," *Washington Post*, July 27, 1993.

27. R. C. Longworth and Sharman Stein, "Middle Class Finds Odd Jobs May Be Only Way to Stay Afloat," *Chicago Tribune*, August 22, 1995; David Hage, "Have Job Prospects Started to Wilt?" *U.S. News & World Report*, May 15, 1995, 65; Nancy Gibbs, "Working Harder, Getting Nowhere," *Time*, July 3, 1995, 16–20. See also Louis Uchitelle, "Recovery? Not in Your Paycheck," *New York Times*, January 8, 1995; Wallace Peterson, *Silent Depression: The Fate of the American Dream* (New York: W. W. Norton, 1994), 9–10; Kurt Eichenwald, "Data Gloomy on Consumers and Retailers," *New York Times*, January 31, 1996; and Thurow, *The Future of Capitalism*, 6, 24, 253, 256–57, 326.

28. Thurow, *The Future of Capitalism*, 2, 261; Lester C. Thurow, "A Surge in Inequality," *Scientific American*, May, 1987, 30–37; Keith Bradsher, "Gap in Wealth in U.S. Called Widest in West," *New York Times*, April 17, 1995; Keith Bradsher, "Widest Gap in Incomes? Research Points to U.S.," *New York Times*, October 27, 1995; and John Cassidy, "Who Killed the Middle Class?" *New Yorker*, October 16, 1995, 113–24. See also Steven V. Roberts, "Workers Take It on the Chin," *U.S. News & World Report*, January 22, 1996, 44–46.

29. Mortimer B. Zuckerman, "Where Have the Good Jobs Gone?" *U.S. News & World*

Report, July 31, 1995, 68; R. C. Longworth and Sharman Stein, "Battered Middle Class Turning Anxious, Angry," *Chicago Tribune,* August 20, 1995; and Longworth and Stein, "Middle Class Finds Odd Jobs May Be Only Way to Stay Afloat."

30. On Buchanan's economic populism, see Malcolm Gladwell, "A Populist From the Past," *Washington Post National Weekly,* February 26–March 1, 1996, 6–7; "Crashing the Party," *U.S. News & World Report,* February 26, 1996, 26–29; and Richard Lacayo, "The Populist Blowup," *Time,* February 26, 1996, 28–30. See also Lisa Anderson and Michael Tackett, "In Switch, GOP Hopefuls Stress Jobs, Jobs, Jobs," *Chicago Tribune,* February 18, 1996. On the issue of CEO pay, see Louis Uchitelle, "1995 Was Good for Companies, and Better for a Lot of C.E.O.'s," *New York Times,* March 29, 1996; Harrison Rainie, "The State of Greed," *U.S. News & World Report,* June 17, 1996, 62, 63; Daniel Kadlec, "How CEO Pay Got Away," *Time,* April 28, 1997, 59–60; David Cay Johnston, "Tracking Executives' Compensation," *New York Times,* April 14, 1997; and Graef Crystal, *In Search of Excess: The Overcompensation of American Executives* (New York: W.W. Norton, 1991).

31. Allan Sloan, "The Hit Men," *Newsweek,* February 26, 1996, 44–48; Dale Russakoff and Steven Pearlstein, "Ma Bell's Changing Tone: In a Reordered Corporate World, It's Employees Who Pay the Toll," *Washington Post National Weekly,* June 3–9, 1996, 6–9; Tim Jones, "Amid Uproar about Layoffs, AT&T Retreats," *Chicago Tribune,* March 16, 1996; and AT&T advertisement, *Chicago Tribune,* March 15, 1996. See also Dale Russakoff and Steven Pearlstein, "Leaving Normal: Layoffs at Companies Like AT&T Pleased Wall Street but Destroyed Main Street," *Washington Post National Weekly,* September 16–22, 1996, 6–7.

32. Louis Uchitelle and N. R. Kleinfield, "The Price of Jobs Lost," in *New York Times, The Downsizing of America* (New York: Random House, 1996), ix, 6–7, 16–17, 19, 309–10.

33. Adam Clymer, "Help Urged for Companies That Treat Workers Well," *New York Times,* February 9, 1996; Alison Mitchell, "Republicans and Democrats Jumping on the Issue of Corporate Responsibility," *New York Times,* February 15, 1996; Henry Allen, "Ha! So Much for Loyalty," *Washington Post National Weekly,* March 4–10, 1996, 11–12; Robert B. Reich, "Companies Are Cutting Their Hearts Out," *New York Times Magazine,* December 19, 1993, 54–55; Nancy Gibbs, "Working Harder, Getting Nowhere," *Time,* July 3, 1995, 16–20; Robert Kuttner, "Ducking 'Class Warfare,'" *Washington Post National Weekly,* March 11–17, 1996, 5; Alison Mitchell, "Clinton Urges Action on Job Initiatives," *New York Times,* March 5, 1996; Bill Saporito, "Good for the Bottom Line," *Time,* May 20, 1996, 40–43; and E. J. Dionne Jr., "The Anxiety Constituency," *Washington Post National Weekly,* March 25–31, 1996, 30. On 1996 economic trends, see Allen R. Myerson, "In Era of Belt-Tightening, Modest Gains for Workers," *New York Times,* February 13, 1997; Louis Uchitelle, "Wages Are Getting Higher, but Prices Hold Their Own," *New York Times,* April 11, 1997, A1, C2; Louis Uchitelle, "Markets Surge as Labor Costs Stay in Check," *New York Times,* April 30, 1997.

34. Clay Chandler, "Budget Turnabout," *Washington Post National Weekly,* February 9, 1998, 6–7; Robert D. Hershey Jr. "U.S. Jobless Rate Declines to 4.9%," *New York Times,* May 3, 1997; John M. Berry, "Consumer Prices at 11-Year Low," *Washington Post,* January 14, 1998; John M. Berry, "Strongest Growth, Lowest Inflation—All in the Same Year," *Washington Post National Weekly,* February 9, 1998, 7; Robert J. Samuelson, "Has Inflation Been Tamed?" *Washington Post National Weekly,* October 27, 1997, 26.

35. Chandler, "Budget Turnabout"; Samuelson, "Has Inflation Been Tamed?"; John M. Berry, "Beating Inflation at the Pay Window," *Washington Post National Weekly,* May 6–12, 1996, 20; Robert D. Hershey Jr., "Wages of U.S. Workers Show Steepest Increase in 5 Years,"

New York Times, May 1, 1996, A1, C3; John Schmelzer, "Economy Eases Up on Output of Angst," *Chicago Tribune*, May 1, 1996; "You're Fired—er, Hired!" *Time*, July 8, 1996, 45; Steven A. Holmes, "Income Disparity between Poorest and Richest Rises," *New York Times*, June 20, 1996; Robert Kuttner, "Tipping the Income Scales," *Washington Post National Weekly*, July 1–7, 1996, 5; Uchitelle, "Markets Surge as Labor Costs Stay in Check"; David E. Sanger, "The Mantra for 1997: It's the Best of Times," *New York Times*, May 3, 1997; "Black-White Income Inequalities," *New York Times*, February 17, 1998; and Michael A. Fletcher, "Race Board's Focus Turns to Economic Gap," *Washington Post*, January 15, 1998. See also Eric Pooley, "Too Good to Be True?" *Time*, May 19, 1997, 29–33; and Steven Pearlstein, "The Mystery of the Booming American Economy . . ." and John M. Berry ". . . And It Keeps On Going and Going and Going," *Washington Post National Weekly*, May 12, 1997, 20–21.

36. Nancy Gibbs, "The Paradox of Prosperity," *Time*, December 29, 1997, 91; and Del Jones, "Workplace Loyalty: Not Dead Yet, but Strained," *USA Today*, February 4, 1998, 4B.

37. Eugene McCarthy and William McGaughey, *Nonfinancial Economics: The Case for Shorter Hours of Work* (New York: Praeger, 1989); Juliet Schor, *The Overworked American: The Unexpected Decline of Leisure* (New York: Basic Books, 1991), 6, 10, 41, 76, 81, 157, 199; and Harper, "Businesses Prefer Buying Equipment to Hiring New Staff."

38. Schor, *The Overworked American*; Arlie Hochschild, *Second Shift: Working Parents and the Revolution at Home* (New York: Viking, 1989); and Linton Weeks, "Stocking Round the Clock," *Washington Post National Weekly*, August 4, 1997: 18.

39. John Marks, "Time Out," *U.S. News & World Report*, December 11, 1995, 85–96; Trip Gabriel, "When Lost Jobs Lead to Found Directions," *New York Times*, September 7, 1995; and Carey Goldberg, "The Simple Life Lures Refugees from Stress," *New York Times*, September 21, 1995.

40. *New York Times, The Downsizing of America*, 307; Amy Saltzman, "When Less Is More," *U.S. News & World Report*, October 27, 1997, 81; and Scott Sullivan, "Life on the Leisure Track," *Newsweek*, June 14, 1993, 48.

41. Stanley Lebergott, *Pursuing Happiness: American Consumers in the Twentieth Century* (Princeton: Princeton University Press, 1993); R. C. Longworth, "The Dream, in Pieces," *Chicago Tribune Magazine*, April 28, 1996, 15–20; Schor, *The Overworked American*, 157, 199; John Schmeltzer, "Day of Debt Reckoning?" *Chicago Tribune*, July 9, 1996; and "When Boomers Become Busted," *Time*, March 31, 1997, 64.

42. Robert D. Putnam, "Bowling Alone: America's Declining Social Capital," *Current* (June 1995): 3–9. Christopher Lasch, *The Revolt of the Elites and the Betrayal of Democracy* (New York: W. W. Norton, 1995).

43. Thurow, *The Future of Capitalism*, 256–57; Eugene McCarthy and William McGaughey, *Nonfinancial Economics*, xi–xii; and Clifford Cobb, Ted Halstead, and Jonathan Rowe, "If the GDP Is Up, Why Is America Down?" *Atlantic Monthly*, October 1995, 59–78.

44. Mike Tharp and William J. Holstein, "Mainstreaming the Militia," *U.S. News & World Report*, April 21, 1997, 25–26; and "Industrial Society and Its Future," a supplement to the *Washington Post*, September 19, 1995, 1–8. See also Stephen Budiansky, "Academic Roots of Paranoia," *U.S. News & World Report*, May 13, 1996, 33–34.

45. Clifford Stoll, *Silicon Snake Oil: Second Thoughts on the Information Highway* (New York: Doubleday, 1995).

46. Kirkpatrick Sale, "Is There Method in His Madness?" *Nation*, September 25, 1996, 305–11; Dirk Johnson, "A Celebration of the Urge to Unplug," *New York Times*, April 15, 1996; Steven Levy, "The Luddites Are Back," *Newsweek*, June 12, 1995, 55; Kirkpatrick Sale, *Rebels against the Future: The Luddites and Their War on the Industrial Revolution: Lessons for the Computer Age* (Reading, Mass.: Addison-Wesley, 1995); and Daniel J. Kevles, "E Pluribus Unabomber," *New Yorker*, August 14, 1995, 2, 4. See also Daniel Grossman, "Neo-Luddites: Don't Just Say Yes to Technology," and Chellis Glendinning, "Notes toward a Neo-Luddite Manifesto," *Utne Reader*, March/April 1990, 44–53.

47. Jeff MacNally, *Des Moines Register*, January 15, 1995; Tom Toles, *Utne Reader*, September/October 1993, 72; and Tom Tomorrow, "This Modern World," *Des Moines Register*, January 8, 1995 (emphasis in original).

48. Wiley Miller, "Non Sequitur," *Chicago Tribune*, February 1, 1996; Art Sansom, "The Born Loser," *Washington Post* January 12, 1993; Sidney Harris, *What's So Funny about Computers?* (Los Altos, Calif.: W. Kaufmann, 1982); Chic Young, "Blondie," *Washington Post*, January 15, 1998; and Mark Cullum and John Marshall,"Walnut Cove," *Chicago Tribune*, March 4, 1996.

49. Joan E. Rigdon, "Retooling Lives: Technological Gains Are Cutting Costs, and Jobs, in Services," *Wall Street Journal*, February 24, 1994. Among writings that highlight the continued complexity of the technological unemployment issue, see General Accounting Office, *Advances in Automation Prompt Concern over Increased U.S. Unemployment* (Washington, D.C.: U.S. Government Printing Office, 1982); Eileen Appelbaum, "The Economics of Technical Progress: Labor Issues Arising from the Spread of Programmable Automation Technologies," in *Automation and the Workplace*, Office of Technology Assessment Technical Memorandum (Washington, D.C.: U.S. Congress, Office of Technology Assessment, March 1983); Marco Vivarelli, *The Economics of Technology and Employment: Theory and Empirical Evidence* (Aldershot, Hants, England: Edward Elgar, 1995); and Sheila McConnell, "The Role of Computers in Reshaping the Work Force," *Monthly Labor Review* 119, no. 8 (1996): 3–5.

50. Thorstein Veblen, *The Engineers and the Price System* (1921; New York: Viking, 1954). See also Joseph Dorfman, *Thorstein Veblen and His America* (New York: Viking, 1935).

51. Harris, *What's So Funny about Computers?*.

52. On complications and costs of computerization, see Edward Tenner, *Why Things Bite Back: Technology and the Revenge of Unintended Consequences* (New York: Vintage Books, 1996).

53. On technological determinism, see John M. Jordan, *Machine-Age Ideology: Social Engineering and American Liberalism, 1911–1939* (Chapel Hill: University of North Carolina Press, 1994); Cecelia Tichi, *Shifting Gears: Technology, Literature and Culture in Modernist America* (Chapel Hill: University of North Carolina Press, 1987); and Merritt Roe Smith and Leo Marx, eds., *Does Technology Drive History? The Dilemma of Technological Determinism* (Cambridge, Mass.: MIT Press, 1995). Most important, see David F. Noble, "Present Tense Technology," Part 1, *democracy* 3, no. 2 (1983): 8–24; Part 2, *democracy* 3, no. 3 (1983): 70–82; and Part 3, *democracy* 3, no. 4 (1983): 71–93.

54. Rigdon, "Retooling Lives"

55. Kuttner, *The Economic Illusion*, 4, 151–52, 168–82; and Kuttner, "The Declining Middle."

ESSAY ON SOURCES

Primary Source Materials

Manuscript Collections

Archival collections of papers from the 1930s served as vital sources of information for this book. Both the Herbert Hoover Presidential Library in West Branch, Iowa, and the Franklin Delano Roosevelt Library in Hyde Park, New York, have strong collections relating to their administration's economic considerations in general and, within that, bits of information about technological unemployment issues. Speeches by various administration officials, press releases, and memos served as especially useful sources. The National Archives in Washington, D.C., houses the records of the WPA National Research Project; though lack of a detailed index can make working with these papers frustrating, they contain a wealth of information about the agency's research program and approach to the technological unemployment question. Other archives that provided source material include the Robert Wagner Labor Archives at New York University, the Hagley Library in Delaware (which has collections of business papers and publications), and the Center for History of Physics at the American Institute of Physics (which has records of the American Physical Society).

Published Government Documents

Published government documents employed in writing this book include *Recent Economic Changes in the United States* (New York: McGraw-Hill, 1929), *Recent Social Trends in the United States* (New York, McGraw-Hill, 1933), and the U.S. National Resources Committee, *Technological Trends and National Policy, Including the Social Implications of New Inventions* (Washington, D.C.: U.S. Government Printing Office, 1937). Among the WPA's National Research Project publications, the most useful include David Weintraub and Harold Posner's *Unemployment and Increasing Productivity* (Philadelphia: U.S. Government

Printing Office, 1937), David Weintraub and Irving Kaplan, *Summary of Findings to Date* (Philadelphia: U.S. Government Printing Office, 1938), and Corrington Gill, *Unemployment and Technological Change* (Washington, D.C.: U.S. Government Printing Office, 1940). The U.S. Bureau of Labor Statistics *Monthly Labor Review* of the 1930s contained a number of useful reports on technological change and employment in various occupations and industries, while the Women's Bureau of the Department of Labor published many useful pieces in the *Bulletin of the Women's Bureau* on the issue of how technical innovation affected female employment. Regular *Congressional Records* of the House and Senate were vital, as were reports of the TNEC investigations, especially the record of the hearings "Technology and Concentration of Economic Power," Part 30, *Investigation of Concentration of Economic Power* (Washington, D.C.: U.S. Government Printing Office, 1940).

Periodicals and Journals

Throughout the 1930s, newspapers such as the *New York Times* provided an excellent barometer of events and ideas in the technological unemployment debate; in addition to regular articles, editorials and letters to the editor were also useful. A good number of popular publications during the Depression published articles on the technological displacement question, either occasionally (for example, *Good Housekeeping* or the *Rotarian*) or on a more regular basis (most notably, *Survey Graphic* and *Harper's Monthly*).

Specialized interest or professional publications provided the key to understanding perspectives that different groups adopted on the issue of machines and jobs. For understanding the mechanical engineering community's involvement in the debate, articles, editorials, and letters in *Mechanical Engineering* proved useful. Especially good sources for business attitudes toward the issue included *Iron Age* and the numerous pamphlets published by the Machinery and Allied Products Institute and the National Industrial Conference Board. In the labor community, the *American Federationist* and the *International Musician* published a number of important articles showing union approaches to technological unemployment.

Popular Books

One of the most influential and hence most useful popular books on technological unemployment in the Depression was, of course, Stuart Chase's *Men and Machines* (New York: Macmillan, 1931). William Ogburn's *You and Machines* (Chicago: University of Chicago Press, 1934) also served as an informative source. Works of fiction discussed here that refer to the technological unemployment debate include John Steinbeck's *The Grapes of Wrath* (New York: Penguin, 1977), Upton Sinclair's *Little Steel* (New York: Farrar and Rinehart, 1938), Theodore Dreiser's *Tragic America* (London: Constable, 1931), Sherwood Anderson's *Perhaps Women* (New York: Horace Liveright, 1931), and, from the realm of children's literature, Virginia Lee Burton's *Mike Mulligan and His Steam Shovel* (Boston: Houghton Mifflin, 1939).

Secondary Sources

Though not concentrating on the issue of technological unemployment, historians and sociologists over recent decades have devoted substantial attention to the subject of how technological change affects the labor process and workplace control, producing material that provides an excellent background for thinking about labor and mechanization. See particularly Harry Braverman, *Labor and Monopoly Capital: The Degradation of Work in the Twentieth Century* (New York: Monthly Review Press, 1974); David Montgomery, *Workers' Control in America: Studies in the History of Work, Technology, and Labor Struggles* (Cambridge: Cambridge University Press, 1979); Richard Edwards, *Contested Terrain: The Transformation of the Workplace in the Twentieth Century* (New York: Basic Books, 1979); Michael Burawoy, *Manufacturing Consent: Changes in the Labor Process under Monopoly Capitalism* (Chicago: University of Chicago Press, 1979) and *The Politics of Production: Factory Regimes under Capitalism and Socialism* (London: Verson, 1984); and David Gordon, Richard Edwards, and Michael Reich, *Segmented Work, Divided Workers: The Historical Transformation of Labor in the United States* (Cambridge: Cambridge University Press, 1982).

Looking at specific industries, some historians have examined how introduction of new machines affected labor in certain instances and in different time periods. For such studies, see Merritt Roe Smith, *Harpers Ferry Armory and the New Technology: The Challenge of Change* (Ithaca, N.Y.: Cornell University Press, 1977), and David Noble, *Forces of Production: A Social History of Industrial Automation* (Oxford: Oxford University Press, 1986). Good articles include Evelyn Nakano Glenn and Roslyn L. Feldberg, "Proletarianizing Clerical Work: Technology and Organizational Control in the Office"; Philip Kraft, "The Industrialization of Computer Programmming: From Programming to 'Software Production'"; Herb Mills, "The San Francisco Waterfront: The Social Consequences of Industrial Modernization"; and Michael Yarrow, "The Labor Process in Coal Mining: Struggle for Control"; all in *Case Studies on the Labor Process*, Andrew Zimbalist, ed. (New York: Monthly Review Press, 1979).

A few historical studies have noted how the question of technological displacement arose in specific occupations, for example: Irwin Yellowitz, "Skilled Workers and Mechanization: The Lasters in the 1890s," *Labor History* 18, no. 2

(1977): 197–213; Patricia Cooper, *Once a Cigar Maker: Men, Women and Work Culture in American Cigar Factories, 1900–1919* (Urbana: University of Illinois Press, 1987); Keith Dix, *What's a Coal Miner to Do? The Mechanization of Coal Mining* (Pittsburgh: University of Pittsburgh Press, 1988); and Grace Palladino, "When Militancy Isn't Enough: The Impact of Automation on New York City Building Service Workers, 1934–1970," *Labor History* 28, no. 2 (1987): 196–220.

In the study of technological unemployment, the literature in the history of technology of course plays a vital role. An excellent overview of American topics is Alan Marcus and Howard P. Segal, *Technology in America: A Brief History* (San Diego: Harcourt Brace Jovanovich, 1989). An invaluable work on the history of technocracy is William Akin's *Technocracy and the American Dream* (Berkeley: University of California Press, 1977). For the history of how the engineering profession established itself and dealt with critical concerns, see Edwin Layton, Jr., *The Revolt of the Engineers* (Baltimore: Johns Hopkins University Press, 1986); and David F. Noble, *America by Design: Science, Technology, and the Rise of Corporate Capitalism* (Oxford: Oxford University Press, 1977). Vital sources in the history of Depression-era science include Peter Kuznick, *Beyond the Laboratory* (Chicago: University of Chicago Press, 1987); and Marcel LaFollette, *Making Science Our Own* (Chicago: University of Chicago Press, 1990).

Two articles that provide excellent discussions of parts of the technological unemployment issue during the Depression are Ester Fano, "A 'Wastage of Men': Technological Progress and Unemployment in the U.S.," *Technology and Culture* 32 (April 1991), and Carroll Pursell, "Government and Technology in the Great Depression," *Technology and Culture* 20 (January 1979).

Since the issue of technological unemployment is so closely related to general political, social, cultural, and labor history during the Depression, a broad range of books supply additional background in these areas. A few useful books include: William Barber, *From New Era to New Deal* (New York: Cambridge University Press, 1985); Irving Bernstein, *The Turbulent Years* (New York: Houghton Mifflin, 1969); David Peeler, *Hope among Us Yet* (Athens, Ga.: University of Georgia Press, 1987); Richard Pells, *Radical Visions and American Dreams* (New York: Harper, 1973); and Warren Susman, *Culture as History* (New York: Pantheon, 1984).

Finally, on some of the "big questions" about technology and progress, see especially John M. Jordan, *Machine-Age Ideology: Social Engineering and American Liberalism, 1911–1939* (Chapel Hill: University of North Carolina Press, 1994); David F. Noble, "Present Tense Technology," Part 1, *democracy* 3, no. 2 (1983): 8–24; Part 2, *democracy* 3, no. 3 (1983): 70–82; and Part 3, *democracy* 3, no. 4 (1983): 71–93. The issue of "technological determinism" is explored well in Merritt Roe Smith and Leo Marx, eds., *Does Technology Drive History? The Dilemma of Technological Determinism* (Cambridge, Mass.: MIT Press, 1995).

INDEX

Adams, Potter, 199
Ad Hoc Committee on the Triple Revolution, 269
advertisements for new machinery, 148, 155–56, 266–68
Advertising Federation of America, 154–55
Advisory Committee on Employment Statistics, 44
African American workers, 258, 291, 295
age discrimination, 34, 36, 258
Agriculture, U.S. Department of, 65–67
agricultural employment and technology, 64–69, 123, 125, 137, 223–24. *See also* cotton-picking machine
Alford, L. P., 216
Alger, George, 106–7
Allen, Robert, 294
Allner, Frederick, 194
All's Fair at the Fair, 225
American Academy of Political and Social Science, 40
American Association for the Advancement of Science (AAAS), 65, 175–77, 192, 197
American Communications Association, 106
American Economic Association, 31
American Engineering Council (AEC), 45–46, 188, 194–95, 199

American Federationist, 98
American Federation of Hosiery Workers, 111
American Federation of Labor (AFL), 111–12; meetings of, 14, 46–47, 104; publications of, 15, 83, 102–3; officials of, 24; statements on technological unemployment, 83, 100–101, 105, 199. *See also* Green, William
American Federation of Musicians (AFM), 93–101, 98, 131–32, 138, 271–72, 285–86.
American Foundation on Automation and Unemployment, 260
American Institute of Physics, 177–81
American Legion, 154
American Machinist, 161
American Management Association, 281
American Philosophical Society, 173
American Rolling Mill Company, 18–19
American Society of Mechanical Engineers (ASME), 51, 170, 184, 189–92, 227; fiftieth anniversary of, 187, 196
American superiority, notion of, 152, 163–64, 173–74, 183, 206, 209, 217, 232
American Statistical Association, 40
American Technotax Society, 76
American Viscose Company, 106
Anderson, Benjamin, Jr., 144–45, 166
Anderson, Dewey, 20

Anderson, Mary, 85
Anderson, Sherwood, 124
Angell, James Rowland, 170
Arkwright, Frank, 122
Armour Foods, 87, 271
Aronowitz, Stanley, 290–91
AT&T (American Telephone & Tele-
    graph): in the 1930s, 22–26, 87, 91, 159,
    224; in the postwar decades, 248–49,
    286, 294–95
Atlantic Monthly, 300
Atoms in Action, 181–82
automation: ideal principles of, 237–42,
    244–47, 306–7; of manufacturing,
    240–43; of office work, 242–43
Automation and Technological Change
    hearings, 243–50
automation funds, 270–72
Automats, 39
automobile manufacture: in the 1930s,
    9–10, 17, 35, 148–53, 229; in the postwar
    decades, 254, 261
automobiles: ownership of, 9–10
Automotive Industries, 149

Babbage, Charles, 41
Baggs, Thomas, 158
Baker, Elizabeth, 32–33
Baltimore *Sun*, 242
banking, 242, 287
Bank of America, 242
Barclay, Hartley, 163
Barnes, Julius, 48
Barnes Company, 237
Barnett, George, 12–13
Barton, Henry, 177–78, 180
Barton, William, 237, 244
Beard, Charles, 44, 206, 209
Behar, M. F., 64
Behind the Scenes in the Machine Age,
    90–91, 137
Beirne, Joseph A., 248–50

Bell System. *See* AT&T
Bent, Silas, 37, 204
Best, Ethel, 24
Bidwell, Percy, 53
Bingay, Malcolm, 164
Black, Hugo, 74
black workers, 258, 291, 295
Blair, John, 231–32
"Blondie" (comic strip), 250, 302
Bone, Homer, 75
books about technological unemploy-
    ment: in the 1930s, 12–14, 53–55, 117–18;
    in the postwar decades, 290–92. *See
    also* fiction, technological unemploy-
    ment discussed in
Borden Company, 223–24
Borsodi, Ralph, 209, 219
Botany Worsted Mill, 124
Bowers, Edison, 197–98
Bowman, Isaiah, 175
Bridges, Harry, 270
Brittain, M. L., 164
Brotherhood of Railroad Trainmen, 17,
    104, 230, 248
Brown, J. J., 237–40, 306
Bruere, Henry, 146
Buchanan, Pat, 294
Budington, Robert, 170, 172
Bureau of Industrial Relations (Univ. of
    Michigan), 107–8
Bureau of Labor Statistics, U.S. 23, 35,
    44–45, 56, 86, 93–94, 149, 286
Burlingame, Roger, 206–7, 209
Burton, Virginia Lee, 128
Bush, George, 284
Bush, Vannevar, 236
business organizations. *See* Chamber of
    Commerce; Machinery and Allied
    Products Institute; National Associa-
    tion of Manufacturers
business public-relations campaigns,
    153–67, 221–22

Business Week, 241, 266, 281

Butler, Nicholas Murray, 120–21

Byrnes, William H., Jr., 77

California Commission on Manpower, Automation, and Technology, 259

Cameron, W. J., 150

Campbell, John, 126–27

Capek, Karel, 126

Carey, James, 244

Caterpillar tractors, 68

Census Bureau, 37–38, 293

Century of Progress exposition, 156–57, 194, 211

Challenge of Leisure, 215

Chamber of Commerce, U.S., 159–60, 163

Changing Technology and Employment in Agriculture, 67

Chaplin, Charlie, 136

Chase, Stuart, 38–39, 41, 121, 151, 200, 218; *Prosperity: Fact or Myth*, 13; *Men and Machines*, 117–18, 146

Chicago Sun-Times, 262

Chicago Tribune, 293

Ching, Cyrus, 165–66

Christensen, C. H., 50

Christian Science Monitor, 61

Chrysler, Walter, 160

Chubb, L. W., 157

Cigarmakers' International Union, 102

cigar- and cigarette-making machinery, 63, 76, 81–82, 92, 100, 223

City, 220–21

Civic Education Service, 26, 87–88

Civilian Conservation Corps, 53

Clague, Ewan, 45

Clark, Birge, 51

Cleveland Plain Dealer, 61

Clinton, Bill, 284, 292–93, 295

coal mining machinery, 80–81, 108–9, 130, 272

Cohn, Fannia, 101

Cold War, 236–37, 249, 255, 257, 297

Collier's, 11

Commerce Department, 45, 47–48, 83, 145, 295

Commercial Telegraphers' Union, 86

Communications Workers of America, 248

Compton, Karl, 173–79, 196, 198–99

computers, 242–43, 251–52, 274–76, 289, 305; as visual symbol, 265, 302

Congress of Industrial Organizations (CIO), 101, 104, 111–12

Congress, U.S.: 1930s debates, 74–78; 1950s debates, 247. *See also* Temporary National Economic Committee hearings; Automation and Technological Change hearings

consumerism: in the 1930s, 72–73, 103, 144–45, 155–56, 161–64, 218–19, 221, 223, 234; in the postwar decades, 236–37, 239, 246, 279, 297–300, 310–11

containerization, 270

continuous-strip steel mills. *See* steel-making

Cooke, Morris L., 198

Coolidge, W. D., 179–80, 189

Cooper, Fred, 56, 138

Cordiner, Ralph, 245–46

cotton-picking machine, 66–67, 70, 130, 135

Coughlin, Charles, 133

Council for Technological Advancement, 241–42

Country Gentleman, 226

Crawford, Ruth, 116–17

cultural lag theory, 40–41, 52, 55, 69, 186, 189, 206, 210, 228, 250, 308–9

Cunliffe, R. B., 212

Damrosch, Walter, 97

Danish, Max, 106–7

Das Kapital (Marx), 29

Daugherty, Carroll, 85

Davis, D. J., 244–45
Davis, Watson, 228
Day, Clarence, 132
Death on the Aisle (Lockridge), 125
Delano, Frederick, 175
"Democracity," 220
deskilling, 35
Desk Set, 251–52
Des Moines Register, 69
Desmond, Thomas, 213–14
determinism, technological. *See* technological determinism
Dewey, Thomas, 73
Diebold, John, 241, 245–46, 262, 306
DiFazio, William, 290–91
Dinner at Eight, 136
Director, Aaron, 30, 38
Dirksen, Everett, 256
Doherty, Robert, 199
Douglas, Paul, 30–32, 37–38, 188, 200
downsizing, 280–82, 284–85
Dreiser, Theodore, 124
Dreyfuss, Henry, 220
Driesen, Daniel, 106
Drucker, Peter, 264, 275
DuBrul, Stephen, 150
Dunn, Gano, 157
Durand, E. Dana, 17

Ecce Homo, 134–35, 233
economic analysis of technological unemployment, 6, 26–32, 36–39
economic conditions: in the 1920s, 9–13, 15–16; in the 1930s, 1, 16–18; in the postwar decades, 236–37, 247, 249, 254–55, 257, 273–74, 277, 279–84, 293–96, 300, 302
Economics for Engineers, 197–98
Edison, Thomas, 11, 41, 178
education, 197–99, 211–12, 267. *See also* retraining
efficiency, 171, 199–202, 305
Einstein, Albert, 178

Elektro (the "moto-man"), 224–26
Elizabeth, New Jersey, *Journal*, 61
Ely, Sumner, 210
End of Work (Rifkin), 291
engineering education, 197–99
engineering organizations. *See* American Engineering Council; American Society of Mechanical Engineers
Engineers and the Price System (Veblen), 306
engineers, on technological unemployment, 182–99
Engines of Democracy (Burlingame), 206
Erickson, Ethel, 89
ERMA banking system, 242, 251
Esquire, 267
Europe, 7
Ezekiel, Mordecai, 38, 72–73

Faddis, Charles, 75
farming. *See* agricultural employment and technology
Fechner, Robert, 53
Federated American Engineering Societies, 47
fiction, technological unemployment discussed in, 122–31, 252–254
Filene, Edward, 155
First to Awaken (Hicks), 127
Fisher, Irving, 121
Flanders, Ralph, 184–85, 188, 190, 210, 254
Fletcher, Leonard, 194
Forbes, 289
Ford, Henry, 9, 117, 149, 166, 210, 221–23, 283
Ford, Henry, II, 256
Ford Motor Company: in the 1930s, 9–10, 35, 135, 150, 229; in the postwar decades, 240, 247, 250–51
Fortune, 237–38, 264, 268, 275–76, 283–85
France, automation in, 257
Frankenstein metaphors, 70, 77, 95, 126
Frankl, Paul, 215

Freeman, Richard, 289
Frey, John, 45
frontier, closing of, 38, 174
Furnas, Clifford, 182–83, 213

Galbraith, John Kenneth, 236
garment making, 92, 106–7
Garvey, John, 211
Gay, Edwin, 51
General Electric, 155–57, 246, 252, 276, 285
General Motors, 10, 149, 156–57, 254, 295
Gephardt, Richard, 295
Germany, automation in, 257
Gernsback, Hugo, 125–27
Gherardi, Bancroft, 182
Gide, Charles, 29
Giese, Henry, 65
Gilbreth, Lillian, 44
Gilfillan, S. Colum, 40–41
Gill, Corrington, 58, 61–62, 230; state-
    ments of, on technological unemploy-
    ment, 57, 60, 167
Gillette, J. M., 68
Givens, Meredith, 52, 120
glass blowing, 12–13, 116–17, 181
Goldberg, Arthur, 256
Good Housekeeping, 117, 152
Gordon, Robert, 289
government reports. *See Recent Economic
    Changes; Recent Social Trends; Tech-
    nological Trends and National Policy;
    National Research Project on Reem-
    ployment Opportunities*
Graham, Rev. Thomas, 214
Grapes of Wrath (Steinbeck), 123, 125
Green, William, 44–45, 112; statements of,
    on technological unemployment,
    14–15, 39, 71, 76, 82, 100–101, 103–4, 169,
    170, 213, 218–19
Guide to Civilized Loafing, 214

Hannon, Margaret, 215–16
Hansen, Alvin, 31–32

Harbord, James, 148
Hare, Michael, 220
Harris, Sidney, 302, 307
Harrison, William, 24
Harper's, 22, 116, 132, 285
Hendrickson, Roy, 70
Hermans, Mabel, 215–16
Hickok, Lorena, 65–66, 110, 115
Hicks, Granville, 127
high-tech industry, 277, 279, 290–92, 299
Hirshfeld, C. F., 186, 192
historians' assessments of technological
    unemployment, 205–9
History of American Civilization (Rugg),
    129
Hobbs, Franklyn, 165
Hobson, John, 29
Hoeppel, John, 77–78
Hook, Charles, 19–21, 146, 153
Hoover, Allan, 51
Hoover, Herbert: engineering career of,
    47; as head of Commerce, 47; Presi-
    dential administration of, 14, 43–51; as
    Republican candidate, 12; statements
    of, on technological unemployment,
    46–48, 51, 153
Hoover, Lou Henry, 51
Hopkins, Harry, 57, 62, 65, 72, 110, 115
Hopkins, John, 66–67
hours of work debate: in the 1930s, 50, 74,
    103–5, 164–65, 212–18; in the postwar
    decades, 267, 297–99
Hunt, Edward Eyre, 45
Hurlin, Ralph, 51–52

IBM (International Business Machines),
    88
illustrations, of technological unemploy-
    ment: in the 1920s, 13; in the 1930s, 2,
    54–56, *56*, 90, 96–98, *98*, 137–41, *139*,
    158–59; in the postwar decades, 250–51,
    258–59, 262–66, *265*, *266*, 281, *288*,
    301–3, *303*, 307

Indianapolis News, 61
Industrial Instruments and Changing Technology, 64
Industrial Revolution, 1, 26, 35
Industrial Workers of the World (IWW), 13
Instruments, 64
International Alliance of Theatrical Stage Hands and Moving Picture Machine Operators, 93
International Association of Machinists (IAM), 257, 260, 277
International Brotherhood of Electrical Workers, 23, 273
International Brotherhood of Paper Makers, 14
International Chamber of Commerce, 159
International Harvester, 67–68
International Ladies' Garment Workers Union, 101, 106–7
International Longshoremen's and Warehousemen's Union, 270
International Longshoremen's Association, 270
International Typographical Union, 92
Introduction to Problems of American Culture (Rugg), 129, 139
Iowa State Federation of Labor, 14
Iron Age, 138, 144, 153–54, 158, 165

Jacks, L. P., 214
Jackson, Dugald, 185–86, 209
Japan, automation in, 257, 276
Javits, Jacob, 257
Jewett, Frank, 25, 156–57, 175, 179
Jobless Future (Aronowitz and DiFazio), 290
Johnson, Lyndon, 273
Jordan, Virgil, 120, 164, 218

Kaempffert, Waldemar, 12, 197
Kaus, Mickey, 283
Keedoozle store, 39

Kendall, Henry, 201
Kenin, Herman, 271
Kennedy, John F., 255–58
Kennedy, Ted, 295
Kennedy, W. P., 247–48, 250
Kettering, Charles, 145, 173, 176, 193, 196, 228
Keynes, John Maynard, 167
Khrushchev, Nikita, 236, 257
Kimball, Dexter, 15, 45–46, 176–77, 183, 190–91, 210
King, Wilford, 121
Kiwanis Magazine, 77
Klein, Julius, 48, 132, 152
Korson, George, 130
Kreps, Theodore, 228
Krugman, Paul, 283
Kuttner, Robert, 277, 299

Labor's Stake in the American Way, 163
labor unions. *See* unions
Labor, U.S. Department of, 140, 242–43, 247, 256, 258, 274, 284. *See also* Bureau of Labor Statistics, U.S.; Women's Bureau, U.S. Department of Labor
Lamont, Robert, 45, 48
Lange, Dorothea, 137
Lasch, Christopher, 299
League for Industrial Democracy, 132
Leaver, Eric W., 237–40, 306
Lee, Joshua, 75
leisure, 212–18, 267, 297–300. *See also* hours of work
Levy, Frank, 283
Levy, Steven, 301
Lewis, J. C., 14
Lewis, John L., 101, 109, 112, 272
Life, 239, 262
Lind, Herman, 153, 160
Linotype. See printing technology
Litchfield, P. W., 44
literature. *See* fiction, technological unemployment discussed in

Littell, Robert, 219

Little Steel (Sinclair), 123, 125

"Little Waters" letter, 198–99

Lockridge, Frances and Richard, 125

Locomotive Engineers' Journal, 139–40

logging, 287–88

longshoring, 270

Lorentz, Pare, 134–35, 233

Lubin, Isador, 36, 63, 67, 72, 230

Luce, Henry, 232

Luddites, 1, 70, 111, 301

Lynd, Robert and Helen, 10

MacDonald, J. A., 13

Machine-Made Jobs, 160

Machinery and Allied Products Institute
    (MAPI), 59, 160–61, 163, 241

*Machinery and the American Standard of
    Living*, 161

Mackay Radio and Telegraph Company,
    106

Macklin, Justin, 145–46, 164

MacNally, Jeff, 301

MacNeil-Lehrer Newshour, 288

Magazine of Wall Street, 137

Malthus, Thomas, 29

Manpower Development and Training
    Act, 258, 273

manufacturing: in the 1930s, 82, 85, 87–90;
    in the postwar decades, 237–41, 252,
    276–77. *See also* automobile manufac-
    ture; steelmaking

Marx, Karl, 29

Mathes, J. M., 155

McCall's, 262

McCarthy, Eugene, 297, 300

McClellan, William, 182–84, 193

McCormick, Cyrus, 157

McCormick, Fowler, 68

McCrory, S. H., 70

McCulloch, J. R., 28

McFall, Robert, 202

McGraw Hill, 57, 161–62

McGraw, James, Jr., 161

Meany, George, 256, 273

meatpacking, 271

Mechanical Engineering, 183–84, 186,
    189–92, 227

media coverage of technological unem-
    ployment: in the 1930s, 116–17; in the
    postwar decades, 250–51, 261–62

Mees, C. E. Kenneth, 217

Mehren, Edward J., 183, 187

Men and Machines (Chase), 117–18, 146

Merriam, Charles, 51

Merriam, J. C., 175

Merrick, F. A., 19, 144, 153

middle class, economic prospects of, 277,
    283, 293–94

*The Middletons Visit the New York World's
    Fair*, 225–26

Mike Mulligan and His Steam Shovel
    (Burton), 128

Mill, John Stuart, 29

Miller, Spencer, 212

Millikan, Robert, 173, 175–76, 179

Milwaukee Federated Trades Council, 14

Milwaukee Journal, 97

mining. *See* coal mining

Mitchell, Wesley, 15, 51

Modern Times, 136

movie-making, 93–100

movies, technological unemployment
    discussed in, 90–91, 134–36, 233–34,
    251–52

Mumford, Lewis, 207–9

Murray, Philip, 19–21, 83–84, 100–101, 106,
    230

Music Defense League, 97–99, 131

musicians: in the 1930s, 93–101, 131–32,
    138, 181; in the postwar decades,
    271–72, 285–86

National Association of Manufacturers
    (NAM), 74, 153–54, 157, 231–32, 245,
    247, 257

National Business Survey Conference, 48
National Committee on Industrial Reha-
bilitation, 148
National Education Association, 260
National Federation of Settlements, 81
National Industrial Conference Board
(NICB), 61, 120, 159, 163–64
National Machine Tool Builders Associa-
tion, 153
National Organization for Taxation of
Labor Displacing Devices, 77
*National Research Project on Reemploy-
ment Opportunities and Recent
Changes in Industrial Techniques,*
57–69, 72, 228–30.
National Resources Committee, 69–71
National Union for Social Justice, 133
National Window Glass Workers, 91
Nation's Business, 144, 149, 158–59, 232
neo-Luddites, 301
New Castle, Pennsylvania, 22
New Deal programs, 43, 112, 216. *See also*
Roosevelt, Franklin Delano
New Principles of Political Economy
(Sismondi), 27
New Republic, 22, 72
Newsweek, 262, 294, 299
"New Technology Bill of Rights" (IAM),
277
New Yorker, 264
New York Museum of Science and Indus-
try, 178
New York State Commission on Old Age
Security, 34
New York Times: 1930s comments on
technological unemployment, 16, 33,
37, 55, 61–62, 82–83, 99, 107, 140–41,
150, 169, 180, 217; postwar comments,
263–66, 286, 289, 294–95
New York World's Fair, 205, 220–27, 236
Nichol, F. W., 88
Nixon, Richard, 236, 255

Norris, George, 78
numerical control machine tools, 276

occupational obsolescence, 31, 306
office work: in the 1930s, 87–89; in the
postwar decades, 242–43, 251–52,
274–75
Ogburn, William Fielding, 40–41, 51–52,
69–71, 171–72, 308–9; *Social Change
with Respect to Culture and Original
Nature,* 40; *You and Machines,* 53–56,
138, 140. *See also* cultural lag theory
Oil, Chemical, and Atomic Workers
Union, 273
older workers, 34, 36, 258
O'Leary, John, 160
Omaha World-Herald, 15
O'Mahoney, Joseph, 78, 228
Ornburn, I. M., 102
Overstreet, H. A., 215
Owens bottle machine. *See* glass blowing

Pack, Arthur Newton, 215
paper manufacture, 14
Parmalee, J. H., 229
Passaic, New Jersey, 124
patents and patent law, 11, 12, 52, 76,
187–88
Patman, Wright, 243–44
Peck, Gustav, 51
Pepper, Claude, 153
Perhaps Women (Anderson), 124
Perkins, Frances, 81
Person, Harlow, 45, 201–2
Petrillo, James, 95–96
Phalen, Clifton, 248
Phillips, Kevin, 299
Pidgeon, Mary, 89
Player Piano (Vonnegut), 252–54
The Plow That Broke the Plains, 134
poetry, technological unemployment dis-
cussed in, 130–31

Popular Science, 203
postal work, 287, 302–3
Potter, Andrey, 199
Prentis, H. W., 74
Presidential Advisory Committee on
    Labor-Management Policy, 256
President's Emergency Committee on
    Employment (PECE), 44, 49–50
President's Organization on Unemploy-
    ment Relief (POUR), 49–50
*Principles of Political Economy and Taxa-
    tion* (Ricardo), 28
Printers' Ink, 158
printing technology, 32–33, 92, 259, 274
Product Engineering, 57
Prosperity: Fact or Myth (Chase), 13
Prosser, C. A., 211
Pupin, Michael, 172–73
Putnam, Robert, 299

Quinn, Jane Bryant, 281

radio, technological unemployment
    discussed on, 132–35
railroad technology: in the 1930s, 17, 83,
    86, 107–8, 229–30; postwar decades,
    248, 259
Railroad Trainman, 108
Rand McNally Bankers Monthly, 148
Randolph, Jennings, 76
RCA (Radio Corporation of America), 106
Readers' Digest, 262
Reagan, Ronald, 279
*Recent Economic Changes in the United
    States*, 15–16, 69, 208
Recent Social Trends in the United States,
    51–52, 69
record-playback technology, 252
re-employment, 36–38, 63, 73, 83–84, 106,
    245
refining technology, 240–41, 273
Reich, Robert, 283, 292

religious comments on machines and
    work, 115–16, 214, 260–61
Research Committee on Social Trends, 51
retraining, 73, 105, 258, 267, 271, 273, 292.
    See also re-employment
Reuther, Walter, 242, 247, 249–50, 301
Ricardo, David, 28
Richmond, Virginia, 86
Riesman, David, 267
Rieve, Emil, 81, 102
Rifkin, Jeremy, 291–92, 300, 305
Ripley, William, 165
River, 134
robots: in the 1930s, 64, 125–27, 131, 141; in
    the postwar decades, 276–78, 285; as
    visual symbol, 2, 54–56, 56, 96, 98,
    137–41, 158, 262, 264–65, 278, 301–3;
    Elektro, 224–26
Rockefeller, Nelson, 259
Roe, Joseph, 193–94
Rogers, H. H., 199
Rogers, Will, 132
Roosevelt, Eleanor, 115
Roosevelt, Franklin Delano: Presidential
    administration of, 53–74; statements
    of, on technological unemployment,
    43, 55, 62, 71, 73, 121, 171, 178; letter
    from, on engineering education,
    198–99
Rotarian, 169
Rowntree, R. Henry, 197–98
Rugg, Harold, 129–30, 137, 139–40, 154, 159,
    207, 209
R.U.R. (Capek), 126
Rust cotton-picker, 67
Ruttenberg, Harold, 22

Sale, Kirkpatrick, 301
Sarnoff, David, 157
Saturday Evening Post, 262
Say, Jean-Baptiste, 27
Science Advisory Board, 174–75

science fiction, technological unemployment discussed in, 125–27
science holiday proposals, 168–69
Science, the Endless Frontier (Bush), 236
Scientific American, 12
scientific management, 171, 199–202
scientific organizations. *See* American Association for the Advancement of Science; American Institute of Physics
scientists' public-relations campaigns, 172–74, 176–82
Scott, Howard, 118–22, 195
Scott, Walter Dill, 157
Scribner's Magazine, 116
secretarial work. *See* office work
service sector, 38–39, 63. *See also* telephone technology
Shaw, Arch, 48
Sheffield Farms, 223–24
Simkin, William, 273
Sinclair, Upton, 123, 125
Sioux City, Iowa, 110
Sismondi, Jean Charles Léonard de, 27–28
Slichter, Sumner, 32, 45, 201
Sloan, Alfred, Jr., 149, 156–57
Smith, A. O., factory of, 129, 137, 150–52, 237, 306
Smith, Elliott Dunlap, 34, 184, 196–97
Smith, L. R., 49, 151–52
Snyder, John, 249, 264
Social Change with Respect to Culture and Original Nature (Ogburn), 40
Society of Arts and Sciences, 171
sociology, 40–41
Soviet Union, automation in, 236, 249, 257
statistics on technological unemployment: in the 1930s, 16–17, 20–24, 32–33, 46, 51, 54, 60–63, 65–67, 81–84, 86–91, 94–95, 129, 133, 145, 162; in the postwar decades, 240, 242–43, 246–49, 254–55, 258–59, 270–71, 274–77, 281, 286, 289, 295

Steel, 19
steelmaking: in the 1930s, 18–22, 81, 123, 125, 135–36; in the postwar decades, 254–55, 261, 287
Steel Workers Organizing Committee (SWOC), 19–22, 83, 101
Steinbeck, John, 123, 125, 135
Steiner, Ralph, 220
Stern, Bernhard, 70
Stern, Boris, 62
Stewart, Ethelbert, 46
Stock, Alfred, 170
Stoll, Clifford, 301
stone cutting, 12–13, 92
Sullivan, Rose, 24–26
Sumners, Hatton, 75–76
Survey Graphic, 130, 149
Sweden and automation, 257, 310
Swope, Gerald, 147
Sylvania, 246

Taylor, Frederick Winslow, 171, 199
Taylor, Paul, 66–68
Taylorism, 171, 199–202
Taylor Society, 199–202
Technics and Civilization (Mumford), 208
Technocracy, 118–22, 195, 305
technocultural unemployment, 33
technological determinism, 11–12, 147, 209–11, 221, 234–35, 250, 308–11
Technological Trends and National Policy, 69–71, 99
Technology and the American Consumer, 161
Technology in Our Economy, 231
technotax proposals, 76–78, 109, 153, 269, 277
telegraph technology, 85–87
telephone technology: in the 1930s, 22–26, 100, 147, 161, 181, 224; in the postwar decades, 248–50, 286. *See also* AT&T
television, development of, 99, 246

Temporary National Economic Committee (TNEC) hearings, 19–21, 68, 205, 227–32
temporary work, 282
Texas Centennial Exposition, 140
textile machinery, 63, 81, 102
Textile Workers Union, 81, 83, 106
"This Modern World" (comic strip), 301–3
Thomas, R. J., 102–3
Thurow, Lester, 293, 299
Tibbits, Clark, 40
Time, 262, 275–76, 278, 284, 295
Times (London), 44
Toles, Tom, 301
Tomorrow, Tom, 301–3
Tragic America (Dreiser), 124
Transport Workers Union, 255
Treatise on Political Economy (Say), 27
Tritle, J. S., 145, 167
Trundle, George F., Jr., 166
Tuggey, J. M., Jr., 87
Tugwell, Rexford, 31
"Twilight" (Campbell), 126

Unabomber, 300
Unemployed, 2, 132
Unemployment and Increasing Productivity, 59–61, 122
Unemployment and the Machine (MacDonald), 13
unions: in the 1920s, 13–15, 81–82, 91–92; in the 1930s, 82–84, 100–113; in the postwar decades, 247–50, 269, 272–73. *See also* American Federation of Labor; Congress of Industrial Organizations; *specific unions*
United Auto Workers (UAW), 102, 108–9, 229, 272–73
United Mine Workers, 108–9, 112, 272
U.S. Industries, 249, 260, 264–65
Using Leisure Time (Hermans and Hannon), 215

U.S. News & World Report, 250, 293
Utne Reader, 303

Valleytown, 135–36, 233
Van Deventer, John, 144, 153–54, 164, 221–22
Van Dresser, Peter, 116
Van Dyke, Willard, 135–36, 220, 233–34
Veblen, Thorstein, 306
vending machines, 39, 252–53, 255, 287
Vonnegut, Kurt, 252–54

Wagner, Robert, 74–75, 132
Wallace, Henry, 65, 197
Warner, Harry, 93
Washington Post, 274
Waste in Industry, 46
Weber, Frank, 14
Weber, Joseph, 96–97, 99
Weeks, Arland D., 40
Weintraub, David, 58–64, 69, 223, 230
Western Union, 87
Westinghouse Electric Company, 19, 224–26
Westinghouse, George, 187
Wheeler, Harry, 49
Whipple, J. V. H., 59
Whitney, A. F., 17, 104
Whitney, W. R., 157
Wickenden, William, 191
Wiener, Norbert, 240, 305
Williams, Aubrey, 55–56
Williams, C. C., 199
Willits, Joseph, 44
Wirtz, W. Willard, 258
Woll, Matthew, 100
Wolman, Leo, 51, 200
Women's Bureau, U.S. Department of Labor, 24, 85, 88–91, 137, 217
women's employment, 84–91, 216–17
Wonder Stories, 125
Worcester, Massachusetts, 24

workers' comments on mechanization: in the 1930s, 21, 33, 80–81, 108–11, 114–15; in the postwar decades, 261, 273–74, 294–95, 309

work, length of. *See* hours of work

Works Progress Administration (WPA), 43, 216. *See also National Research Project on Re-employment Opportunities*

world's fairs, 219–20. *See also* Century of Progress exposition; New York World's Fair

World War II, 232–37

Wright, Carroll D., 29

Wright, Roy, 191

You and Machines (Ogburn), 53–56, 138

Young, Owen, 171, 180

Library of Congress Cataloging-in-Publication Data

Bix, Amy Sue.
    Inventing ourselves out of jobs? : America's debate over
technological unemployment, 1929–1981 / Amy Sue Bix.
       p.   cm. — (Studies in industry and society)
    Includes bibliographical references and index.
    ISBN 0-8018-6244-2 (alk. paper)
     1. Technological unemployment—United States—History. I. Title.
II. Series.
HD6331.2.U5B59 2000
331.13'7042'0973—dc21   99-32829
                                                          CIP

Printed in the United States
3911